Polymorphism in Pharmaceutical Solids

DRUGS AND THE PHARMACEUTICAL SCIENCES

Executive Editor
James Swarbrick
AAI, Inc.
Wilmington, North Carolina

Advisory Board

Larry L. Augsburger
University of Maryland
Baltimore, Maryland

David E. Nichols
Purdue University
West Lafayette, Indiana

Douwe D. Breimer
Gorlaeus Laboratories
Leiden, The Netherlands

Stephen G. Schulman
University of Florida
Gainesville, Florida

Trevor M. Jones
The Association of the
British Pharmaceutical Industry
London, United Kingdom

Jerome P. Skelly
Copley Pharmaceutical, Inc.
Canton, Massachusetts

Hans E. Junginger
Leiden/Amsterdam Center
for Drug Research
Leiden, The Netherlands

Felix Theeuwes
Alza Corporation
Palo Alto, California

Vincent H. L. Lee
University of Southern California
Los Angeles, California

Geoffrey T. Tucker
University of Sheffield
Royal Hallamshire Hospital
Sheffield, United Kingdom

Peter G. Welling
Institut de Recherche Jouveinal
Fresnes, France

DRUGS AND THE PHARMACEUTICAL SCIENCES

A Series of Textbooks and Monographs

Polymorphism in Pharmaceutical Solids

edited by
Harry G. Brittain
Discovery Laboratories, Inc.
Milford, New Jersey

MARCEL DEKKER, INC.

NEW YORK · BASEL

1999

ISBN: 0-8247-0237-9

This book is printed on acid-free paper.

Headquarters
Marcel Dekker, Inc.
270 Madison Avenue, New York, NY 10016
tel: 212-696-9000; fax: 212-685-4540

Eastern Hemisphere Distribution
Marcel Dekker AG
Hutgasse 4, Postfach 812, CH-4001 Basel, Switzerland
tel: 44-61-261-8482; fax: 44-61-261-8896

World Wide Web
http://www.dekker.com

The publisher offers discounts on this book when ordered in bulk quantities. For more information, write to Special Sales/Professional Marketing at the headquarters address above.

Current printing (last digit):
10 9 8 7 6 5 4 3 2 1

PRINTED IN THE UNITED STATES OF AMERICA

Preface

Since the middle of the last century, it has been noted that organic molecules can be obtained in more than one distinct crystal form, a property that became known as polymorphism. Once experimental methods based on the diffraction of x-rays were developed to determine the structures of crystalline substances, it was quickly learned that an extremely large number of molecules were capable of exhibiting the phenomenon. In addition, numerous compounds were shown to form other nonequivalent crystalline structures through the inclusion of solvent molecules in the lattice.

It was also established that the structure adopted by a given compound upon crystallization would exert a profound effect on the solid-state properties of that system. For a given material, the heat capacity, conductivity, volume, density, viscosity, surface tension, diffusivity, crystal hardness, crystal shape and color, refractive index, electrolytic conductivity, melting or sublimation properties, latent heat of fusion, heat of solution, solubility, dissolution rate, enthalpy of transitions, phase diagrams, stability, hygroscopicity, and rates of reactions are all determined primarily by the nature of the crystal structure.

Workers in pharmaceutical-related fields realized that the solid-state property differences derived from the existence of alternate crystal forms could translate into measurable differences in properties of pharmaceutical importance. In early works, it was found that the various polymorphs of phenylbutazone exhibited different dissolution rates. The existence of nonequivalent dissolution rates and solubilities for the different polymorphs of chloramphenicol led to the observation that the various forms were not bioequivalent. It has therefore been recognized that the determination of possible polymorphism in a new drug entity must be thoroughly investigated during the early stages of development. Fortunately, a wide variety of experimental techniques are now available for the characterization of polymorphic solids, with the most important of these being x-ray diffraction methodologies.

This book represents an attempt to summarize the major issues pertaining to the pharmaceutical aspects of polymorphism, as well as the effects of solvate formation. The book is subdivided into five main sections, the first consisting of two chapters defining the phenomenon. Chapter 1 presents the theory and origin of polymorphism in solids, and Chapter 2 examines the phase characteristics of polymorphic systems in the systematic manner permitted by the Phase Rule. The second section covers the crystallographic considerations that define different polymorphic and solvate species, with Chapters 3 and 4 providing detailed summaries of the structural aspects.

The fundamentals of the effect having been established, the next section contains two chapters that present some highly practical subjects. Chapter 5 lays out all the various procedures one would use to generate all the possible polymorphs and hydrate species that might exist for a given substance. Chapter 6 summarizes the principal analysis methods routinely used for the characterization and determination of different crystal forms. The fourth section of the book presents two chapters that address the consequences associated with the existence of polymorphic forms. The topic of solubility and dissolution rate is of utmost importance, in that these quantities directly affect the relative bioavailabilities of different crystalline forms. Different polymorphic and/or hydrate species can exert definite effects on both the processing and stability of drug dosage forms, and these effects are explored.

The final section contains two chapters that are obliquely related

to the area of polymorphism. Chapter 9 concerns the structural variations that can arise from the existence of molecular dissymmetry, manifested primarily in marked differences in solid-state properties between
solids composed of racemates relative to solids composed of separated
enantiomers. The amorphous condition can be considered one polymorphic state available to all compounds, and the final chapter deals with
the impact of polymorphism on the quality of amorphous materials
produced by lyophilization.

There is no doubt that an awareness of solid-state pharmaceutics
continues to develop as the sophistication of drug development advances. The technology available to workers in this area was developed
in our earlier volume, *Physical Characterization of Pharmaceutical
Solids*. In the present work, we have sought to provide in a single volume a comprehensive view of the principles, practical concerns, and
consequences of the existence of polymorphism and solvate formation.
As with the earlier, almost companion, volume we hope to suggest
approaches that will in turn stimulate work and encourage additional
growth in this area.

Harry G. Brittain

Contents

Contributors

Harry G. Brittain Discovery Laboratories, Inc., Milford, New Jersey

Stephen R. Bryn Department of Medicinal Chemistry and Phamacognosy, Purdue University, West Lafayette, Indiana

Eugene F. Fiese Pharmaceutical Research and Development, Pfizer Central Research, Groton, Connecticut

David J. W. Grant Department of Pharmaceutics, College of Pharmacy, University of Minnesota, Minneapolis, Minnesota

J. Keith Guillory College of Pharmacy, The University of Iowa, Iowa City, Iowa

Kenneth R. Morris Department of Industrial and Physical Pharmacy, Purdue University, West Lafayette, Indiana

Michael J. Pikal School of Pharmacy, University of Connecticut, Storrs, Connecticut

1

Theory and Origin of Polymorphism

David J. W. Grant

University of Minnesota
Minneapolis, Minnesota

I. INTRODUCTION

Many pharmaceutical solids exhibit *polymorphism*, which is frequently defined as the ability of a substance to exist as two or more crystalline phases that have different arrangements and/or conformations of the mol-

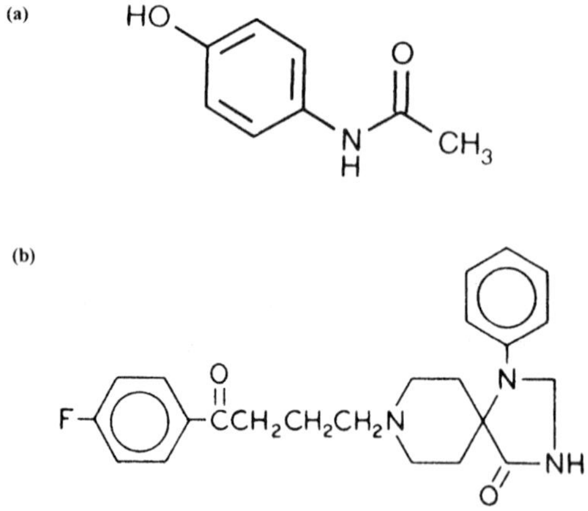

(a)

(b)

Fig. 1   Molecular structure of (a) acetaminophen and (b) spiperone.

ecules in the crystal lattice [1–3]. Thus, in the strictest sense, polymorphs are different crystalline forms of the same pure substance in which the molecules have different arrangements and/or different conformations of the molecules. As a result, the polymorphic solids have different unit cells and hence display different physical properties, including those due to packing, and various thermodynamic, spectroscopic, interfacial, and mechanical properties, as discussed below [1–3].

For example, acetaminophen (paracetamol, 4-acetamidophenol, 4-hydroxyacetanilide, shown in Fig. 1a) can exist as a monoclinic form, of space group $P2_1/n$ [4], which is thermodynamically stable under ambient conditions. The compound can also be obtained as a less stable orthorhombic form, of space group $Pbca$, and which has a higher density indicative of closer packing [5–7]. The unit cells of these two forms are compared in Fig. 2 and Table 1. The molecule of acetaminophen is rigid on account of resonance due to conjugation involving the hy-

Fig. 2   View of the unit cell contents for two polymorphs of acetaminophen: (a) orthorhombic form (b) monoclinic form [4,5,7]. (Reproduced with permission of the copyright owner, the American Crystallographic Association, Washington, DC.)

(a)

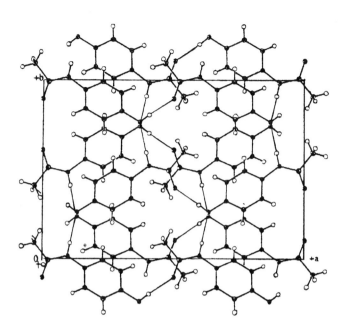

(b)

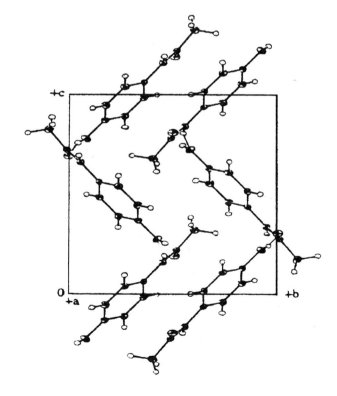

Table 1 Crystal Data for Two Polymorphs of Acetaminophen

Crystal data and structure refinement	Orthorhombic phase	Monoclinic phase
Empirical formula	$C_8H_9NO_2$	$C_8H_9NO_2$
Formula weight	151.16	151.16
Crystal system	Orthorhombic	Monoclinic
Space group	*Pbca*	*P2$_1$/n*
Unit cell dimensions	a = 17.1657(12) Å	a = 7.0941(12) Å
	b = 11.7773(11) Å	b = 9.2322(11) Å
	c = 7.212(2) Å	c = 11.6196(10) Å
	α = 90.000°	α = 90.000°
	β = 90.000°	β = 97.821(10)°
	γ = 90.000°	γ = 90.000°
Volume	1458.1(4) Å^3	753.9(2) Å^3
Z	8	4
Density (calculated)	1.377 g/cm^3	1.332 g/cm^3
Crystal size	0.28 × 0.25 × 0.15 mm	0.30 × 0.30 × 0.15 mm
Refinement method	Full-matrix least-squares on F^2	Full-matrix least-squares on F^2
Hydrogen bond lengths and angles		
H(5)O(2)	1.852(26) Å	1.772(20) Å
H(6)O(1)	2.072(28) Å	2.007(18) Å
O(1)—H(5)O(2)	170.80(2.35)°	166.15(1.75)°
N(1)—H(6)O(1)	163.52(2.19)°	163.93(1.51)°

Source: Refs. 4, 5, and 7. Reproduced with permission of the copyright owner, the American Crystallographic Association, Washington, DC.

droxyl group, the benzene ring, and the amido group. Therefore the conformation of the molecule is virtually identical in the two polymorphs of acetaminophen. On the other hand, the spiperone molecule (8-[3-(p-fluorobenzoyl)-propyl]-1-phenyl-1,3,8-triazaspiro[4,5]decan-4-one, shown in Fig. 1b) contains a flexible -CH_2-CH_2-CH_2- chain and is therefore capable of existing in different molecular conformations [8]. Two such conformations, shown in Fig. 3, give rise to two different conformational polymorphs (denoted Forms I and II), which have different unit cells (one of which is shown in Fig. 4) and densities, even

Form I

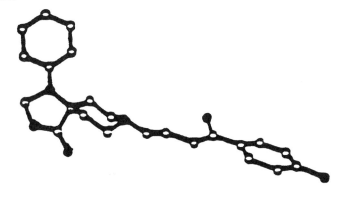

Form II

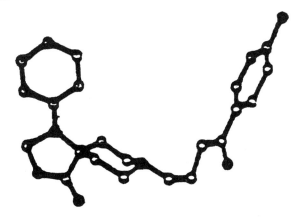

Fig. 3 The molecular conformations of the spiperone molecule in polymorphic forms I and II [8]. (Reproduced with permission of the copyright owner, the American Pharmaceutical Association, Washington, DC.)

though their space groups are the same, both being $P2_1/n$, monoclinic, as shown in Table 2 [8].

As mentioned above, the various polymorphs of a substance can exhibit a variety of different physical properties. Table 3 lists some of the many properties that differ among different polymorphs [1–3,9]. Because of differences in the dimensions, shape, symmetry, capacity

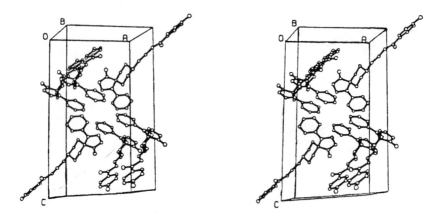

Fig. 4 View of the unit cell contents for the form I polymorph of spiperone [8]. (Reproduced with permission of the copyright owner, the American Pharmaceutical Association, Washington, DC.)

Table 2 Crystal Data for Two Polymorphs of Spiperone

	Form I	Form II
Empirical formula	$C_{23}H_{26}FN_3O_2$	$C_{23}H_{26}FN_3O_2$
Molecular weight	395.46	395.46
Crystal system	Monoclinic	Monoclinic
Space group	$P2_1/a$	$P2_1/c$
Unit cell dimensions	$a = 12.722$ Å	$a = 18.571$ Å
	$b = 7.510$ Å	$b = 6.072$ Å
	$c = 21.910$ Å	$c = 20.681$ Å
	$\alpha = 90.00°$	$\alpha = 90.00°$
	$\beta = 95.08°$	$\beta = 118.69°$
	$\gamma = 90.00°$	$\gamma = 90.00°$
Unit cell volume	2085.1 Å^3	2045.7 Å^3
Z	4	4

Source: Ref. 8. Reproduced with permission of the copyright owner, the American Pharmaceutical Association, Washington, DC.

Table 3 List of Physical Properties that Differ Among Various Polymorphs

1. Packing properties
    a. Molar volume and density
    b. Refractive index
    c. Conductivity, electrical and thermal
    d. Hygroscopicity
2. Thermodynamic properties
    a. Melting and sublimation temperatures
    b. Internal energy (i.e., Structural energy)
    c. Enthalpy (i.e., Heat content)
    d. Heat capacity
    e. Entropy
    f. Free energy and chemical potential
    g. Thermodynamic activity
    h. Vapor pressure
    i. Solubility
3. Spectroscopic properties
    a. Electronic transitions (i.e., ultraviolet–visible absorption spectra)
    b. Vibrational transitions (i.e., infrared absorption spectra and Raman spectra)
    c. Rotational transitions (i.e., far infrared or microwave absorption spectra)
    d. Nuclear spin transitions (i.e., nuclear magnetic resonance spectra)
4. Kinetic properties
    a. Dissolution rate
    b. Rates of solid state reactions
    c. Stability
5. Surface properties
    a. Surface free energy
    b. Interfacial tensions
    c. Habit (i.e., shape)
6. Mechanical properties
    a. Hardness
    b. Tensile strength
    c. Compactibility, tableting
    d. Handling, flow, and blending

(number of molecules), and void volumes of their unit cells, the different polymorphs of a given substance have different physical properties arising from differences in molecular packing. Such properties include molecular volume, molar volume (which equals the molecular volume multiplied by Avogadro's number), density (which equals the molar mass divided by the molar volume), refractive index in a given direction (as a result of the interactions of light quanta with the vibrations of the electrons in that direction), thermal conductivity (as a result of the interaction of infrared quanta with the intramolecular and intermolecular vibrations and rotations of the molecules), electrical conductivity (as a result of movement of the electrons in an electric field), and hygroscopicity (as a result of access of water molecules into the crystal and their interactions with the molecules of the substance). Differences in melting point of the various polymorphs arise from differences of the cooperative interactions of the molecules in the solid state as compared with the liquid state. Differences in the other thermodynamic properties among the various polymorphs of a given substance are discussed below. Also involved are differences in spectroscopic properties, kinetic properties, and some surface properties. Differences in packing properties and in the energetics of the intermolecular interactions (thermodynamic properties) among polymorphs give rise to differences in mechanical properties.

Many pharmaceutical solids can exist in an amorphous form, which, because of its distinctive properties, is sometimes regarded as a polymorph. However, unlike true polymorphs, amorphous forms are not crystalline [1,2,10]. In fact, amorphous solids consist of disordered arrangements of molecules and therefore possess no distinguishable crystal lattice nor unit cell and consequently have zero crystallinity. In amorphous forms, the molecules display no long-range order, although the short-range intermolecular forces give rise to the short-range order typical of that between nearest neighbors (see Fig. 5). Thermodynamically, the absence of stabilizing lattice energy causes the molar internal energy or molar enthalpy of the amorphous form to exceed that of the crystalline state. The absence of long-range order causes the molar entropy of the amorphous form to exceed that of the crystalline state. Furthermore, the lower stability and greater reactivity of the amorphous form indicates that its molar Gibbs free energy exceeds that of the crys-

(a)

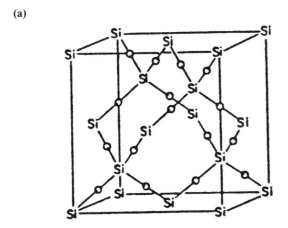

(b)

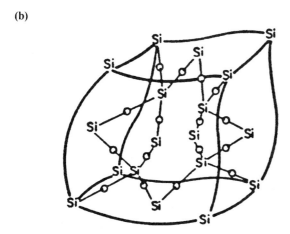

Fig. 5 Schematic diagram showing the difference in long-range order of silicon dioxide in (a) the crystalline state (crystobalite) and (b) the amorphous state (silica glass) [2]. The two forms have the same short-range order. (Reproduced with permission of the copyright owner, the American Pharmaceutical Association, Washington, DC.)

talline state. This observation implies that the increased molar enthalpy of the amorphous form outweighs the $T\Delta S$ term that arises from its increased molar entropy.

II. THERMODYNAMICS OF POLYMORPHS

The energy of interaction between a pair of molecules in a solid, liquid, or real gas depends on the mean intermolecular distance of separation according to the Morse potential energy curve shown in Fig. 6 [11,12]. For a given pair of molecules, each polymorph, liquid or real gas has its own characteristic interaction energies and Morse curve. These intermolecular Morse curves are similar in shape but have smaller energies and greater distances than the Morse potential energy curve for the interaction between two atoms linked by a covalent bond in a diatomic molecule or within a functional group of a polyatomic molecule. The Morse potential energy curve in Fig. 6 is itself the algebraic sum of a curve for intermolecular attraction due to van der Waals forces or hydrogen bonding and a curve for intermolecular electron–electron and nucleus–nucleus repulsion at closer approach. The convention employed is that attraction causes a decrease in potential energy, whereas repulsion causes an increase in potential energy. At the absolute zero of temperature, the pair of molecules would occupy the lowest or zero point energy level. The Heisenberg uncertainty principle requires that the molecules have an indeterminate position at a defined momentum or energy. This indeterminate position corresponds to the familiar vibration of the molecules about the mean positions that define the mean intermolecular distance. At a temperature T above the absolute zero, a proportion of the molecules will occupy higher energy levels according to the Boltzmann equation:

$$\frac{N_1}{N_0} = \exp\left(\frac{-\Delta\varepsilon_1}{kT}\right) \qquad (1)$$

where N_1 is the number of molecules occupying energy level 1 (for which the potential energy exceeds the zero point level by the energy difference $\Delta\varepsilon_1$), N_0 is the number of molecules occupying the zero point

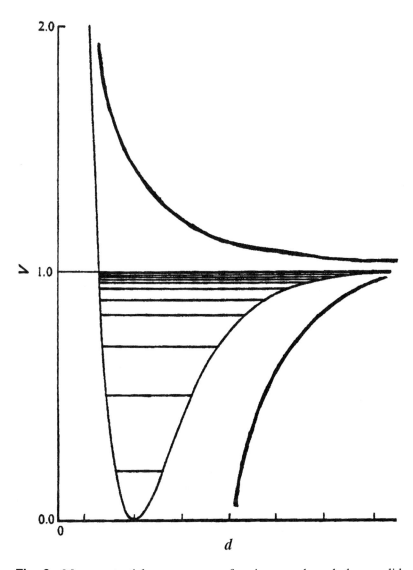

Fig. 6 Morse potential energy curve of a given condensed phase, solid or liquid [11]. The potential energy of interaction *V* is plotted against the mean intermolecular distance *d*. (Reproduced with permission of the copyright owner, Oxford University Press, Oxford, UK.)

level, and k is the Boltzmann constant (1.381×10^{-23} J/K, or 3.300×10^{-26} cal/K, i.e. the gas constant per molecule).

With increasing temperature, increasing numbers of molecules occupy the higher energy levels so that the distribution of the molecules among the various energy levels (known as the Boltzmann distribution) becomes broader, as shown in Fig. 7. At any given temperature, the number of distinguishable arrangements of the molecules of the system among the various energy levels (and positions in space) available to them is termed the thermodynamic probability Ω. With increasing temperature, Ω increases astronomically. According to the Boltzmann equation,

$$S = k \cdot \ln \Omega \tag{2}$$

where the entropy S is a logarithmic function of Ω, so increasing temperature causes a steady rise, though not an astronomical rise, in the entropy. In a macroscopic system, such as a given polymorph, the product $T \cdot S$ represents the energy of the system that is associated with

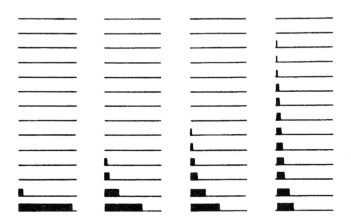

Fig. 7 Populations of molecular states at various temperatures [11]. The temperature is increasing from left to right. (Reproduced with permission of the copyright owner, Oxford University Press, Oxford, UK.)

the disorder of the molecules. This energy is the bound energy of the system that is unavailable for doing work.

The sum of the individual energies of interaction between nearest neighbors, next nearest neighbors, and so on, throughout the entire crystal lattice, liquid, or real gas can be used to define the internal energy E (i.e., the intermolecular structural energy) of the phase. Normally the interactions beyond next nearest neighbors are weak enough to be approximated or even ignored. For quantitative convenience one mole of substance is considered, corresponding to molar thermodynamic quantities. At constant pressure P (usually equal to atmospheric pressure), the total energy of a phase is represented by the enthalpy H:

$$H = E + P \cdot V \tag{3}$$

where V is the volume of the phase (the other quantities have already been defined). With increasing temperature, E, V, and H tend to increase.

Figure 8 shows that the enthalpy H and the entropy S of a phase

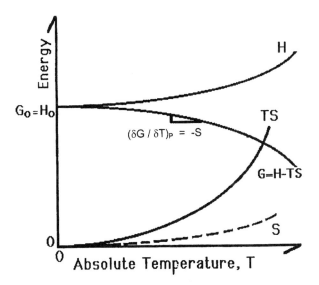

Fig. 8 Plots of various thermodynamic quantities against the absolute temperature T of a given solid phase (polymorph) or liquid phase at constant pressure. H = enthalpy, S = entropy, and G = Gibbs free energy.

tend to increase with increasing absolute temperature T. According to the third law of thermodynamics, the entropy of a perfect, pure crystalline solid is zero at the absolute zero of temperature. The product $T \cdot S$ increases more rapidly with increasing temperature than does H. Hence the Gibbs free energy G, which is defined by

$$G = H - T \cdot S \tag{4}$$

tends to decrease with increasing temperature (Fig. 8). This decrease also corresponds to the fact that the slope $(\delta G/\delta T)$ of the plot of G against T is negative according to the equation

$$\left(\frac{\delta G}{\delta T} \right)_p = -S \tag{5}$$

As already stated, the entropy of a perfect, pure crystalline solid is zero at the absolute zero of temperature. Hence the value of G at $T = 0$ (termed G_0) is equal to the value of H at $T = 0$, termed H_0 (Fig. 8). Each polymorph yields an energy diagram similar to that of Fig. 6, although the values of G, H, and the slopes of the curves at a given temperature are expected to differ between different polymorphs.

Because each polymorph has its own distinctive crystal lattice, it has its own distinctive Morse potential energy curve for the dependence of the intermolecular interaction energies with intermolecular distance. The liquid state has a Morse curve with greater intermolecular energies and distances, because the liquid state has a higher energy and molar volume (lower density) than does the solid state. Figure 9 presents a series of Morse curves, one for each polymorph (A, B, and C) and for the liquid state of a typical substance of pharmaceutical interest. The composite curve in Fig. 9 is the algebraic sum of the Morse curves for each phase (polymorph or liquid). The dashed line corresponds to the potential energy of the separated, noninteracting molecules in the gaseous state. The increase in potential energy from the zero point value of a given polymorph to the dashed line corresponds to the lattice energy of that polymorph or energy of sublimation (if at constant pressure, the enthalpy of vaporization). For the liquid state the increase in potential energy from the average value in the liquid state to the dashed line for the gaseous molecules corresponds to the energy of vaporiza-

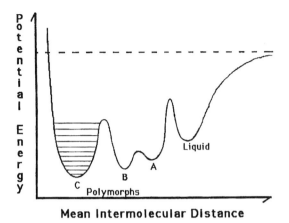

Fig. 9 Composite Morse potential energy curve of a series of polymorphs, A, B, and C, and of the corresponding liquid phase.

tion (if at constant pressure, the enthalpy of vaporization). The increase in potential energy from the zero point value of a given polymorph to the average value for the liquid state corresponds to the energy of fusion (if at constant pressure, the enthalpy of fusion).

When comparing the thermodynamic properties of polymorph 1 and polymorph 2 (or of one polymorph 1 and the liquid state 2) the difference notation is used:

$$\Delta G = G_2 - G_1 \tag{6}$$

$$\Delta S = S_2 - S_1 \tag{7}$$

$$\Delta H = H_2 - H_1 \tag{8}$$

$$\Delta V = V_2 - V_1 \tag{9}$$

In discussions of the relative stability of polymorphs and the driving force for polymorphic transformation at constant temperature and pressure (usually ambient conditions), the difference in Gibbs free energy is the decisive factor and is given by

$$\Delta G = \Delta H - T\,\Delta S \tag{10}$$

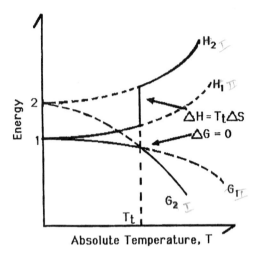

Fig. 10   Plots of the Gibbs free energy G and the enthalpy H at constant pressure against the absolute temperature T for a system consisting of two polymorphs, 1 and 2 (or a solid, 1, and a liquid, 2). T_t is the transition temperature (or melting temperature) and S is the entropy.

   Figure 10 shows the temperature dependence of G and H for two different polymorphs 1 and 2 (or for a solid 1, corresponding to any polymorph, and a liquid 2) [13]. In Fig. 10 the free energy curves cross. At the point of intersection, known as the transition temperature T_t (or the melting point for a solid and a liquid), the Gibbs free energies of the two phases are equal, meaning that the phases 1 and 2 are in equilibrium (i.e., $\Delta G = 0$). However, at T_t Fig. 10 shows that polymorph 2 (or the liquid) has an enthalpy H_2 that is higher than that of polymorph 1 (or the solid), so that $H_2 > H_1$. Equations 10 and 6 show that, if $\Delta G = 0$, polymorph 2 (or the liquid) also has a higher entropy S_2 than does polymorph 1 (or the solid), so that $S_2 > S_1$. Therefore according to Equation 10, at T_t,

$$\Delta H_t = T_t \Delta S_t \tag{11}$$

where $\Delta H_t = H_2 - H_1$ and $\Delta S_t = S_2 - S_1$ at T_t. By means of differential scanning calorimetry, the enthalpy transition ΔH_t (or the enthalpy of fusion ΔH_f) may be determined. For a polymorphic transition, the rate of temperature increase must be slow enough to allow polymorph 1 to change completely to polymorph 2 over a few degrees. Because in Fig. 10, $H_2 > H_1$, ΔH is positive and the transition is endothermic in nature.

Figure 10 shows that, below T_t, polymorph 1 (or the solid) has the lower Gibbs free energy and is therefore more stable (i.e., $G_2 > G_1$). On the other hand, above T_t, polymorph 2 (or the liquid) has the lower Gibbs free energy and is therefore more stable (i.e., $G_2 < G_1$). Under defined conditions of temperature and pressure, only one polymorph can be stable, and the other polymorph(s) are unstable. If a phase is unstable but transforms at an imperceptibly low rate, then it is sometimes said to be metastable.

The Gibbs free energy difference ΔG between two phases reflects the ratio of "escaping tendencies" of the two phases. The escaping tendency is termed the fugacity f and is approximated by the saturated vapor pressure, p. Therefore

$$\Delta G = RT \ln \left(\frac{f_2}{f_1} \right) \tag{12}$$

$$\sim RT \ln \left(\frac{p_2}{p_1} \right) \tag{13}$$

where the subscripts 1 and 2 refer to the respective phases, R is the universal gas constant, and T is the absolute temperature. The fugacity is proportional to the thermodynamic activity a (where the constant of proportionality is defined by the standard state), while thermodynamic activity is approximately proportional to the solubility s (in any given solvent) provided the laws of dilute solution apply. Therefore

$$\Delta G = RT \ln \left(\frac{a_2}{a_1} \right) \tag{14}$$

$$\sim RT \ln \left(\frac{s_2}{s_1} \right) \tag{15}$$

in which the symbols have been defined above. Hence, because the most stable polymorph under defined conditions of temperature and pressure has the lowest Gibbs free energy, it also has the lowest values of fugacity, vapor pressure, thermodynamic activity, and solubility in any given solvent. During the dissolution process, if transport-controlled under sink conditions and under constant conditions of hydrodynamic flow, the dissolution rate per unit surface area J is proportional to the solubility according to the Noyes–Whitney [14] equation; therefore

$$\Delta G = RT \ln \left(\frac{J_2}{J_1} \right) \tag{16}$$

According to the law of mass action, the rate r of a chemical reaction (including the decomposition rate) is proportional to the thermodynamic activity of the reacting substance. Therefore

$$\Delta G = RT \ln \left(\frac{r_2}{r_1} \right) \tag{17}$$

To summarize, the most stable polymorph has the lowest Gibbs free energy, fugacity, vapor pressure, thermodynamic activity, solubility, and dissolution rate per unit surface area in any solvent, and rate of reaction, including decomposition rate.

III. ENANTIOTROPY AND MONOTROPY

If as shown in Fig. 10 one polymorph is stable (i.e., has the lower free energy content and solubility over a certain temperature range and pressure), while another polymorph is stable (has a lower free energy and solubility over a different temperature range and pressure), the two polymorphs are said to be enantiotropes, and the system of the two solid phases is said to be enantiotropic. For an enantiotropic system a reversible transition can be observed at a definite transition temperature, at which the free energy curves cross before the melting point is reached. Examples showing such behavior include acetazolamide, carbamazepine, metochlopramide, and tolbutamide [9,14,15].

Sometimes only one polymorph is stable at all temperatures below the melting point, with all other polymorphs being therefore unstable. These polymorphs are said to be monotropes, and the system of the two solid phases is said to be monotropic. For a monotropic system the free energy curves do not cross, so no reversible transition can be observed below the melting point. The polymorph with the higher free energy curve and solubility at any given temperature is, of course, always the unstable polymorph. Examples of this type of system include chloramphenicol palmitate and metolazone [9,14,15].

To help decide whether two polymorphs are enantiotropes or monotropes, Burger and Ramberger developed four thermodynamic rules [14]. The application of these rules was extended by Yu [15]. The most useful and applicable of the thermodynamic rules of Burger and Ramberger are the heat of transition rule and the heat of fusion rule. Figure 11, which includes the liquid phase as well as the two polymorphs, illustrates the use of these rules. The heat of fusion rule states that, if an endothermic polymorphic transition is observed, the two forms are enantiotropes. Conversely, if an exothermic polymorphic transition is observed, the two forms are monotropes.

The heat of fusion rule states that, if the higher melting polymorph has the lower heat of fusion, the two forms are enantiotropes. Conversely, if the higher melting polymorph has the higher heat of fusion, the two forms are monotropes. Figure 11, which includes the liquid phase as well as the two polymorphs, is necessary to illustrate the heat of fusion rule.

The above conditions, that are implicit in the thermodynamic rules, are summarized in Table 4. The last two rules in Table 4, the infrared rule and the density rule, were found by Burger and Ramberger [14] to be significantly less reliable than the heat of transition rule and the heat of fusion rule and are therefore not discussed here.

IV. KINETICS OF CRYSTALLIZATION

Among the various methods for preparing different polymorphs are sublimation, crystallization from the melt, crystallization from supercritical fluids, and crystallization from liquid solutions. In the pharmaceutical sciences, different polymorphs are usually prepared by crystallization from solution employing various solvents and various temperature regimes, such as initial supersaturation, rate of de-supersaturation, or final supersaturation. The supersaturation of the solution

(a)

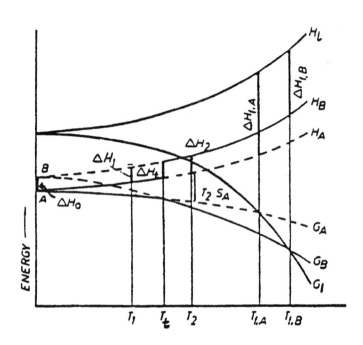

(b)

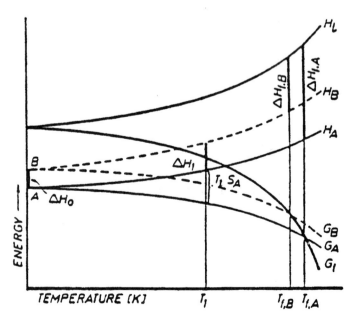

Table 4 Thermodynamic Rules for Polymorphic Transitions According to Burger and Ramberger [14], Where Form I is the Higher-Melting Form

Enantiotropy	Monotropy
Transition < melting I	Transition > melting I
I Stable > transition	I always stable
II Stable < transition	
Transition reversible	Transition irreversible
Solubility I higher < transition	Solubility I always lower than II
Solubility I lower > transition	
Transition II → I is endothermic	Transition II → I is exothermic
$\Delta H_f^I < \Delta H_f^{II}$	$\Delta H_f^I > \Delta H_f^{II}$
IR peak I before II	IR peak I after II
Density I < density II	Density I > density II

Source: Reproduced from Refer. 9 with permission of the copyright owner, Elsevier, Amsterdam, The Netherlands.

that is necessary for crystallization may be achieved by evaporation of the solvent (although any impurities will be concentrated), cooling the solution from a known initial supersaturation (or heating the solution if the heat of solution is exothermic), addition of a poor solvent (sometimes termed a precipitant), chemical reaction between two or more soluble species, or variation of pH to produce a less soluble acid or base from a salt or vice versa (while minimizing other changes in composition).

During the 19th century, Gay Lussac observed that, during crystallization, an unstable form is frequently obtained first that subsequently transforms into a stable form [13]. This observation was later explained thermodynamically by Ostwald [13,16–19], who formulated the law of successive reactions, also known as Ostwald's step rule. This

Fig. 11 Plots of the Gibbs free energy G and the enthalpy H at constant pressure against the absolute temperature T for a system consisting of two polymorphs, A and B, and a liquid phase, 1 [14]. T_t is the transition temperature, T_f is the melting temperature, and S is the entropy for (a) an enantiotropic system and (b) a monotropic system. (Reproduced with permission of the copyright owner, Springer Verlag, Vienna, Austria.)

rule may be stated as, "In all processes, it is not the most stable state with the lowest amount of free energy that is initially formed, but the least stable state lying nearest in free energy to the original state [13]."

Ostwald's step rule [13,16–19] is illustrated by Fig. 12. Let an enantiotropic system (Fig. 12a) be initially in a state represented by point X, corresponding to an unstable vapor or liquid or to a supersaturated solution. If this system is cooled, the Gibbs free energy will de-

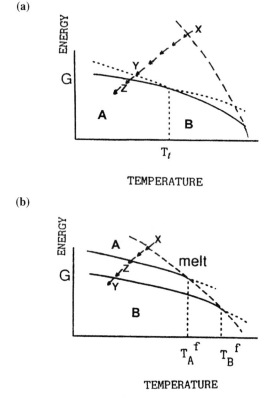

Fig. 12 Relationship between the Gibbs free energy G and the temperature T for two polymorphs for (a) an enantiotropic system and (b) a monotropic system in which the system is cooled from point X [9]. The arrows indicate the direction of change. (Reproduced with permission of the copyright owner, Elsevier, Amsterdam, The Netherlands.)

crease as the temperature decreases. When the state of the system reaches point Y, form B will tend to be formed instead of form A, because according to Ostwald's step rule Y (not Z) is the least stable state lying nearest in free energy to the original state. Similarly, let a monotropic system (Fig. 12b) be initially in a state represented by point X, corresponding to an unstable vapor or liquid or to a supersaturated solution. If this system is cooled, the Gibbs free energy will decrease as the temperature decreases. When the state of the system reaches point Z, form A will tend to be formed instead of form B, because according to Ostwald's step rule Z (not Y) is now the least stable state lying nearest in free energy to the original state. This rule is not an invariable thermodynamic law but a useful practical rule that is based on kinetics, and it is not always obeyed.

An understanding of the kinetics of the crystallization process involves consideration of the various steps involved. In the first step (termed nucleation) tiny crystallites of the smallest size capable of independent existence (termed nuclei) are formed in the supersaturated phase. Molecules of the crystallizing phase then progressively attach themselves to the nuclei, which then grow to form macroscopic crystals in the process known as crystal growth, until the crystallization medium is no longer supersaturated because saturation equilibrium has now been achieved. If the crystals are now allowed to remain in the saturated medium, the smaller crystals, which have a slightly greater solubility according to the Thomson (Kelvin) equation [11,20], tend to dissolve. At the same time, the larger crystals, which consequently have a lower solubility, tend to grow. This process of the growth of larger crystals at the expense of smaller crystals is sometimes termed Ostwald ripening.

The nucleation step is the most critical for the production of different polymorphs and is therefore discussed in some detail below. Nucleation may be primary (which does not require preexisting crystals of the substance that crystallizes) or secondary (in which nucleation is induced by preexisting crystals of the substance). Primary nucleation may be homogeneous, whereby the nuclei of the crystallizing substance arise spontaneously in the medium in which crystallization occurs, or heterogeneous, whereby the nuclei comprise foreign solid matter, such as particulate contaminants (including dust particles or the walls of the container).

Heterogeneous (i.e., spontaneous) nucleation is a stochastic process that is governed by the algebraic opposition of a volume term that favors the accretion of additional molecules from the supersaturated medium and a surface term that favors the dissolution of the molecular aggregates that would otherwise form nuclei. The resulting curve (Fig. 13) resembles an inverted Morse curve [21]. The molecules of the crystallizing substance tend to aggregate in the supersaturated medium under the influence of the volume term that tends to reduce the Gibbs free energy of the system. The prenuclear aggregates, termed embryos, are relatively small and have a high ratio of surface area to volume. The smaller the embryos, the larger will be the surface-to-volume ratio, and the more effective is the surface term in causing the embryos to

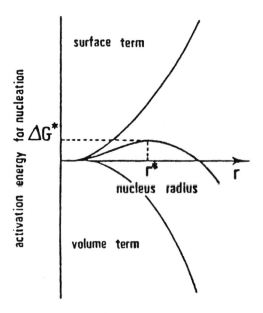

Fig. 13 Plot of the Gibbs free energy G of molecular aggregates (embryos) that are capable of forming nuclei against the size (mean radius r) of the aggregates [21]. ΔG^* is the activation energy for the formation of a nucleus of critical size r^* at which the nucleus can spontaneously grow (G decreases as r increases) or dissolve (G decreases as r decreases) by addition or removal of a single molecule. (Reproduced with permission of the copyright owner, Academic Press, New York, NY.)

dissolve. The resultant free energy curve in Fig. 13 has a maximum corresponding to the critical nuclear aggregate of critical radius r^* and representing an activation energy barrier ΔG^*. Embryos of smaller radius than r^* tend to dissolve, whereas those larger than r^* are true nuclei that tend to grow to form macroscopic crystals [21].

V. NUCLEATION OF POLYMORPHS

For a substance capable of existing in two or more polymorphic forms, each polymorph has its own characteristic curve typified by Fig. 13, each with its own characteristic value of r^* and ΔG^*. Within the limits imposed by their characteristic curves, the aggregates or embryos of the various polymorphs compete for molecules as depicted in Fig. 14 [22]. Depending on the nature of its curve, the aggregate present at the highest concentration (or for which the critical activation energy is the lowest) will form the first nucleus leading to the crystallization of that particular polymorph [22]. This mechanism explains the usual situation in which one polymorph crystallizes depending on the conditions that exist. However, examples are known in which more than one polymorph is obtained in the crystallization process. In these cases, conditions presumably exist whereby more than one type of nucleus is formed in the supersaturated medium at about the same time.

The formation of prenuclear molecular aggregates or embryos in a supersaturated solution can be studied by various physical methods, such as laser Raman spectroscopy [23], a technique that is especially

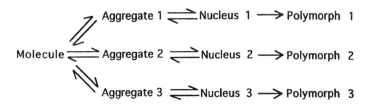

Fig. 14 Nucleation of polymorphs. The aggregate present at the highest concentration, or for which the critical activation energy is lowest, will form the first nucleus leading to the crystallization of that particular polymorph. (Reproduced with permission from Ref. 22.)

useful for aqueous solutions. The vibrational spectra of the aggregates contain peaks that are characteristic of some of the intermolecular interactions (such as hydrogen bonding) that are present in the solid phase that ultimately crystallizes. By appropriate examination of the supersaturated solution, perhaps by spectroscopic methods such as laser Raman spectroscopy [23], it may be possible to identify the intermolecular interaction in the aggregates and hence to identify the nature of the polymorph that will form before it actually crystallizes.

The foregoing theoretical discussion on nucleation, and on the factors that influence nucleation, readily explains why and how the following factors determine the polymorph that crystallizes out: solvent medium, supersaturation, temperature, impurities or additives dissolved, surface of the crystallization vessel, suspended particles, and seed crystals.

Under appropriate thermodynamic conditions discussed at the beginning of this chapter, a less stable polymorph may be converted into a more stable polymorph. The rate of conversion to the more stable polymorph is often rapid, if mediated by the solution phase or vapor phase. In these phases the less stable polymorph (having the greater solubility or vapor pressure) dissolves or sublimes, while the more stable polymorph (having the lower solubility or vapor pressure) crystallizes out. The rate of conversion to the more stable polymorph is usually slower, if the transformation proceeds directly from one solid phase to another. In this case, the mechanism of interconversion is likely to involve the following three steps: (1) loosening and breaking of the intermolecular forces (not covalent bonds) in the less stable polymorph, (2) formation of a disordered solid, similar to a localized amorphous form, and (3) formation of new intermolecular forces leading to crystallization of the more stable polymorph as the product phase [24].

We have seen earlier in this chapter that for an enantiotropic system, one polymorph may transform to another polymorph on the appropriate side of the transition temperature. Figure 15 [9] shows a plot of the rate of polymorphic change as a function of temperature. Close to the transition temperature, the rate is minimal but increases at higher temperatures, at which I $\rightarrow$ II, or at lower temperatures, at which II $\rightarrow$ I. If the temperature is lower than a certain optimal value, the rate of polymorphic change of II $\rightarrow$ I decreases based on the rules of chemical

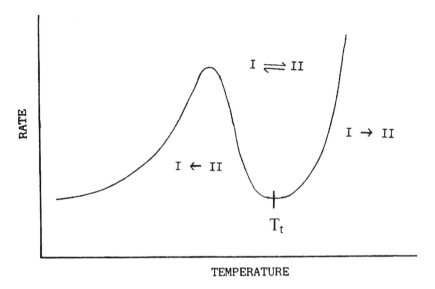

Fig. 15 Temperature dependence of the rates of transformation for a typical first-order transition between a low-temperature polymorph (I) and a high temperature polymorph (II) in an enantiotropic system for which T_t is the transition temperature [9]. (Reproduced with permission of the copyright owner, Elsevier, Amsterdam, The Netherlands.)

kinetics. At temperatures much lower than the transition temperature, the rate of change II $\rightarrow$ I may be negligible, explaining the observation that the higher temperature polymorph II is metastable at sufficiently low temperatures [9].

VI. NEW OR DISAPPEARING POLYMORPHS

We have seen that the nature of the polymorph that crystallizes depends on the relative rates of nucleation of the polymorphs. These kinetic factors also explain why solid state transformations in molecular crystals often display pronounced hysteresis [25]. For example, to induce the transition of the low-temperature enantiotrope to the high-temperature form, the former may have to be heated well above the transition temperature. Analogously, the absence of a solid state transformation of the lower melting form below the melting point may not necessarily indicate monotropy but could merely arise from slow nucleation of an enantiotropic transition. Similarly, on cooling the high-temperature form, transformations to the low-temperature form are usually associated with hysteresis. Thus X-ray diffraction studies of crystals have been achieved at 100K, well below (by more than 200K) the temperature range of thermodynamic stability. For example, single-crystal X-ray structural analysis was performed at 98K on the white high-temperature polymorph of dimethyl-3,6-dichloro-2,5-dihydroxyterephthalate, although this polymorph is thermodynamically unstable below 340K [26,27]. Thus a metastable high-temperature form can sometimes remain kinetically stable well below the transition point.

There are several documented examples of the inability to obtain a previously prepared crystal form [27,28]. Dunitz and Bernstein [25] quoted the following passage by Webb and Anderson [29], "Within the fraternity of crystallographers anecdotes abound about crystalline compounds which, like legendary beasts, are observed once and then never seen again." Similar anecdotes have been recounted by some industrial pharmaceutical scientists prior to 1970, but published reports relating to drugs and excipients are exceedingly difficult to find, undoubtedly because they would indicate a lack of process control. Most crystallographers and preformulation scientists recognize the role of seeding in initiating nucleation, and many consider the disappearance of a metastable form to be a local and temporary phenomenon. Jacewicz and Nayler [30] concluded that "any authentic crystal form should be capable of being re-prepared, although selection of the right conditions may require some time and trouble."

The chemical and pharmaceutical literature documents a number of examples of crystal forms that were apparently displaced by a more stable polymorph. One example is benzocaine picrate, for which a crystal form melting at 129–132°C was referred to in the 1968 edition of the *Pharmacopoeia Nordica* as one of the identification tests for the local anesthetic. The 8th edition of the *Merck Index* (1968) gives 134°C as the melting point [31]. Nielsen and Borka [32] described a more stable polymorph melting at 162–163°C, which can be obtained by drying the original lower-melting form at 105°C for two or more hours or by vacuum drying at 100°C and 0.1 mmHg with or without sublimation. On a hot stage under a microscope, the phase-pure polymorphs melt at 132–133°C or 162–163°C, respectively. A partially transformed sample that contains both polymorphs partially melts at 132°C, whereupon the molten benzocaine picrate resolidifies within seconds [32]. The new resolidified crystals that grow from the liquid phase are found to melt at 162–163°C, typical of the more stable polymorph. The infrared spectra of the two polymorphs differ mainly around 3500 cm^{-1} and in the 1500–1700 cm^{-1} region [32]. The authors report [32] that, once the stable (higher-melting) form had been obtained in either of the two laboratories, the metastable (lower-melting) polymorph could no longer be isolated. Most significantly, it was reported that the lower-melting polymorph could be isolated again after discarding all the samples, washing the equipment and laboratory benches, and waiting for 8–12 days. This cleansing procedure had been repeated several times in the laboratories of the above authors, who commented "Obviously, the seeding effect during the formation of the primary crystals (or during the very procedure of determination of the melting point) is exceptionally strong" [32]. After these findings the monograph in the 1973 edition of the *Pharmacopoeia Nordica* was modified, stating that benzocaine picrate has a melting point between 161°C and 164°C and may be formed as a metastable modification with melting point between 129°C and 132°C, which will not in every case be transformed into the higher-melting modification during the determination of the melting point [32].

Another example of the displacement of a metastable polymorph by a stable polymorph is xylitol (the RS or meso form), which is used as a sweetening agent in tablets, syrups, and coatings and as an alternative to sucrose in foods, confectionery, and toiletries [33–35]. Xylitol is also described in a NF monograph [36]. In the early 1940s, two polymorphs of xylitol were described. One of these is a metastable, hygroscopic, monoclinic form, melting at 61–61.5°C [37] and the other a stable orthorhombic form melting at 93–94.5°C [38]. After a sample of the orthorhombic form was introduced into a laboratory in which the monoclinic polymorph had been prepared, the latter "changed in a few days into the high-melting and stable form on exposure to the air of the laboratory" [38]. Later, Kim and Jeffrey determined the crystal structure of the stable orthorhombic polymorph [39]. These authors stated, "Attempts to obtain the lower melting monoclinic form from alcoholic solutions either at room temperature or close to 0°C have hitherto been unsuccessful. We invariably grow the orthorhombic crystals. It is interesting to note that although xylitol was first prepared as a syrup in 1891 there was no report of crystallization until fifty years later, when it was the metastable hygroscopic form that was prepared first. Having now obtained the stable form, it is difficult to recover the metastable crystals . . . The availability of appropriate nuclei in the laboratory is clearly a determining factor, as is well known to carbohydrate chemists [39]."

Since the late 1980s, solid state chemists and pharmaceutical scientists have increasingly recognized the possibility of regulating the processes of nucleation and growth of different polymorphs by careful control of the environmental conditions. One interesting approach is to suppress the growth of a particular crystal form and thereby to promote the growth of the other forms, or at least to present them with a competitive advantage, by addition of "tailor-made" additives or impurities [40]. In this example, certain polypeptides can preferentially induce the crystallization of one of the homochiral crystals of histidine hydrochloride (R- or S-His·HCl·H$_2$O) at 25°C instead of the racemic compound (R, S-His·HCl·2H$_2$O), which is the thermodynamically more stable form below 45°C. In the absence of the additives, the racemic compound crystallizes below 45°C, whereas the homochiral crystals form above 45°C. Other examples of the use of tailor-made additives to direct the crystallization of one crystal form at the expense of another crystal form have been reported, and the number of examples is increasing. In many of these examples the additive preferentially blocks the growth of certain faces of the crystal form that is being suppressed, as in the just-discussed case of histidine hydrochloride [40].

Dunitz and Bernstein [25] pointed out that their examples of disappearing polymorphs involve molecules that can adopt different shapes (i.e., conformational polymorphism). These molecules often possess conformational freedom, or different configurations (epimers, such as α and β sugars), or different arrangements of their parts (e.g., benzocaine picrate) [25]. When present in solution or in the liquid state, the different conformations will exist in a dynamic equilibrium. The most stable conformer in the solution may not necessarily be that present in the thermodynamically most stable crystal form. Dunitz and Bernstein, supporting the scheme presented in Fig. 14 by Etter [22], argue that the rate of formation of nuclei of a stable polymorph could be significantly reduced by a low concentration of the required conformer, while another conformer could be incorporated into the nuclei of a less stable polymorph, which then grows rapidly leading to a metastable crystal [25]. Of course, a polymorph that is metastable at or above ambient temperature might be obtained as the thermodynamically stable form at a lower temperature below the transition point. In preformulation studies of pharmaceutical compounds it is usually, if not always, important to resolve these kinetic and thermodynamic issues. Dunitz and Bernstein [25], echoing Jacewicz and Nayler [30], state that it should always be possible to prepare a previously known polymorph again, although the repreparation will require the appropriate experimental conditions, which might be found quickly or only after some effort. This statement probably expresses the prevailing view and emphasizes the importance of initiating and carrying out a comprehensive screening procedure for polymorphic forms appropriate to the drug substance under consideration [41].

REFERENCES

1. J. Haleblian, and W. McCrone, *J. Pharm. Sci.*, *58*, 911–929 (1969).
2. J. K. Haleblian, *J. Pharm. Sci.*, *64*, 1269–1288 (1975).
3. T. L. Threlfall, *Analyst*, *120*, 2435–2460 (1995).
4. M. Haisa, S. Kashino, R. Kawai, and H. Maeda, *Acta Cryst.*, *B32*, 1283–1285 (1976).
5. M. Haisa, S. Kashino, and H. Maeda, *Acta Cryst.*, *B30*, 2510–2512 (1974).
6. P. Di Martino, P. Conflant, M. Drache, J.-P. Huvenne, and A.-M. Guyot-Hermann, *J. Therm. Anal.*, *48*, 447–458 (1997).
7. G. Nichols, and C. Frampton, Poster No. 96-83, presented at the Annual Meeting of the British Crystallographic Association, Leeds, England, April 1997.
8. M. Azibi, M. Draguet-Brughmans, R. Bouche, B. Tinant, G. Germain, J. DeClercq, and M. Van Meersshe, *J. Pharm. Sci.*, *72*, 232–235 (1983).
9. D. Giron, *Thermochim. Acta*, *248*, 1–59 (1995).

10. B. C. Hancock, and G. Zografl, *J. Pharm. Sci.*, *86*, 1–12 (1997).

11. P. W. Atkins, *Physical Chemistry*, Oxford Univ. Press, Oxford, UK, 1978, pp. 194, 563–566, 658–660.

12. J. D. Wright, *Molecular Crystals*, 2d edition, Cambridge Univ. Press, Cambridge, UK, 1995.

13. A. R. Verma, and P. Krishna, *Polymorphism and Polytypism in Crystals*, John Wiley, New York, 1966, pp. 15–30.

14. A. Burger, and R. Ramberger, *Mikrochim. Acta [Wien]*, *II*, 259–271, 273–316 (1979).

15. L. Yu, *J. Pharm. Sci.*, *84*, 966–974 (1995).

16. W. Ostwald, *Lehrbuch der Allgemeinen Chemie, 2*, W. Engelmann, Leipzig, Germany, 1896, p. 444.

17. W. Ostwald, *Z. Physik. Chem.*, *22*, 289–330 (1897).

18. W. Ostwald, *Grundriss der Allgemeinen Chemie*, W. Engelmann, Leipzig, Germany, 1899.

19. J. W. Mullin. *Crystallization*, 3d ed. Butterworth Heinemann, London, UK, 1993, pp. 172–201.

20. D. J. W. Grant, and H. G. Brittain, "Solubility of Pharmaceutical Solids", Chapter 11 in *Physical Characterization of Pharmaceutical Solids* (H. G. Brittain, ed.), Marcel Dekker, New York, 1995, pp. 321–386.

21. R. Boistelle, *Advances in Nephrology*, *15*, 173–217 (1986).

22. M. C. Etter, "Hydrogen Bonding in Organic Solids," in Polymorphs and Solvates of Drugs, S. R. Byrn, organizer, Short Course, Purdue University, West Lafayette, IN, 1990.

23. D. E. Bugay, and A. C. Williams, "Vibrational Spectroscopy," Chapter 3 in *Physical Characterization of Pharmaceutical Solids* (H. G. Brittain, ed.), Marcel Dekker, New York, 1995, pp. 59–91.

24. S. R. Byrn, *Solid State Chemistry of Drugs*, Academic Press, New York, 1982.

25. J. D. Dunitz, and J. Bernstein, *Acc. Chem. Res.*, *28*, 193–200 (1995).

26. Q.-C. Yang, M. F. Richardson, and J. D. Dunitz, *Acta Cryst.*, *B45*, 312–323 (1989).

27. M. F. Richardson, Q.-C. Yang, E. Novotny-Bregger, and J. D. Dunitz, *Acta Cryst.*, *B46*, 653–660 (1990).

28. G. D. Woodward, and W. C. McCrone, *J. Appl. Cryst.*, **8**, 342 (1975).

29. J. Webb, and B. Anderson, *J. Chem. Ed.*, *55*, 644–645 (1978).

30. V. W. Jacewicz, and J. H. C. Nayler, *J. Appl. Cryst.*, *12*, 396–397 (1979).

31. P. G. Stecher, M. Windholz, D. S. Leahy, D. M. Bolton, and L. G. Eaton, *The Merck Index*, 8th edition, Merck and Co., Rahway, NJ, 1968, p. 431.

32. T. K. Nielsen, and L. Borka, *Acta Pharm. Suecica*, *9*, 503–505 (1972).

33. H. E. C. Worthington and P. M. Olinger in *Handbook of Pharmaceutical Excipients*, 2nd edition (A. Wade and P. J. Weller, eds.) American Pharmaceutical Association, Washington, DC; The Pharmaceutical Press, London, UK, 1994, pp. 564–567.

34. J. E. F. Reynolds, K. Parfitt, A. V. Parsons, and S. C. Sweetman, *Martindale: The Extra Pharmacopeia*, 31st edition, The Pharmaceutical Press, London, UK, 1996, p. 1395.

35. S. Budavari, M. J. O'Neil, A. Smith, P. E. Heckelman, and J. F. Kinneary, *The Merck Index*, 12th edition, Merck and Co., Whitehouse Station, NJ, 1996, No. 10218, p. 1723–1724.

36. Xylitol, *USP23/NF18*, United States Pharmacopeial Convention, Inc., Rockville, MD, 1995, pp. 2319–2320.

37. M. L. Wolfrom, and E. J. Kohn, *J. Am. Chem. Soc.*, *64*, 1739 (1942).

38. J. F. Carson, S. W. Waisbrot, and F. T. Jones, *J. Am. Chem. Soc.*, *65*, 1777–1778 (1943).

39. H. S. Kim, and G. A. Jeffrey, *Acta Cryst.*, *B25*, 2607–2613 (1969).

40. I. Weissbuch, D. Zbaida, L. Addidi, L. Leiserowitz, and M. Lahav, *J. Am. Chem. Soc.*, *109*, 1869–1871 (1987).

41. S. Byrn, R. Pfeiffer, M. Ganey, C. Hoiberg, and G. Poochikian, *Pharm. Res.*, *12*, 945–954 (1995).

2

Application of the Phase Rule to the Characterization of Polymorphic Systems

Harry G. Brittain

Discovery Laboratories, Inc.
Milford, New Jersey

I. INTRODUCTION TO THE PHASE RULE

When considering questions of equilibria, one ordinarily thinks of chemical reactions taking place in a suitable medium. However, it is well known that a variety of physical equilibria are also possible, and thermodynamics is a powerful tool for the characterization of such equilibria. The existence of alternate crystal structures for a given compound can be successfully examined from an equilibrium viewpoint, and this approach is especially useful when establishing the relative stability of such polymorphic systems and their possible ability to interconvert.

Consider the situation presented by elemental sulfur, which can be obtained in either a rhombic or a monoclinic crystalline state. Each of these melts at a different temperature and is stable under certain well-defined environmental conditions. What are the conditions under which these two forms can equilibrate with liquid sulfur (either singly or together), and what are the conditions under which the two equilibrate in the absence of a liquid phase? These questions can be answered with the aid of chemical thermodynamics, the modern practice of which can be considered as beginning with publication of the seminal papers of J. Willard Gibbs [1].

Almost immediately after the law of conservation of mass was established, Gibbs showed that all cases of equilibria could be categorized into general class types. His work was perfectly general in that it was free from hypothetical assumptions, and it immediately served to show how different types of chemical and physical changes actually could be explained in a similar fashion. Gibbs began with a system that needed only three independent variables for its complete specification, these being temperature, pressure, and the concentration of species in the system. From these considerations, he defined a general theorem known as the phase rule, where the conditions of equilibrium could be specified according to the composition of that system.

The following discussion of the phase rule, and its application to systems of polymorphic interest, has primarily been distilled from the several classic accounts published in the first half of this century [2–8]. It may be noted in passing that one of the most serious disagreements in the history of physical chemistry was between the proponents of computational thermodynamics and those interested in the more qualitative phase rule. Ultimately the school of exact calculations prevailed [9], and consequently their view has dominated the modern science of physical chemistry. Nevertheless, it is still true that one may obtain a great deal of understanding about phase transformation equilibria through considerations of the phase rule, even if its approach is considered out of vogue at the present time.

A. Phases

A heterogeneous system is composed of various distinct portions, each of which is in itself homogenous in composition, but which are separated from each other by distinct boundary surfaces. These physically distinct and mechanically separable domains are termed phases. A single phase must be chemically and physically homogeneous and may consist of single chemical substance or a mixture of substances.

Theoretically, an infinite number of solid or liquid phases can exist side by side, but there can never be more than one vapor phase. This is because all gases are completely miscible with each other in all proportions and will therefore never undergo a spontaneous separation into component materials. It is important to remember, however, that equilibrium is independent of the relative amounts of the phases present in a system. For instance, once equilibrium is reached the vapor pressure of a liquid does not depend on either the volume of the liquid or vapor phases.

In a discussion of polymorphic systems, one would encounter the vapor and liquid phases of the compound under study as separate phases. In addition, each polymorph would constitute a separate phase. Once the general rule is deduced and stated, the phase rule can be used to deduce the conditions under which these forms can be in an equilibrium condition.

B. Components

A component is defined as a species whose concentration can undergo independent variation in the different phases. Another way to state this definition is to say that a component is a constituent that takes part in the equilibrium processes. For instance, in the phase diagram of pure water, there is only one component (water), although this compound is formed by the chemical reaction of hydrogen and oxygen. Since hydrogen and oxygen are combined in definite proportions, their concentration cannot be varied independently, so they cannot be considered as being separate components.

For the specific instance of polymorphic systems, the substance itself will ordinarily be the only component present. The situation complicates for solvates or hydrates since the lattice solvent will comprise a second component. Hydrate/solvate systems cannot be systems of one component, since the different phases will not have the same composition. The general rule is that the number of components present in an equilibrium situation are to be chosen as the smallest number of the species necessary to express the concentration of each phase participating in the equilibrium.

C. Degrees of Freedom

The number of degrees of freedom of a system is defined as the number of variable factors that must be arbitrarily fixed to define completely the condition of the system at equilibrium. The variables normally specified for phase equilibria are temperature, pressure, or component concentrations. One may also speak of the variance of a system, which is defined by the number of degrees of freedom required to specify the system.

For instance, consider the situation of a substance forming an ideal gas in its vapor phase. The equation of state for an ideal gas is given by

$$PV = nRT \tag{1}$$

where P is the pressure, V is the volume, n is the number of moles present, T is the absolute temperature, and R is the gas constant. For a given amount of gas, if two out of the three independent parameters are specified, then the third is determined. The system is then said to be bivariant, or exhibits two degrees of freedom. If the gaseous substance is then brought into a state of equilibrium with its condensed phase, then empirically one finds that the condition of equilibrium can be specified by only one variable. The system exhibits only one degree of freedom and is now termed univariant. If this system is cooled down until the solid phase forms, one empirically finds that this equilibrium condition can only be attained if all independent parameters are specified. This latter system exhibits no degrees of freedom and is said to be invariant.

D. The Phase Rule

If one considers a substance capable of existing in two different phases, then equilibrium between these can only occur when the intensity factor of chemical energy is the same in each phase. Gibbs termed this chemical intensity factor the chemical potential and defined equilibrium as the situation where the chemical potential of each component in a phase is the same as the chemical potential of that component in the other phase.

Consider the system which consists of C components present in P phases. In order to specify the composition of each phase, it is necessary to know the concentrations of $(C - 1)$ components in each of the phases. Another way to state this is that each phase possesses $(C - 1)$ variables. Besides the concentration terms, there are two other variables (temperature and pressure), so that altogether the number of variables existing in a system of C components in P phases is given by

$$\text{variables} = P(C - 1) + 2 \text{ variables} \tag{2}$$

In order to define the system completely one requires as many equations as variables. If for some reason there are fewer equations than variables, then values must be assigned to the variables until the number of unknown variables equals the number of equations. Alternatively, one must assign values to undefined variables or else the system must remain unspecified. The number of these variables that must be defined or assigned to specify a system is the variability or the degree of freedom of the system.

The equations by which the system is to be defined are obtained from the relationship between the potential of a component and its phase composition, temperature, and pressure. If one chooses as a standard state one of the phases in which all components are found, then the chemical potential of any component in another phase must equal the chemical potential of that component in the standard state. It follows that for each phase in equilibrium with the standard phase, there will be a definite equation of state for each component in that phase. One concludes that if there are P phases, then each component will be specified by $(P - 1)$ equations. Then for C components, we deduce that the maximum number of available equations is given by

$$\text{equations} = C(P - 1) \tag{3}$$

The variance (degrees of freedom) in a system is given by the difference between the number of variables and the number of equations available to specify these. Denoting the number of degrees of freedom as F, this can be stated as

$$F = \text{variables} - \text{equations} \tag{4}$$

Substituting Eqs. (2) and (3) into Eq. (4), and simplifying, yields

$$F = C + 2 - P \tag{5}$$

which is often rearranged to yield the popular statement of the phase rule:

$$P + F = C + 2 \tag{6}$$

One can immediately deduce from Eq. (5) that for a given number of components, an increase in the number of phases must lead to a concomitant decrease in the number of degrees of freedom. Another way to state this is that with an increase in the number of phases at equilibrium, the condition of the system must become more defined and less variable. Thus for polymorphic systems where one can encounter additional solid-state phases, the constraints imposed by the phase rule can be exploited to obtain a greater understanding of the equilibria involved.

II. SYSTEMS OF ONE COMPONENT

Setting aside any consideration of solvate species or considerations of chemical reaction, systems of polymorphic interest consist of only one component. The complete phase diagram of a polymorphic system would provide the boundary conditions for the vapor state, the liquid phase, and for each and every true polymorph possible. From the phase rule, it is concluded that the maximum amount of variance (two degrees of freedom) is only possible when the component is present in a single phase. All systems of one component can therefore be perfectly defined by assigning values to a maximum of two variable factors. However, this bivariant system is not of interest to our discussion.

When a single component is in equilibrium between two phases, the phase rule predicts that it must be a univariant system exhibiting only one degree of freedom. It is worthwhile to consider several univariant possibilities, since the most complicated phase diagram of a polymorphic system can be broken down into its component univariant systems. The phase rule applies equally to all of these systems, and all need to be understood for the entire phase diagram to be most useful.

A. Characteristics of Univariant Systems

When a single component exists in a state of equilibrium between two phases, the system is characterized by only one degree of freedom. The types of observable equilibria can be of the liquid/vapor, solid/vapor, solid/liquid, and (specifically for components that exhibit polymorphism) solid/solid types. We will consider the important features of each in turn.

1. Liquid/Vapor Equilibria

A volatile substance in equilibrium with its vapor constitutes a univariant system, which will be defined if one of the variables (pressure or temperature) is fixed. The implications of this deduction are that the vapor pressure of the substance will have a definite value at a given temperature. Alternatively, if a certain vapor pressure is maintained, then equilibrium between the liquid and vapor phase can only exist at a single definite temperature. Each pressure therefore corresponds to a definite temperature, and a plot of pressure against temperature will yield a continuous line defining the position of equilibrium. Relations of this type define the *vaporization curve* and are ordinarily plotted to illustrate the trends in vapor pressure as a function of system temperature. It is generally found that vaporization curves exhibit the same general shape, being upwardly convex in the usual pressure–temperature diagram.

As an example, consider the system formed by liquid water in equilibrium with its own vapor. The pressure–temperature diagram for this system has been constructed over the range of 1–99°C [10] and is shown in Fig. 1. The characteristics of a univariant system (one degree of freedom) are evident in that for each definite temperature value, water exhibits a fixed and definite pressure value.

In a closed vessel, the volume becomes fixed. An input of heat (i.e., an increase in temperature) into a system consisting of liquid and vapor in equilibrium must result in an increase in the vapor pressure. It must happen that with the increase of pressure, the density of the vapor increases, while with the corresponding increase in temperature, the density of the liquid decreases. At some temperature value, the densities of the liquid and vapor will become identical, and at that point the heterogeneous system becomes homogeneous. At this critical point (defined by a critical temperature and a critical pressure), the entire system passes into one homogeneous phase. The vaporization curve must terminate at the critical point, unless there are accessible metasta-

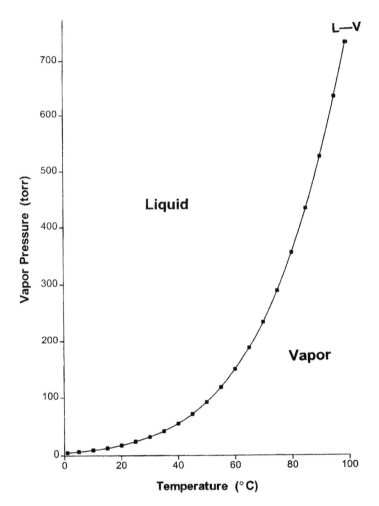

Fig. 1 Vapor pressure of water as a function of temperature. (The data were plotted from published values; from Ref. 10.)

ble states that merely delay the inevitable. As is evident in Fig. 1, the vapor pressure of a liquid approaches that of the ambient atmospheric pressure as the boiling point is reached.

It is useful to consider the principle of Le Chatelier, which states that if an equilibrium system is stressed by a force that shifts the position of equilibrium, then a reaction to the stress that opposes the force will take place. Consider a liquid/vapor system that is sufficiently isolated from its surroundings so that heat transfer is prevented (adiabatic process). An increase in the volume of this system results in a decrease in the pressure of the system, causing liquid to pass into the vapor state. This process requires the input of heat, but since none is available from the surroundings, it follows that the temperature of the system must fall.

While qualitative changes in the position of liquid/vapor equilibrium can be predicted by Le Chatelier's principle, the quantitative specification of the system is given by the Clausius–Clapeyron equation:

$$\frac{dP}{dT} = \frac{q}{T(v_2 - v_1)} \tag{7}$$

where q is the quantity of heat absorbed during the transformation of one phase to the other, v_2 and v_1 are the specific volumes of the two phases, and T is the absolute temperature at which the change occurs. Integration of Eq. (7) leads to useful relations that permit the calculation of individual points along the vaporization curve.

2. Solid/Vapor Equilibria

A solid substance in equilibrium with its vapor phase will exhibit a well-defined vapor pressure for a given temperature, which will be independent of the relative amounts of solid and vapor present. The curve representing the solid/vapor equilibrium conditions is termed a *sublimation curve*, which generally takes a form similar to that of a vaporization curve. Although the sublimation pressure of a solid is often exceedingly small, for many substances it can be considerable.

One example of a solid that exhibits significant vapor pressure is camphor, for which a portion of its sublimation curve is shown in Fig. 2. This compound exhibits the classic pressure–temperature profile [11], finally attaining a vapor pressure of 422.5 torr at its melting point (179.5°C). When heated above the fusion temperature, only a short vaporization curve is possible since the boiling point of camphor is reached at 207.4°C.

The sublimation curve of every substance has an upper limit at

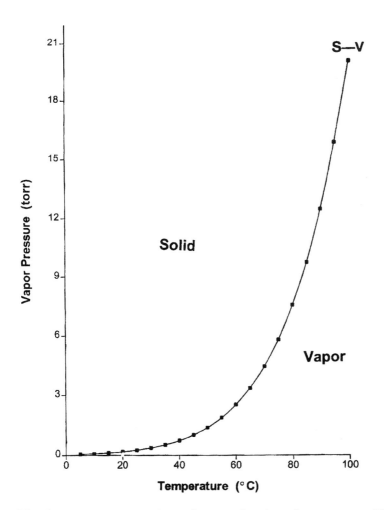

Fig. 2 Vapor pressure of camphor as a function of temperature. (The data were plotted from published values; from Ref. 11.)

the melting point and a theoretical lower limit of absolute zero. However, since low-temperature polymorphic transitions can be encountered, there can be considerable complexity in the phase diagrams. One need only consider the example of water, of which at least seven crystalline forms are known.

If the sublimation pressure of a solid exceeds that of the atmospheric pressure at any temperature below its melting point, then the solid will pass directly into the vapor state (sublime) without melting when stored in an open vessel at that temperature. In such instances, melting of the solid can only take place at pressures exceeding ambient. Carbon dioxide is one of the best known materials that exhibits sublimation. At the usual room temperature conditions, solid "dry ice" sublimes easily. Liquid carbon dioxide can only be maintained between its critical point (temperature of $+31.0°C$ and pressure of 75.28 atm) and its triple point (temperature of $-56.6°C$ and pressure of 4.97 atm) [12].

The direction of changes in sublimation pressure with temperature can be qualitatively predicted using Le Chatelier's principle and quantitatively calculated by means of the Clausius–Clapeyron equation.

3.   Solid/Liquid Equilibria

When a crystalline solid is heated to the temperature at which it melts and passes into the liquid state, the solid/liquid system is univariant. Consequently, for a given pressure value, there will be a definite temperature (independent of the quantities of the two phases present) at which the equilibrium can exist. As with any univariant system, a curve representing the equilibrium temperature and pressure data can be plotted, and this is termed the melting point curve or *fusion curve*. Since both phases in a solid/liquid equilibrium are condensed (and difficult to compress), the effect of pressure on the melting point of a solid is relatively minor unless the applied pressures are quite large.

Using Le Chatelier's principle, one can qualitatively predict the effect of pressure on an equilibrium melting point. The increase in pressure results in a decrease in the volume of the system. For most materials, the specific volume of the liquid phase is less than that of the solid phase, so that an increase in pressure would have the effect of shifting the equilibria to favor the solid phase. This shift will have the observable effect of raising the melting point. For those unusual systems where the specific volume of the liquid exceeds that of the solid phase, then the melting point will be decreased by an increase in pressure.

An example of a fusion curve is provided in Fig. 3, which uses

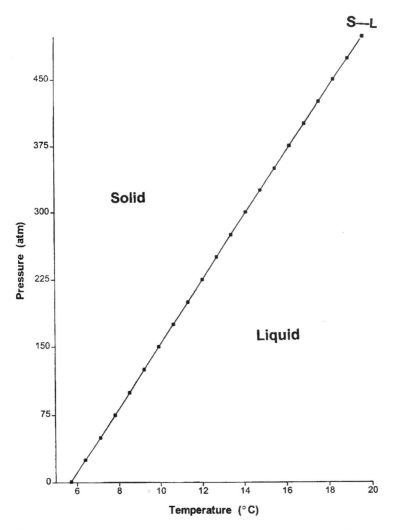

Fig. 3 Effect of pressure on the melting point of benzene. (The data were plotted from published values; from Ref. 13.)

benzene as the example [13]. It can be seen that to double the melting point requires an increase in pressure from 1 atm to approximately 250 atm. The fusion curve of Fig. 3 is fairly typical in that in the absence of any pressure-induced polymorphic transformations, the curve is essentially a straight line.

The quantitative effect of pressure on the melting point can be calculated using the inverse of the Clausius–Clapeyron equation:

$$\frac{dT}{dP} = \frac{T(v_2 - v_1)}{q} \tag{8}$$

However, the magnitude of such shifts of the melting point with pressure are relatively minor, since the differences in specific volumes between the liquid and solid phases are ordinarily not great.

According to Eq. (8), for a fusion curve to exhibit a positive slope (like the one in Fig. 3), the specific volume of the liquid must be greater than the specific volume of the solid. In such systems, the substance would expand upon melting. Other systems are known where the specific volume of the liquid is less than the specific volume of the solid, so that these substances contract upon fusion. The classic example of the latter behavior is that of ice, which is known to contract upon melting. For example, while the melting point of water is 0°C at a pressure of 1 atm, the melting point decreases to −9.0°C at a pressure of 9870 atm [14].

B. The Triple Point

When one component is present in three phases at equilibrium, the phase rule states that the system is invariant and possesses no degrees of freedom. This implies that such a system at equilibrium can only exist at one definite temperature and one definite pressure, which is termed the triple point. For instance, the solid/liquid/vapor triple point of water is found at a temperature of 273.16K and a pressure of 4.58 torr.

Although the solid/liquid/vapor triple point is the most commonly encountered, the existence of other solid phases yields additional triple points. The number of triple points possible to a polymorphic system increases very rapidly with the number of potential solid phases.

It has been shown that the number of triple points in a one-component system is given by [15]

$$\#TP = \frac{P(P-1)(P-2)}{6} \tag{9}$$

Thus for a system capable of existing in two solid-state polymorphs, a total of four phases would be possible, which would then imply that a total of four triple points would be theoretically accessible. Denoting the liquid phase as L, the vapor phase as V, and the two solid phases as S_1 and S_2, the triple points correspond to

$$\begin{align*}
&S_1-L-V \\
&S_2-L-V \\
&S_1-S_2-L \\
&S_1-S_2-V
\end{align*} \tag{10}$$

The S_1-S_2-V point is the *transition point* of the substance, the S_1-L-V and S_2-L-V points are *melting points*, and the S_1-S_2-L point is a *condensed transition point* (7). Whether all these points can be experimentally attained depends on the empirical details of the system itself.

It has already been established that each S_i-V curve ends at the melting point. At this point, liquid and solid are each in equilibrium with vapor at the same pressure, so they must also be in equilibrium with each other. It follows that the particular value of temperature and vapor pressure must lie on each S_i-V curve(s) as well as on the L–V curve. Applying the Clausius–Clapeyron equation to both transitions, one concludes that a discontinuity must take place on passing from the S_i-V curve(s) to the L–V curve. Since the change in specific volume for each transition is essentially the same, and since the heat required to transform solid to vapor must necessarily exceed the heat required to transform liquid to vapor, it must follow that the value of dP/dT for the solid/vapor transition must exceed that for the liquid/vapor transition. Therefore the S_i-V curve(s) must increase more rapidly than does the L–V curve, with the curves intersecting at a triple point. Using a similar argument, it can be deduced that each S_i-L curve must also pass through a triple point. One therefore deduces that the triple point is a point of intersection of three univariant curves. These relationships are illustrated in Fig. 4, which provides the phase behavior for a typical substance for which the specific volume of the liquid exceeds that of the solid.

The triple point differs from the ordinary melting point, since the latter is ordinarily determined at atmospheric pressure. At the triple point, the solid and the liquid are in a state of equilibrium under a pressure that equals their vapor pressure.

Interesting conclusions can be reached if one considers the consequences of applying Le Chatelier's principle to a system in equilibrium at its triple point. Stressing the system through a change in either pressure or temperature must result in an opposing effect that restores the equilibrium. However, since the system is invariant, the position of equilibrium cannot be shifted. Therefore as long as the system remains in equilibrium at the triple point, the only changes that can take place

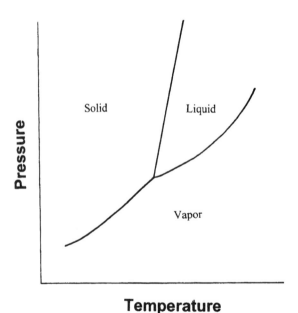

Temperature

Fig. 4 Phase diagram of a hypothetical substance for which the specific volume of the liquid exceeds that of the solid. The triple point is defined by the intersection of the three univariant curves describing the solid–vapor, liquid–vapor, and solid–liquid equilibria.

are changes in the relative amounts of the phases present. For the specific instance of polymorphic solids, this deduction must apply to all other triple points of the system. It should be emphasized that at the triple point, all three phases must be involved in the phase transformations.

III. SOLID-STATE POLYMORPHISM AND THE PHASE RULE

That a given solid can exist in more than one crystalline form was first established by Mitscherlich for the specific instance of sodium phosphate [16]. The phenomenon has been shown to be widespread for both inorganic and organic systems, with various compilations of polymorphic systems having been published for compounds of pharmaceutical interest [17–19]. For the purpose of the present discussion, one must remember that the structural phase differences exist only in the solid state, and that the liquid and vapor phases of all polymorphs of a given component must necessarily be identical.

According to the definition of Section I.A, each solid-state polymorphic form constitutes a separate phase of the component. The phase rule can be used to predict the conditions under which each form can coexist, either along or in the presence of the liquid or vapor phases. One immediate deduction is that since no stable equilibrium can exist when four phases are simultaneously present, two polymorphic forms cannot be in equilibrium with each other and be in equilibrium with both their solid and vapor phases. When the two crystalline forms (denoted S_1 and S_2) are in equilibrium with each other, then the two triple points (S_1–S_2–V and S_1–S_2–L) become exceedingly important.

A. The Transition Point

The S_1–S_2–V triple point is obtained as the intersection of the two univariant sublimation curves, S_1–V and S_2–V. Below this triple point only one of the solid phases can exist in stable equilibrium with the vapor (i.e., being the stable solid phase), and above the triple point only the other phase can be stable. The S_1–S_2–V triple point therefore provides the pressure and temperature conditions at which the relative stability of the two phases inverts and hence is referred to as the *transition point*.

The S_1–S_2–V triple point is also the point of intersection for the S_1–S_2 curve, which of course delineates the conditions of equilibrium for the two polymorphic forms with each other. Since the S_1–S_2 curve defines a univariant system, it follows that the temperature at which the two phases can be in equilibrium will depend on the pressure. An ordinary transition point is often defined as the temperature of equal phase stability at atmospheric pressure, but this point in the phase diagram must be distinguished from the S_1–S_2–V triple point. The ordinary transition point bears the same relationship to the S_1–S_2–V triple point that the ordinary melting point bears to the S–L–V triple point.

The transition point, like the melting point, is affected by pressure. Depending on the relative values of the specific volumes of the two polymorphs, an increase in pressure can either raise or lower the transition temperature. However, since this difference in specific volumes is ordinarily very small, the Clausius–Clapeyron equation predicts that the magnitude of dT/dP will not be great.

To illustrate the phase behavior of a substance at the S_1–S_2–V triple point, we will return to the example of camphor, whose sublimation curve was shown in Fig. 2. The pressure dependence of the S_1–S_2 (Form 1/Form 2) phase transformation is known [20], and the phase diagram resulting from the addition of these data to the sublimation curve is shown in Fig. 5. Since the data used to construct the S_1–S_2 curve were obtained at pressure values ranging up to 2000 atm, the location of the triple point must be deduced from an extrapolation of the S_1–S_2 curve to its intersection with the S–V curve. One finds that the triple point is located at a temperature of 87°C and a pressure of 0.017 atm (13 torr). This finding would imply that the S–V sublimation curve reported for camphor actually represents the composite equilibrium of the two phases with their common vapor phase.

B. The Condensed Transition Point

For the sake of this argument, let us assume that phase S_1 is more stable than is phase S_2 at ordinary ambient conditions. If one increases the

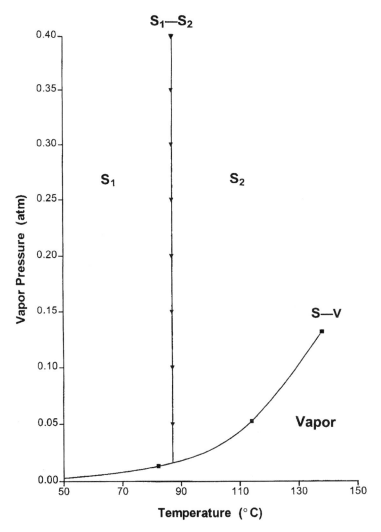

Fig. 5 Location of the Form 1/Form 2/vapor triple point in the phase diagram of camphor. The triple point is deduced from the extrapolated intersection of the S_1–S_2 transition curve with that of the S–V sublimation curve. (The data were plotted from published values; from Refs. 11, 20.)

pressure on the system, the position of equilibrium will be displaced along the S_1–S_2 transition curve, which will have the effect of raising the transition temperature. At some point, the univariant S_1–S_2 curve will intersect with the univariant S_1–L fusion curve, producing a new triple point S_1–S_2–L, which is denoted the *condensed transition point*. The S_1 phase ceases to exist in a stable condition above this triple point, and the S_2 phase will be the only stable solid phase possible.

When observable, the S_1–S_2–L triple point is encountered at extremely high pressures. For this reason, workers rarely determine the position of this triple point in a phase diagram but focus instead on the S_1–S_2–V triple point for discussions of relative phase stability.

To illustrate a determination of an S_1–S_2–L triple point, we will return to the example of benzene, for which the low-pressure portion of the S–L fusion curve was shown in Fig. 3. When the full range of pressure–temperature melting point data is plotted [13], one finds that the pressure-induced volume differential causes a definite nonlinearity to appear in the data. Adding the S_1–S_2 transition data [21] generates the phase diagram of benzene, which is shown in Fig. 6, where the triple point is obtained as the intersection of the S–L fusion curve and the S_1–S_2 transition curve. The S_1–S_2–L triple point is deduced to exist at a temperature of 215°C and a pressure of 11,500 atm. Such pressures are only attainable through the use of sophisticated systems, which explains why the S_1–S_2–L triple point is only rarely determined during the course of ordinary investigations.

C. Enantiotropy and Monotropy

The S_1–S_2–V triple point is one at which the reversible transformation of the crystalline polymorphs can take place. If both S_1 and S_2 are capable of existing in stable equilibrium with their vapor phase, then the relation is termed *enantiotropy*, and the two polymorphs are said to bear an enantiotropic relationship to each other. For such systems, the S_1–S_2–V triple point will be a stable and attainable value on the pressure–temperature phase diagram. A phase diagram of a hypothetical enantiotropic system is shown in Fig. 7. Each of the two polymorphs exhibits a S–V sublimation curve, and they cross at the same temperature at which they meet the S_1–S_2 transition curve. The S_2–V curve

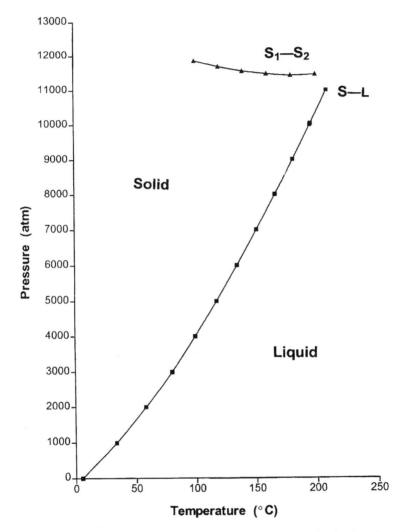

Fig. 6 Location of the Form 1/Form 2/liquid triple point in the phase diagram of benzene. The triple point is deduced from the extrapolated intersection of the S_1–S_2 transition curve with that of the S–L fusion curve. (The data were plotted from published values; from Refs. 13, 21.)

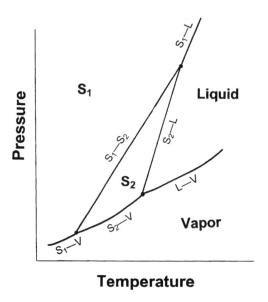

Fig. 7 Idealized phase diagram of a substance whose two polymorphs exhibit an enantiotropic relationship.

crosses the stable L–V fusion curve at an attainable temperature that
is the melting point of the S_2 phase. The S_1–S_2 transition and the S_2–
L fusion curves eventually intersect with the S_1–L fusion curve, forming the condensed transition point.

It should be noted that the ordinary transition point of enantiotropic systems (which is measured at atmospheric pressure) will be
less than the melting point of either solid phase. Each polymorph will
therefore be characterized by a definite range of conditions under which
it will be the most stable phase, and each form is capable of undergoing
a reversible transformation into the other.

The melting behavior of an enantiotropic system is often interesting to observe. If one begins with the polymorph that is less stable at
room temperature and heats the solid up to its melting point, the S_2–
L melting phase transformation is first observed. As the temperature
is raised further, the melt is observed to resolidify because the liquid
is metastable with respect to the most stable polymorph, S_1. Continued
heating will then result in the S_1–L phase transformation. If one allows

this latter melt to resolidify and cool back to room temperature, only the S_1–L melting transition will be observed.

Other systems exist where the second polymorph (S_2) has no region of stability anywhere on a pressure–temperature diagram. This type of behavior is termed *monotropy*, and such polymorphs bear a monotropic relationship to each other. The melting point of the metastable S_2 polymorph will invariably be less (in terms of both pressure and temperature) than the melting point of the stable form (S_1). This in turn has the effect of causing the S_1–S_2–V triple point to exceed the melting point of the stable S_1 phase. Monotropy therefore differs from enantiotropy in that the melting points of an enantiotropic pair are higher than the S_1–S_2–V triple point, while for monotropic systems one or both of the melting points is less than the S_1–S_2–V triple point.

The phase diagram of a hypothetical monotropic system is illustrated in Fig. 8. The S_1–S_2–V triple point (transition point) point is clearly virtual in that fusion of all solid phases takes place before the thermodynamic point of phase stability can be attained. The phase diagram indicates that only one of the polymorphs can be stable at all temperatures up to the melting point, and the other polymorph must

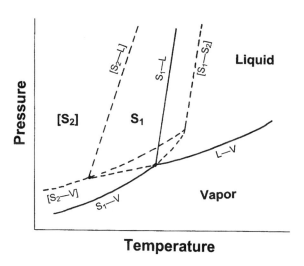

Fig. 8 Idealized phase diagram of a substance whose two polymorphs exhibit a monotropic relationship.

be considered as being a metastable phase. For such systems, there is no transition point attainable at atmospheric pressure, and the transformation of polymorphs can take place irreversibly in one direction only.

Very complicated phase diagrams can arise when substances can exist in more than two crystalline polymorphs. In certain cases, some of the forms may be enantiotropic to each other, and monotropic to yet others. For instance, of the eight polymorphs of elemental sulfur, only the monoclinic and rhombic modifications exhibit enantiotropy and the possibility of reversible interconversion. All of the other forms are monotropic with respect to the monoclinic and rhombic forms and remain as metastable phases up to the melting point.

IV. KINETICALLY IMPAIRED EQUILIBRIA

Using the computational tools of quantitative thermodynamics, one can predict the course of an equilibrium process and determine what will be the favored product. Unfortunately, classical thermodynamics has nothing to say about the velocity of reactions, so a short discussion as to the possible kinetics associated with phase transformation reactions is appropriate.

A. Suspended Phase Transformations

It is well established that certain phase transformations, predicted to be spontaneous on the basis of favorable thermodynamics, do not take place as anticipated. For instance, the diamond phase of carbon is certainly less stable than the graphite phase, but under ordinary conditions (i.e., in a gemstone setting) one does not observe any evidence for phase transformation. The diamond polymorph of carbon is metastable with respect to the graphite phase of carbon, but the phase interconversion can only take place if appropriate energy is added to the system.

Fahrenheit found that pure liquid water, free from suspended particles, could be cooled down to a temperature of $-9.4°C$ without formation of a solid ice phase [22]. If the temperature of the supercooled water was decreased below $-9.4°C$, solidification was observed to take place spontaneously. However, if a crystal of solid ice was added to supercooled water whose temperature was between 0 and −9.4°C, crystallization was found to take place immediately. Fixing the system pressure as that of the atmosphere, one can define the metastable region of stability for supercooled water as 0 to −9.4°C. Supercooled water is unstable at temperatures less than −9.4°C.

Suspended phase transformations are those phase conversions that are predicted to take place at a defined S_1–S_2–V triple point but do not, owing to some nonideality in the system. One can immediately see that only through the occurrence of a suspended transformation could a metastable polymorph be obtained in the first place. In the case of two solids, slow conversion kinetics can permit the transition point to be exceeded when moving in either direction along the S_1–S_2 transition curve, permitting the isolation of the otherwise unobtainable metastable phase.

One of the best known examples of suspended transformation is found with the polymorphs formed by quartz [23]. The three principal polymorphic forms are quartz, tridymite, and cristobalite, which are enantiotropically related to each other. The ordinary transition point for the quartz/tridymite transition is 870°C, while the ordinary transition point for the tridymite/cristobalite transition is 1470°C. The melting point of cristobalite is at 1705°C, which exceeds all of the solid phase transition points. However, the phase transformations of these forms are extremely sluggish, and consequently each mineral form can be found in nature existing in a metastable form.

Ordinarily, the rate-determining step during phase conversion is the formation of nuclei of the new phase. If suitable nuclei cannot be formed at the conditions of study, then the phase transformation is effectively suspended until the nuclei either form spontaneously or are added by the experimenter. Synthetic chemists have long used "seed" crystals of their desired phase to obtain a sufficient crop of that material and to suppress the formation of unwanted by-products. This procedure is especially important during the resolution of enantiomers and diastereomers by direct crystallization.

For example, the inclusion of seed crystals of chloramphenicol palmitate Form A to a mass of Form B was found to lead to accelerated phase transformation during a simple grinding process [24]. The same type of grinding-induced conversion was obtained when seed crystals of form B were added to bulk Form C prior to milling. In this study, the conversion kinetics were best fitted to a two-dimensional nuclear growth equation, but the parameters in the fitting were found to depend drastically on the quantity of seeds present in the bulk material. The practical import of this study was that Form A was the least desirable from a bioavailability viewpoint, and that milling of phase-impure chloramphenicol palmitate could yield problems with drug products manufactured from overly processed material.

B. Pressure–Temperature Relations Between Stable and Metastable Phases

It has already been mentioned that in the vicinity of the S–L–V triple point, the S–V sublimation curve increases more rapidly than does the L–V vaporization curve. If follows that if the L–V curve is to be extended below the triple point (as would have to happen for a super-cooled liquid), the continuation of the curve must lie above the S–V curve. This implies that the vapor pressure of a supercooled liquid (a metastable phase) must always exceed the vapor pressure of the solid (the stable phase) at the same temperature.

For solids capable of exhibiting polymorphism, in the vicinity of the S_1–S_2–V triple point, the sublimation curve for the metastable phase (S_2–V) will always lie above the sublimation curve for the stable phase (S_1–V). It follows that the vapor pressure of a metastable solid phase will always exceed the vapor pressure of the stable phase at a given temperature. This generalization was first deduced by Ostwald, who proved that for a given temperature of a one-component system, the vapor pressure of any metastable phase must exceed that of the stable phase [25]. This behavior was verified for the rhombic and mo-noclinic polymorphs of elemental sulfur, where it was found that the ordinary transition point of the enantiotropic conversion was 95.5°C [26]. The vapor pressure curve of the rhombic phase was found invari-ably to exceed that of the monoclinic phase at all temperature values above 95.5°C, while the vapor pressure of the monoclinic phase was higher than that of the rhombic phase below 95.5°C. This behavior provided direct evidence that the rhombic phase was the most stable phase below the transition point and that the monoclinic phase was more stable above the transition point.

Owing to the experimental difficulties associated with measurement of the families of S_1–V sublimation curves required for the use of Ostwald's rule of relative phase stability, a variety of empirical rules (not based on the phase rule) have been advanced for the deduction of relative phase stabilities. However, when the pertinent data can be measured, application of the rule can yield unequivocal results. The pressure–temperature diagram for the α-, β-, and γ-phases of sulfanilamide was constructed using crystallographic and thermodynamic data, and by assigning the temperatures of the experimentally observed phase transitions to triple points involving the vapor phase [26]. At temperatures below 108°C, the order of vapor pressures was β < α < γ, which indicated that the β-phase was more stable than the α-phase, which is itself more stable than the γ-phase. Between 108 and 118°C, the order of vapor pressures was determined to be β < γ < α, so that within this range the β-phase remained the most stable, and that the γ-phase was more stable than the α-phase. At temperatures exceeding 118°C, the order of vapor pressures was γ < β < α, indicating that the γ-phase became the most stable, and that the α-phase remained the least stable. The data clearly indicate that the β- and γ-phases are enantiotropically related, having a transition point of 118°C. It was further concluded that since no stability region could be identified for the α-phase (it only became less metastable as the temperature increased), it bore a monotropic relationship to the other two phases.

V. SYSTEMS OF TWO COMPONENTS

When the substance under study is capable of forming a hydrate or solvate system, the number of components must necessarily increase to two. The two components are the substance itself and the solvent of solvation, since any other compound can be described as some combination of these. The various phases that can be in equilibrium will generally not exhibit the same composition, so that the usual variables of pressure, volume, and temperature must be augmented by the inclusion of the additional variable of concentration. In fact, it is a general rule that if the composition of different phases in equilibrium varies, then the system must contain more than one component.

Two components present in a single phase constitute a tervariant system, characterized by three degrees of freedom. The equilibrium condition between two phases is a bivariant system, while three phases in equilibrium would be univariant. For a system of two components to be invariant, there must be four phases in equilibrium. From the phase rule, one immediately concludes that there cannot be more than four phases in equilibrium under any set of environmental conditions. The graphical expression of phase relationships on a two-dimensional surface will require the a priori specification of a number of conditions. Fortunately, for the two component systems of greatest interest to pharmaceutical scientists (hydrates and their anhydrates), the normal studies are conducted at atmospheric pressure, which immediately fixes one of the variables. For phase equilibria, this allows the construction of the usual planar diagrams.

Since this discussion will specifically focus on the phase equilibria of solvate species, only two classes of two-component systems will be discussed. One of these will be those systems where the molecule of solvation is thermally evolved prior to melting, and the other will be where the molecule of solvation is thermally evolved subsequent to melting.

A. Solid/Vapor Equilibria

The first example concerns systems where one or more solid phases exists in a state of equilibrium with a single vapor phase. This type of situation would exist for solvation/desolvation equilibria whose transition temperatures are substantially less than the fusion point corresponding to generation of a liquid phase, and it is certainly the most commonly encountered type of solvate system of pharmaceutical interest. For most compounds, the solid substance in question has no appreciable vapor pressure, so that the sole component of the vapor phase will be the volatile solvent. The usual occurrence where the evolved solvent passes entirely into the vapor phase will be assumed, where it does not form a discrete liquid phase of its own.

Upon heating, the solvate species can dissociate either into a solvate of lower solvation or into an anhydrous phase. Each stage of such equilibria represents a system of two components (substance and solvent) present in three phases (initial solvate, solvate product, and solvent vapor). According to the phase rule, this constitutes a univariant system, so a definite vapor pressure must correspond to each temperature. This is termed the dissociation pressure and will be independent of the relative or absolute amounts of phases present.

1. Single Solvation State Systems

Copper chloride dihydrate is an example of simple dehydration, which upon simple heating below the melting point is capable of losing its water of hydration:

$$CuCl_2 \cdot 2H_2O \rightarrow CuCl_2 + 2H_2O \tag{11}$$

At atmospheric pressure, the dehydration of the dihydrate is essentially complete by 75°C [28]. The pressure–temperature curve of the dihydrate consists of a simple dissociation curve having the form illustrated in Fig. 9.

When dehydrating the dihydrate phase at constant temperature, the pressure would be maintained at the value corresponding to the dissociation pressure of the dihydrate until the complete disappearance of that phase. At that point, the pressure would fall to that characteristic (and negligible) vapor pressure of the anhydrate phase. If the external pressure on the dihydrate is reduced below its dissociation pressure at a given temperature, then the solid will undergo spontaneous efflorescence and will lose the requisite water of hydration to the atmosphere.

Conversely, if one begins with the anhydrous phase and exposes the solid to water vapor, as long as the vapor pressure is less than that of the dissociation pressure at that temperature, no hydrate phase will form. This does not imply that adventitious water will not be absorbed, however, but simply that the crystalline dihydrate cannot be formed. This situation arises since according to the phase diagram, only the anhydrate phase is stable below the lowest dissociation pressure. At the dissociation pressure, however, a univariant system is obtained, since with formation of the hydrate phase there are now three phases in equilibrium. With the experiment being conducted at constant labo-

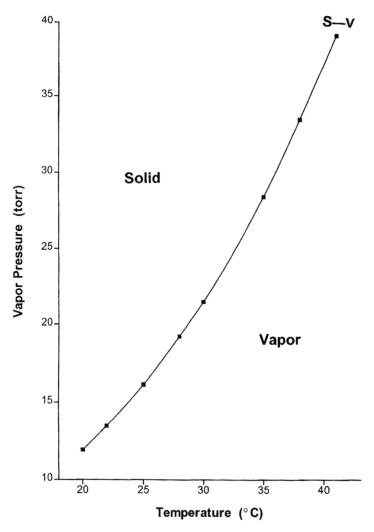

Fig. 9 Vapor pressure of water over copper chloride dihydrate as a function of temperature. (The data were plotted from published values; from Ref. 28.)

ratory temperature, the pressure must also be constant. Continued addition of water vapor can only result in an increase in the amount of dihydrate phase and a decrease in the amount of anhydrate phase present. When the anhydrate is completely converted, the system again becomes bivariant, and the pressure increases again with the amount of water added. Since no higher hydrate forms are possible for copper chloride, only adventitious water can be absorbed. Of course, if sufficient water is absorbed, the solid can presumably dissolve in the extra water, a phenomenon known as *deliquescence*.

2. Multiple Solvation State Systems

When substances are capable of forming multiple solvated forms, it is observed that the different solvates will exhibit different regions of stability, and the pressure–temperature phase diagram becomes much more complicated. Each solvate will be characterized by its own dissociation curve, and these families of curves mutually terminate at points of intersection. Each dissociation curve will exhibit first an initial increase and then a plateau as conversion to another solvation state begins, and then a decrease as the vapor pressure of the solvate product becomes established. At temperature values slightly above the intersection point of two dissociation curves, the solvate product would have a higher vapor pressure than the solvate reactant and would therefore be metastable with respect to the higher solvate. However, once the temperature is allowed to rise beyond the plateau value, the solvate product becomes the stable phase.

The hydrate system formed by lithium iodide will be used to illustrate the stepwise dehydration process. When heated at temperature values below the melting point of anhydrous lithium iodide (446°C), the trihydrate is capable of losing its water of hydration to form a dihydrate and a monohydrate on the way to the anhydrate phase:

$$LiI \cdot 3H_2O \rightarrow LiI \cdot 2H_2O + H_2O$$
$$LiI \cdot 2H_2O \rightarrow LiI \cdot 1H_2O + H_2O$$

(12)

At atmospheric pressure, the transition point for the trihydrate/dihydrate conversion is 72°C, and the transition point for the dihydrate/

monohydrate conversion is 87°C [29]. As illustrated in Fig. 10, the pressure–temperature phase diagram of the system consists of three discrete dissociation curves that intersect at the ordinary transition points.

When dehydrating the trihydrate phase at constant temperature, the pressure would be maintained at the value corresponding to the dissociation pressure of the trihydrate until the complete disappearance of that phase. At that point, the pressure would fall to that characteristic pressure of the dihydrate phase. Continued dehydration would take place at the dissociation pressure of the dihydrate phase until it was completely transformed to the monohydrate, whereupon the pressure would immediately fall to the dissociation pressure of the monohydrate. As previously discussed, once the external pressure on a given hydrate is reduced below its characteristic dissociation pressure, the solid will undergo spontaneous efflorescence to a lower hydration state and will evolve the associated water of hydration.

Conversely, if one begins with the anhydrous lithium iodide and exposes the solid to water vapor, as long as the vapor pressure is less than any of the dissociation pressures, no hydrate phase can form. At the lowest dissociation pressure a univariant system is obtained, since upon formation of the hydrate phase there must be three phases in equilibrium. Since the experiment is being conducted at constant laboratory temperature, the pressure must also be constant. Continued addition of water vapor can only result in an increase in the amount of hydrate phase and a decrease in the amount of anhydrate phase present. When the anhydrate is completely converted, the system again becomes bivariant, and the pressure increases again with the amount of water added. The higher hydrate forms are in turn produced at their characteristic conversion pressures in an equivalent manner.

3. Desolvated Solvates

A desolvated solvate is the species formed upon removal of the solvent from a solvate. Depending on the empirical details of the system, the desolvated solvate may be produced as either a crystalline or an amorphous phase. These materials are not equivalent, possessing different free energies, and the amorphous phase will ordinarily be the less stable

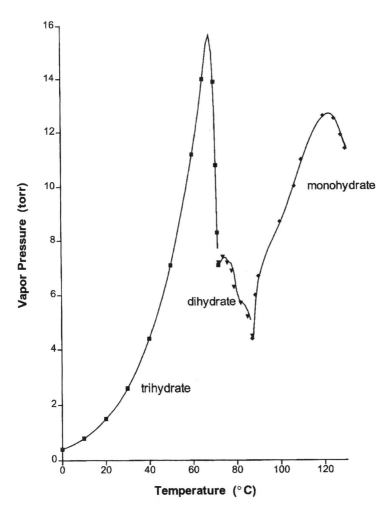

Fig. 10 Vapor pressure of water over the trihydrate, dihydrate, and monohy-drate phases of lithium iodide as a function of temperature. (The data were plotted from published values; from Ref. 29.)

of any of the crystalline forms. For example, the thermal dehydration product of theophylline monohydrate could be formulated into tablets, which then exhibited different dissolution rates from tablets formed from either the monohydrate or the anhydrate phase [30].

However, from a phase rule viewpoint, a completely desolvated solvate, from which the solvent vapor has been totally removed from the residual solid, is simply a system of one component. The characteristics and phase equilibria of such systems have been amply described earlier in Section II. All of the deductions reached about systems of one component must necessarily hold for solids produced by the desolvation of a solvate species.

B. Solid/Liquid/Vapor Equilibria

The second example concerns those situations where the solid phase containing the solvated compound is in equilibrium with both its liquid and its vapor phase. This system would result from the congruent melting of the solid phase, which was in turn accompanied by the simultaneous volatilization of the included solvation molecules. The equilibrium therefore consists of two components (substance and solvent) present in three phases (initial solvate, fused liquid, and solvent vapor). According to the phase rule, this constitutes a univariant system, so just as for the system described in Section A, for each temperature there will correspond a definite vapor pressure. This is still a dissociation pressure and will be independent of the relative or absolute amounts of phases present.

Examples of this type of behavior are not commonly encountered for compounds of pharmaceutical interest, since the melting points of drug substances generally lie at considerably higher temperatures than do the dehydration points. Even for excipients characterized by low melting points, the dehydration steps take place at lower temperatures than do the fusion transitions. One of the closest pairs of dehydration and melting temperatures was noted for the crystalline dihydrates of magnesium stearate and palmitate, but even here the melting transition occurred approximately 20°C higher than the dehydration transition [31]. As a result, the crystalline hydrates could be completely dehydrated prior to the onset of any melting.

Nevertheless, the phase rule can be used to deduce some conclusions about systems where a congruently melting solid remained in equilibrium with the vapor phase. One deduction is that one would not expect to encounter a condition where in addition to being in equilibrium with liquid and vapor phases, the solvate phase was in equilibrium with any other type of solid phase. Such a system would constitute an invariant system and could only exist at a characteristic quadrupole point. Since it is hardly likely to encounter a quadrupole point at ambient temperature or pressure, the possibility can be effectively discounted from ordinary experience.

The power of the phase rule is immediately evident in that the solid/liquid/vapor system is characterized by the same amount of variance as was the solid/vapor system. As a result, the arguments made regarding the pressure–temperature curves of the former system can be extended to apply to the latter system, except that the liquid phase takes the place of the anhydrate phase.

C. Kinetically Impaired Equilibria

It goes without saying that suspended transformations may also accompany the phase interconversions of anhydrates and solvates. Efflorescence may not occur immediately once the pressure is reduced below the dissociation pressure, but such reactions will always take place upon formation of a suitable nucleus. For instance, it has been known since the time of Michael Faraday that the decahydrate phase of sodium sulfate is unstable with respect to open air, since the vapor pressure of the salt exceeds the vapor pressure of water vapor at room temperature. However, the system only dehydrates upon contact with the anhydrate phase, demonstrating the metastable nature of the decahydrate phase.

The instability of many anhydrate phases with respect to water has been long known. For instance, it was shown by Shefter and Higuchi that hydrate phases of cholesterol, theophylline, caffeine, glutethimide, and succinyl sulfathiazole would spontaneously form during dissolution studies [32]. Similar behavior has been reported for metronidazole benzoate [33] and carbamazepine [34]. In each of these systems, the integrity of the anhydrous phases can be maintained only as long as the relative humidity is kept below the dissociation pressure of the hydrate species.

Ampicillin is known to form crystalline anhydrate and trihydrate phases, which exhibit an ordinary transition point of 42°C when in contact with bulk water [35]. The anhydrate phase is found to be the stable phase below the transition point, and the trihydrate is the stable phase above this temperature. The trihydrate is the phase of pharmaceutical interest, but fortunately can be maintained in a stable condition as long as contact with other phases is suppressed. When milled in contact with anhydrate phase, or when placed in contact with bulk water at room temperature, the anhydrate phase forms from the trihydrate with great velocity.

VI. SUMMARY

Owing to the qualitative nature of conclusions that can be reached through its use, the phase rule has not been a topic of research study for quite some time. Most scientists are more enamored with the hard deductions permitted by the application of quantitative thermodynamics, and they routinely use the signs and magnitudes of free energy changes to deduce questions of phase stability. Nevertheless, the phase rule is extremely useful for yielding a physical understanding of polymorphic systems and for providing a physical interpretation of phase transformation phenomena. Its greatest power is in its ability to rule out the existence of simultaneous multiple equilibria that violate its fundamental equation, permitting more quantitative investigations to focus on the possible aspects of such systems.

REFERENCES

1. J. Willard Gibbs, *Collected Works*, Vol. 1, Longmans, Green, New York, 1928, pp. 96–144.
2. A. C. D. Rivett, *The Phase Rule*, Oxford Univ. Press, London, 1923.
3. J. E. W. Rhodes, *Phase Rule Studies*, Oxford Univ. Press, London, 1933.

4. J. S. Marsh, *Principles of Phase Diagrams*, McGraw-Hill, New York, 1935.

5. S. T. Bowden, *The Phase Rule and Phase Reactions*, Macmillan, London, 1938.

6. A. Findlay, and A. N. Campbell, *The Phase Rule and its Applications*, Dover, New York, 1938.

7. J. E. Ricci, *The Phase Rule and Heterogeneous Equilibrium*, Van Nostrand, New York, 1951.

8. K. Denbigh, *The Principles of Chemical Equilibrium*, Cambridge Univ. Press, Cambridge, 1955, pp. 180–210.

9. G. N. Lewis, and M. Randall, *Thermodynamics and the Free Energy of Chemical Substances*, McGraw-Hill, New York, 1923, pp. 185–186.

10. E. W. Washburn, *International Critical Tables*, Vol. 3 (E. W. Washburn, ed.), McGraw Hill, New York, 1928, pp. 210–212.

11. J. A. Dean, ed., *Lange's Handbook of Chemistry*, 12th ed., McGraw Hill, New York, 1979, pp. 10–38.

12. R. C. Weast, ed., *Handbook of Chemistry and Physics*, 50th ed., Chemical Rubber Company, Cleveland, OH, 1969, p. D-139.

13. P. W. Bridgeman, *Proc. Am. Acad., 47*, 441 (1912).

14. L. Merrill, *J. Phys. Chem. Ref. Data, 6*, 1205 (1977).

15. E. Riecke, *Z. Phys. Chem., 6*, 411 (1890).

16. E. Mitscherlich, *Ann. Chim. Phys., 19*, 414 (1821).

17. M. Kuhnert-Brandstätter, *Thermomicroscopy in the Analysis of Pharmaceuticals*, Pergamon Press, Oxford, 1971.

18. L. Borka, and J. K. Haleblian, *Acta Pharm. Jugosl., 40*, 71 (1990).

19. L. Borka, *Pharm. Acta Helv., 66*, 16 (1991).

20. P. W. Bridgeman, in *International Critical Tables*, Vol. 4 (E.W. Washburn, ed.), McGraw Hill, New York, 1928, pp. 16.

21. P. W. Bridgeman, in *International Critical Tables*, Vol. 4 (E.W. Washburn, ed.), McGraw Hill, New York, 1928, pp. 15.

22. G. D. Fahrenheit, *Phil. Trans., 39*, 78 (1724).

23. A. N. Winchell, and H. Winchell, *The Microscopic Characters of Artificial Inorganic Solid Substances*, Academic Press, New York, 1964, pp. 63–64.

24. M. Otsuka, and N. Kaneniwa, *J. Pharm. Sci., 75*, 506 (1986).

25. W. Ostwald, *Z. Phys. Chem., 22*, 313 (1897).

26. G. Ruff, and R. Graf, *Z. Anorg. Chem., 58*, 209 (1908).

27. S. Toscani, A. Dzyabchenko, V. Agafonov, J. Dugué, and R. Céolin, *Pharm. Res., 13*, 151 (1996).

28. F. C. Kracek, *International Critical Tables*, Vol. 3 (E.W. Washburn, ed.), McGraw: Hill, New York, 1928, p. 366.
29. C. Kracek, *International Critical Tables*, Vol. 3 (E.W. Washburn, ed.), McGraw Hill, New York, 1928, p. 369.
30. N. V. Phadnis, and R. Suryanarayanan, *J. Pharm. Sci., 86*, 1256 (1997).
31. S. A. Sharpe, M. Celik, A. W. Newman, and H. G. Brittain, *Struct. Chem., 8*, 73 (1997).
32. E. Shefter, and T. Higuchi, *J. Pharm. Sci., 52*, 781 (1963).
33. A. Hoelgaard, and N. Møller, *Int. J. Pharm., 15*, 213 (1983).
34. E. Laine, V. Tuominen, P. Ilvessalo, and P. Kahela, *Int. J. Pharm., 20*, 307 (1984).
35. J. W. Poole, and C. K. Bahal, *J. Pharm. Sci., 57*, 1945 (1968).

3

Structural Aspects of Polymorphism

Harry G. Brittain

Discovery Laboratories, Inc.
Milford, New Jersey

Stephen R. Byrn

Purdue University
West Lafayette, Indiana

I. INTRODUCTION

A century ago, the study of crystals was only concerned with examinations of their external form and with the physical properties as these varied along the different crystal directions. The regularity of the appearance and of the external form of naturally occurring crystals led observers to believe that these materials were formed by the regular repetition of identical building blocks. It was deduced that when a crystal grows in a constant environment, the shape remains unchanged while the mass is built up from the continuous addition of fundamental units. Mineralogists made the important discovery that the faces of a crystal could be indexed to exact integers, and in 1784 Haüy showed that only the arrangement of identical particles in a three-dimensional periodic array could account for the law of rational indices [1]. It was subsequently suggested by Seeber that the elementary building blocks of crystals were small spheres, held together in a lattice array through the balance of attractive and repulsive forces [2]. Different crystal structures were believed to arise from the various ways spheres could be packed into alternate motifs. Shrewd guesses about the structures of a number of crystals were made by Barlow, who argued from considerations of symmetry and the packing of spheres [3].

The modern study of the physical properties of the solid state essentially began in 1912 with the publication of a paper by Laue and coworkers on the diffraction of Röntgen rays (now known as x-rays) [4]. In the first part of the paper, Laue developed an elementary theory for the diffraction of x-rays by a periodic array of atoms, while in the second part Friedrich and Knipping reported the first experimental observations of x-ray diffraction by crystals. This work demonstrated simultaneously that x-rays must possess a wave character (since they could be diffracted) and that solids were composed of a regular network of atoms. These experimental proofs are considered to mark the beginning of the field of solid state physics. W. L. Bragg reported the first determinations of crystal structures in 1913, deducing the structures of KCl, NaCl, KBr, and KI [5].

II. LATTICE THEORY OF CRYSTAL STRUCTURE

An ideal crystal is constructed by the infinite regular repetition in space of identical structural units. In the simplest crystals formed by monatomic elements, the basic structural unit consists of a single atom. For the organic molecules of pharmaceutical interest, the structural unit will contain one or more molecules. One can describe the structure of all crystals in terms of a single periodic lattice, which represents the translational repetition of the fundamental structural unit. For elemental or ionic crystal structures, each point in the lattice may be a single atom. For organic molecules, a group of atoms is often attached to a lattice point or situated in an elementary parallelepiped.

A *lattice* is defined as a regular periodic arrangement of points in space and is by definition a purely mathematical abstraction. The nomenclature refers to the fact that a three-dimensional grid of lines can be used to connect the lattice points. It is important to note that the points in a lattice may be connected in various ways to form an infinite number of different lattice structures. The crystal structure is formed only when a fundamental unit is attached identically to each lattice point and extended along each crystal axis through translational repetition.

The points on a lattice are defined by three fundamental translation vectors, a, b, and c, such that the atomic arrangement looks the same in every respect when viewed from any point r as it does when viewed at point r':

$$r' = r + n_1 a + n_2 b + n_3 c \tag{1}$$

where n_1, n_2, and n_3 are arbitrary integers. The lattice and translation vectors are said to be *primitive* if any two points from which an identical atomic arrangement is obtained through the satisfaction of Eq. (1) with a suitable choice of the n_1, n_2, and n_3 integers. It is common practice to define the primitive translation vectors as the axes of the crystal, although other nonprimitive crystal axes can be used for the sake of convenience. A lattice translation operation is defined as the displacement within a lattice, with the vector describing the operation being given by

$$T = n_1a + n_2b + n_3c \tag{2}$$

A. The Unit Cell

The crystal axes a, b, and c form three adjacent edges of a parallelepiped. The smallest parallelepiped built upon the three unit translations is known as the *unit cell*. Although the unit cell is an imaginary construct, it has an actual shape and a definite volume. The entire crystal structure is generated through the periodic repetition, by the three unit translations, of matter contained within the volume of the unit cell. A unit cell does not necessarily have a definite absolute origin or position, but it does have the definite orientation and shape defined by the translation vectors A cell by repetition will fill all space under the action of suitable crystal translation operations, and in itself it will occupy the minimum volume permissible. By elementary vector analysis, the volume of a unit cell is defined by the magnitude of the cell axes as

$$V_c = |a \times b \cdot c| \tag{3}$$

The unit cell is defined by the lengths of the crystal axes a, b, and c and by the angles α, β, and γ between these. The convention is that α defines the angle between the b and c axes, β defines the angle between the a and c axes, and γ defines the angle between the a and b axes. There are seven fundamental types of primitive unit cell, whose characteristics are provided in Table 1. These same unit cell characteristics define the seven *crystal classes*. As stated earlier, each unit cell will occupy one lattice point in the structure.

Table 1 The Seven Crystal Classes, Defined from the Fundamental Unit Cells

System	Relationship between cell edges	Relationship between cell angles
Cubic	$a = b = c$	$\alpha = \beta = \gamma = 90°$
Tetragonal	$a = b \neq c$	$\alpha = \beta = \gamma = 90°$
Orthorhombic	$a \neq b \neq c$	$\alpha = \beta = \gamma = 90°$
Monoclinic	$a \neq b \neq c$	$\alpha = \gamma = 90°$ $\beta \neq 90°$
Triclinic	$a \neq b \neq c$	$\alpha \neq \beta \neq \gamma \neq 90°$
Hexagonal	$a = b \neq c$	$\alpha = \beta = 90°$ $\gamma = 120°$
Trigonal	$a = b = c$	$\alpha = \beta = 90°$ $\gamma \neq 90°$

B. Unit Cell Symmetry Operations

Crystal lattices can be carried into themselves not only by the lattice translation defined in Eq. (2) but also by the performance of various point symmetry operations. A *symmetry operation* is defined as an operation that moves the system into a new configuration that is equivalent to and indistinguishable from the original one. A *symmetry element* is a point, line, or plane with respect to which a symmetry operation is performed. There are a total of five symmetry operations of import to the solid state.

The simplest operation is the *identity*, which leaves the system unchanged and hence in an orientation identical with that of the original. Although the performance of the identity operation does not yield any translation, it is a mathematical requirement of the theory of groups necessary to handle the ensemble of symmetry operations. Its presence also serves to distinguish between identical and equivalent orientations.

The second symmetry operation is that of *reflection* through a plane; it is denoted by the symbol σ. Using Cartesian coordinates, the effect of reflection is to change the sign of the coordinates perpendicular to the plane while leaving unchanged the coordinates parallel to the plane. The third symmetry operation is that of *inversion* through a point and is denoted by the symbol i. In Cartesian coordinates, the effect of inversion is to change the sign of all three coordinates that define a lattice point in space.

The fourth type of symmetry operation is the *proper rotation*, a simple rotation about an axis that passes through a lattice point. Only rotations by angles of 2π (360°), $2\pi/2$ (180°), $2\pi/3$ (120°), $2\pi/4$ (90°), and $2\pi/6$ (60°) radians are permissible. These operations are referred to as one-fold (symbol C_1), two-fold (symbol C_2), three-fold (symbol C_3), four-fold (symbol C_4), and six-fold (symbol C_6) proper rotation axes. Rotations by other angles will not bring a three-dimensional lattice system into an equivalent configuration and are therefore not permissible symmetry operations in the solid state.

The final type of symmetry operation is the *improper rotation*, a proper rotation followed by a reflection. Once again, only improper rotations by angles of 2π (360°), $2\pi/2$ (180°), $2\pi/3$ (120°), $2\pi/4$ (90°), and $2\pi/6$ (60°) radians are permissible and are denoted as one-fold (symbol S_1), two-fold (symbol S_2), three-fold (symbol S_3), four-fold (symbol S_4), and six-fold (symbol S_6) proper rotation axes.

C. The 14 Bravais Lattices

Bravais proved that if the contents of a unit cell have symmetry, the number of distinct types of space lattices becomes fourteen. These are the only lattices that can fill all space and are commonly termed the 14 Bravais lattices. Since there are seven crystal systems, it might be thought that by combining the seven crystal systems with the idea of a primitive lattice a total of seven distinct Bravais lattices would be obtained. However, it turns out that the trigonal and hexagonal lattices so constructed are equivalent, and therefore only six lattices can be formed in this way. These lattices, which are given the label P, define the primitive unit cells (or P-cells) in each case.

The other eight Bravais lattices are derived by taking each of the six P-lattices and considering what happens when other lattice points are added through the use of certain centering conditions. This operation yields eight centered lattices, seven of which are given a name (body-centered, all-face-centered, one-face-centered) and a new symbol (*I, F, and A, B, or C*). The last new lattice is a specially centered hexagonal lattice that can be regarded after appropriate redefinition of axes of reference as a primitive rhombohedral lattice.

One crystal system can have all four space lattices (P, I, F, and C), while some crystal systems can have only the P-lattice. For each crystal system, the I-, F-, or C-lattices have unit cells that contain more than one lattice point, since we have added lattice points in various centered positions. Cells with more than one lattice point are sometimes known as multiple-primitive unit cells. The same axes refer to all the space lattices in a given crystal system, so all of these conventional, centered unit cells will display exactly the same rotational symmetry as does the corresponding P-cell in a given crystal system.

The assignation of axes of reference in relation to the rotational symmetry of the crystal systems defines six lattices that, by definition, are primitive or P-lattices. To determine if new lattices can be formed from these P-lattices, one must determine if more points can be added so that the lattice condition is still maintained, and whether this addition of points alters the crystal system. For example, if one starts with a simple cubic primitive lattice and adds other lattice points in such a way that a lattice still exists, it must happen that the resulting new lattice still possesses cubic symmetry. Since the lattice condition must be maintained when new points are added, the points must be added to highly symmetric positions of the P-lattice. These types of positions are (a) a single point at the body center of each unit cell, (b) a point at the center of each independent face of the unit cell, (c) a point at the center of one face of the unit cell, and (d) the special centering positions in the trigonal system that give a rhombohedral lattice.

In the triclinic system, there are no restrictions on the magnitudes of the lengths of the unit cell axes or on their interaxial angles. One can therefore always take a triclinic lattice and center it, to produce a new lattice that will be compatible with the conditions of the triclinic crystal system. However there is nothing new about this lattice, since a smaller primitive cell can be determined with the same complete arbitrariness of the cell edges and angles. Thus for the triclinic crystal system there can be only one Bravais lattice, the primitive or P-lattice.

For the monoclinic system, a new lattice can be obtained by centering a monoclinic P-cell along the face defined by the plane formed by the intersection of the a and b crystal axes. This is because it is not possible to maintain the fundamental monoclinic conditions and yet describe this lattice as a P-lattice. There are therefore two unique monoclinic Bravais lattices.

The primitive orthorhombic lattice can be thought of as arising from a primitive monoclinic lattice with the added restriction that the third angle is also 90°. In that case, all the unit cell translation vectors are 90° to one another but have different lengths. In the orthorhombic system, one can construct a C-centered cell, which can also be described as an A- or B-lattice by an interchange of the orthogonal axes. In addition, there can also be an all-face-centered F-lattice structure and a body-centered I-lattice. Thus for the orthorhombic crystal system there are four unique Bravais lattices, P, I, F, and C.

In the tetragonal system, centering any of the P-cell faces results only in a lattice equivalent to the P-lattice itself. It happens that the only new lattice that can be generated in the tetragonal system is the one formed through body-centering of the P-cell. The surroundings of every point in this cell are still identical, with each point having eight nearest neighbors at the same distance and in the same directions, which preserves the fourfold symmetry. Therefore for the tetragonal crystal system there are only two distinct Bravais lattices.

Placing a lattice point at the body-center position of a cubic P-cell yields a new lattice. Each point is surrounded by eight other nearest-neighbor points, all in the same relative positions regardless of the observation point. The I-cell type remains cubic since the four threefold axes are preserved. This structure is frequently called the body-centered cubic cell. One can also add points at the face centers of the primitive cubic unit cell to obtain another new lattice. Twelve nearest-neighbor points surround every lattice point in the F-cell, still preserving the four threefold axes. This F-lattice is often called the face-centered cubic cell. It is clear that the cubic crystal system cannot have a single face-centered lattice, because centering only one face would destroy the four threefold axes of symmetry. Thus there are three cubic Bravais lattices.

The hexagonal and trigonal systems yield special complications owing to the relationship between crystal axes and angles. One finds that face-centering or body-centering the primitive lattice types with simultaneous preservation of the threefold or sixfold axes is not possible. It suffices to say that only the two Bravais lattices of Fig. 1 are

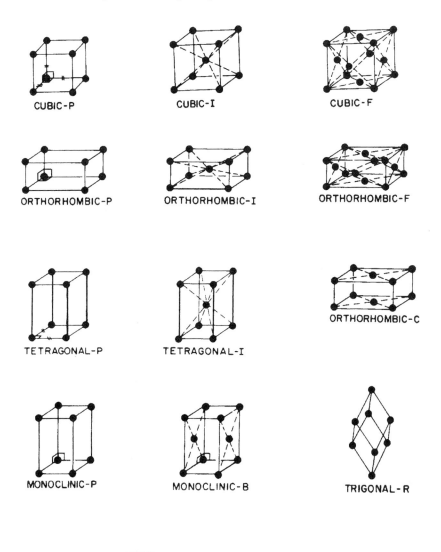

Fig. 1 The fourteen Bravais lattices.

possible, and readers seeking additional information on this question are referred to the literature [7].

D. The 230 Space Groups

The structure of a given crystal might be assigned to one of the seven crystal systems or to one of the 14 Bravais lattices. In those classification schemes, one can state with certainty the crystal system as well as the type of lattice. On a deeper level, one could deduce the molecular point group of the molecules in the unit cell, which would then pinpoint to which of the 32 point groups the crystal belongs. This latter classification would yield both the crystal system and the relationships between the tensors that describe the macroscopic behavior and the properties of the crystal.

In a further development of detail, one can take into account how the atoms of the solid are distributed spatially. The issue of symmetry in context with a fixed point in the crystal, and the symmetry of Bravais lattices, has been addressed, but in order to describe the entire crystal the effects of two new types of symmetry operation must be included. A *space group* determined in this way describes the spatial symmetry of the crystal. By definition, a crystallographic space group is the set of geometrical symmetry operations that take a three-dimensional periodic crystal into itself. The set of operations that make up the space group must form a group in the mathematical sense and must include the primitive lattice translations as well as other symmetry operations.

Bravais also showed that when the full range of lattice symmetry operations is taken into account, there are a maximum of 230 distinct varieties of crystal symmetry. The ensemble of these 230 configurations is the space groups, and the structure of any given crystal must be described by one of these. A detailed description of all 230 space groups is beyond the scope of this discussion, but such treatments are available [7].

The set of symmetry operations is ordinarily divided into two main classes. The product of the symmetry operations with the identity of the translation group (zero translation) yields the category termed essential space group operations. The other products of the symmetry operations with primitive lattice translations are termed nonessential space group operations. Operations obtained by the product of the identity of the symmetry operations with the primitive lattice translations are simply the translation symmetry operations.

Space group symmetry operations may involve a translation, which is smaller than a primitive lattice translation, and which is coupled with a rotation or reflection operation. These symmetry operations are known as glide or screw operations. This leads to an important classification of space groups as being either symmorphic or nonsymmorphic. A symmorphic space group is one that is entirely specified by symmetry operations acting at a common point and that do not involve one of the glide or screw translations. The symmorphic space groups are obtained by combining the 32 point groups with the 14 Bravais lattices. This generates the ensemble of symmorphic space groups, which account for 73 space groups out of the total of 230.

When it is necessary to specify a space group by at least one operation involving a glide or screw translation, the space group is said to be nonsymmorphic. The *screw operation* is a symmetry operation derived from the coupling of a proper rotation with a nonprimitive translation parallel to the axis of rotation, which is called the screw axis. The combination of a reflection operation plus a translation operation yields a *glide plane*. Three types of glide plane are known in crystals, and these are termed the axial glide, the diagonal glide, and the diamond glide. In all of these glides, one reflects across a plane and translates by some distance that is a fraction of the unit cell. In the *axial glide*, the magnitude of the translation vector is one half of a unit-cell translation parallel to the reflection plane. One refers to the axial glide as an a-, b-, or c-glide according to the axis along which the translation is carried out. The reflection that accompanies this translation can be across any of the planes defined by the intersection of two of the translation axes. The *diagonal glide* involves translations along two or three directions, which are usually $(a + b)/2$, $(b + c)/2$, or $(c + a)/2$. In the case of tetragonal and cubic crystal systems the diagonal can also include a translation of $(a + b + c)/2$. The *diamond glide* operation consists of a translation having a magnitude of $(a \pm b)/4$ or $(b \pm c)/4$ or $(a \pm c)/4$ and $(a \pm b \pm c)/4$ in tetragonal and cubic crystals.

The nonsymmorphic space groups are specified by at least one operation that involves a nonprimitive translation. Taking such operations into account, there are a total of 157 nonsymmorphic space groups. These taken together with the 73 symmorphic space groups produce the entire 230 space groups originally derived by Bravais.

The best description of the properties of individual space groups uses the ensemble of symmetry operations that give rise to the fundamental structure. One must explain the coordinates of the general equivalent positions, and the positions and occurrence of all the symmetry elements (both essential and nonessential) with respect to any particular choice of origin. The procedure is to consider the point group of any space group and to ascertain the number of essential symmetry operations. This will automatically yield the number of general equivalent positions in the primitive unit cell. Crystallographers use matrix operator methods to determine the coordinates of the entire general equivalent positions and the presence and positions of all symmetry elements.

III. CRYSTAL STRUCTURES DERIVED FROM THE CLOSE-PACKING OF SPHERES

The first crystal structures to be solved were those of inorganic compounds, for which great help was found through considerations of how one could pack spheres into structures that minimized the magnitude of interstitial space [8]. In this approach, the crystal is built up from spherical structural units that make contact with one another. Alternatively, one could consider the points of a lattice network to be inflated into spheres. As a result, all individual symmetry is removed from the individual particles, and the symmetry of the lattice structure follows from the fashion in which the spheres are arranged.

The model of sphere packing requires no definition as to the nature of the bonding holding the spheres together, and in fact it requires that no directionality be exhibited in the bonding. This model does not work well when applied to structures built up of covalently bonded organic molecules, but it does work reasonably well for the crystals formed by ionic or metallic solids.

A. Spheres of Equivalent Size

If spheres of equal size are packed together in a plane, they will arrange themselves in one of the patterns shown in Fig. 2. The sphere centers will lie at the corners of equilateral triangles, and each sphere will be in contact with six others. The three-dimensional structure of closely packed spheres of equal sizes will consist of such. The conditions that must be satisfied are that the space occupied per sphere is a minimum and that the spheres all have the same environment. The relative orientation of the spheres is disregarded if it is assumed that they possess

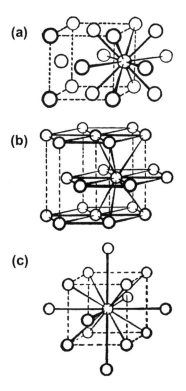

(a)

(b)

(c)

Fig. 2 Close-packing arrangements of spheres of equal sizes. Shown are (a) cubic close-packing, (b) hexagonal close-packing, and (c) body-centered cubic packing.

full spherical symmetry, and that there is no way of determining their orientations.

There are two ways to arrange equivalent spheres in a regular array that minimizes the interstitial volume. If the positions of the atoms in a layer are labeled A, then an exactly similar layer can be superimposed on the first so that the centers of the atoms in the upper layer are vertically placed above the positions B. When the third layer is placed above the second layer, the centers of the atoms may lie above either the C or the A positions. In the latter, case one obtains the layer sequence of ABABAB etc., which is the simplest form and is called hexagonal close packing. In the former case, the sequence of layers is ABCABC etc. and is called cubic close packing. The terms hexagonal and cubic refer to the symmetry of the resulting aggregate.

The highest axial symmetry of hexagonal close packing consists of sixfold axes perpendicular to the layers. In cubic close packing, there are threefold axes of symmetry in four directions, inclined to one another as are the body-diagonals of a cube. This structure requires that the close-packed layers of the planes be perpendicular to any body-diagonal of the cube. The unit cell of cubic close packing is a group of atoms situated at the corners and midpoints of the faces of a cube, which leads to the alternative name of face-centered cubic for a cubic close-packed structure.

In both of these close-packed arrangements each atom has twelve equidistant nearest neighbors, six in its own plane and three in each adjacent layer. The next nearest neighbors are, however, differently arranged in the two cases. It is apparent that any sequence of close-packed layers must lead to a close-packed structure, for any atom will always have twelve equidistant nearest neighbors. Any other sequences of layers, however, differ from the sequences ABABAB . . . and ABCABC . . . in that the environment of all the atoms is not the same. Although the arrangement of the nearest neighbors is identical, in the arrangement of the next nearest neighbors the atoms fall into two or more sets. For example, in the sequence ABACABAC . . . (''double'' hexagonal close packing) there are two types of nonequivalent spheres.

There are numerous examples of cubic and hexagonal close packing. Metallic solids are known to exhibit both cubic and hexagonal structures, while the inert gases are found to form solids characterized by cubic packing. The structures of many simple diatomic molecules can also be described in terms of cubic close packing.

It is evident that even the simplest model for the packing of spheres of equivalent size permits the possibility of polymorphism. Most of the metals of the lanthanide series exist in the hexagonal close-packed structure at room temperature but transform into the body-centered cubic form at temperatures exceeding 800°C [9]. For the most part, these transitions appear to be enantiotropic in nature.

B. Spheres of Differing Sizes

As soon as spheres of different sizes (i.e., different species) are considered, the number of possibilities for packing arrangements increases significantly. Such situations arise most often for ionic solids, formed by the interactions of positively charged (cations) and negatively charged (anions) spheres.

For this approach, monatomic ions may be considered as incompressible spheres that are only slightly polarized by the ions of opposite charge around them. The two conditions for an arrangement of ions in space to have minimum potential energy are that (a) the larger ions (usually the anions) around a smaller ion of opposite sign must all be in contact, and (b) the coordination number of an ion must be as large as possible (subject to the first condition being observed). Thus coordination numbers of 5, 7, 9, 10, and 11 are excluded by geometry if the ionic arrangement is to form a regular spatial pattern. In addition, the coordination number of 12 is excluded by the requirement that positive and negative charges balance one another.

For instance, in the well-known sodium chloride structure (body-centered cubic packing), only six chloride ions can be accommodated around a sodium ion so that they all "touch" the cation, even though the available space would accommodate more sodium ions. But since electrical neutrality must be achieved, and as the system is stable only when the cation touches all the anions surrounding it, the coordination in sodium chloride must be six. Thus any single sodium ion has six octahedrally arranged chlorides as its nearest neighbors, while each chloride ion similarly has a corresponding arrangement of sodium ions around it.

For a compound having a $1:1$ stoichiometry of cations and anions, the number of anions surrounding a cation in a given structure will depend on the ratio of the cationic (r_c) and anionic (r_a) radii. For instance, the smallest radius ratio r_c/r_a that permits four anions to surround a single cation in the same plane is 0.414. This value therefore represents the smallest radius ratio for which six-coordination of the cation is possible. In addition to the four anions within the plane, one anion can be accommodated above and one below the cation. For lower values of r_c/r_a, six-coordination of the cation is not possible because the anions touch one another but not the cation. The limiting values of radius ratios for various coordination numbers are given in Table 2.

The situation becomes complicated for stoichiometries other than $1:1$. In crystals of an AB_2 compound, the coordination number of A is twice that of B and the structure is determined by the coordination number of the smaller ion. The radius ratio rule is clear-cut only with simple ions, and polymorphism is much more prevalent when complex ions are involved. For example, the commonly employed pigment titanium dioxide is known to exist in two polymorphs, anatase and rutile, whose structures are each built up from different arrangements of octahedral coordination. As shown in Fig. 3, the coordination polyhedron of rutile is essentially ideal, while for anatase there are very distorted octahedra of oxygen atoms about each titanium ion. Although rutile is

Table 2 Limiting Radius Ratios, Cation Coordination Numbers, and Ionic Arrangements in Ionic Solids

Radius ratio (r_c/r_a)	Cation coordination number	Arrangement
$1 \rightarrow 0.73$	8	at cube corners
$0.73 \rightarrow 0.41$	6	octahedral
$0.41 \rightarrow 0.22$	4	tetrahedral
For anisodesmic crystals		
$0.73 \rightarrow 0.41$	4	square
$0.22 \rightarrow 0.16$	3	triangular
0.16	2	linear

(a) (b)

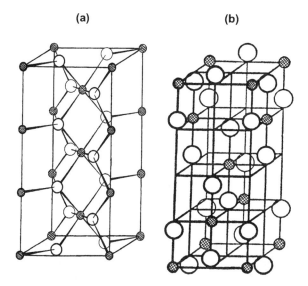

Fig. 3 Solid-state structure adopted by titanium dioxide in the (a) rutile and (b) anatase structures. The titanium atoms are indicated by the cross-hatched circles, and the oxygen atoms are indicated by the open circles.

ordinarily assumed to be the most stable form owing to its common occurrence, anatase is actually more stable by approximately 8–12 kJ/mol [10].

IV. STRUCTURAL THEORY OF POLYMORPHISM

The problem with thermodynamics is that the rate of formation of individual polymorphic modifications is not determined solely on the basis of energetics through a reduction in their free energy. Entering into the picture are other structural factors that play a part and that, under certain circumstances, may even prove to be decisive. One often encounters apparent kinetic obstacles to thermodynamic predictions, which lead to the observation of metastable states. This situation has been pointed out in the first two chapters of this book and indicates the need for supplementing the thermodynamic approach by structural considerations based on atomic and molecular theory.

The classification of polymorphic substances into monotropic and enantiotropic classes from the standpoint of observed phenomena is not appropriate as a basis for the lattice theory of polymorphism. In the lattice theory, one begins by considering whether the lattices of the polymorphic forms are related in structure or not. Only in the case of lattices that are structurally related will it be possible for mutual transformation to take place at a transition point fixed by the temperature and pressure of the system. If the lattices differ in such a way that atoms or molecules must be completely regrouped during the transformation (changing their state of bonding), no point of contact for mutual reversible transformation will exist.

Generally speaking, the concepts of monotropy and enantiotropy in phase theory appear to coincide with the structural concepts of unrelated and related lattices. Nevertheless, one must avoid equating the two, for it is certainly possible that one of two related lattices of the same substance is less stable than the other under all conditions of temperature and pressure. This would indicate the existence of monotropy in spite of the existence of related lattices. This situation becomes especially important for polymorphic organic compounds, which form molecular lattices.

A. Polymorphism Among Structurally Unrelated Lattices

The most striking differences in properties are exhibited by those polymorphic modifications for which the lattices have a completely different arrangement. Frequently, a complete change in the arrangement of the lattice particles is also accompanied by a change in the state of bonding. For example, completely different chemical bonding exists for the diamond/graphite and white/red phosphorus polymorphic pairs. The difference in energy between polymorphs may be quite considerable (as with the modifications of phosphorus and arsenic), but it also can be surprisingly small (as with diamond/graphite). In almost all instances, polymorphic changes that require the breaking of primary bonds and the rejoining of atoms into a different arrangement necessitate the existence of monotropy.

The modifications of various nonmetals usually differ drastically in their structure and bonding. Systems are known where one or more nonmetallic modifications form molecular lattices in which a definite small number of elementary atoms are closely united to form a structural grouping, while other semi metallic modifications exist that do not exhibit such a grouping in the lattice. A classic example of this type of behavior is provided by white phosphorus (with a P_4 structural grouping) and black phosphorus (which exhibits a layered structure) [11].

Even where there is not such a marked difference in the kind of bonding, pronounced differences in the structural arrangement are able to stipulate very different physical and chemical properties. This type of behavior is typified by elemental sulfur. The ''plastic'' modification consists of long disordered chains of sulfur atoms, whereas the two crystalline modifications contain the structural group S_8 as a molecule in the lattice [12].

An explanation for the existence of the very different kinds of lattices that are observed for the various nonmetals demands a profound knowledge of the possible ways in which the valence electrons of these elements can be arranged. This information is only available from the performance of suitable molecular orbital calculations, which yield detailed information about the distribution of electron density in the components of a unit cell. For instance, only such computational methods can show why phosphorus forms P_4 molecules in the vapor phase and why these same molecular species are again encountered in the lattice of white phosphorus.

B. Polymorphism Among Structurally Related Lattices

Any structural theory of polymorphism as applied to related lattices must deal with three issues. One must first establish the structural relationships between the different lattice types on the basis of the lattice theory. Next one must explain why a particular substance is able to arrange its structural units in two closely related lattices. Last one must find the manner and conditions under which a rearrangement of the units from one lattice type to another can occur.

The structural relations existing between the 230 space groups have been dealt with adequately, and a very detailed summary of these relationships is available [7]. Hückel has provided an exhaustive summary of the structural causes of polymorphism among related crystal lattice types and has provided also the entire range of mechanisms whereby a polymorphic transition can take place [13].

Three-dimensional lattices of the same symmetry class characterize the two polymorphic forms, but the unit cells exist with different coordination numbers and different coordination polyhedra. The classic example of this type of behavior is given by cesium chloride, which undergoes a reversible transformation from a cubic body-centered lattice to a cubic face-centered lattice at 445°C [14]. On the body-centered modification, two simple primitive cubic lattices (one of Cs cations and one of Cl anions) are placed inside one another so that the corners of one kind of cube are situated at the centers of cubes of the other kind. The face-centered modification is built up from face-centered ionic lattices situated inside one another.

The two polymorphs are characterized by three-dimensional lattices with the same coordination scheme but with different arrangements of the more remote lattice points. The structures are related to one another in a manner that finds expression in twin-formation at the plane of twinning. Examples of this category are the two polymorphs of zinc sulfide, zinc blende and wurtzite [15]. At ambient temperatures, zinc blende is the stable modification, but it changes enantiotropically into wurtzite at 1024°C. For both forms, the zinc occupies a site of tetrahedral symmetry. In zinc blende, all the tetrahedra are situated inversely with respect to each other, so that all the Zn and S atoms respectively are structurally equivalent. In wurtzite, three tetrahedra are situated inversely with respect to each other and the fourth as a mirror image. The one structure follows from the other by rotating alternate layers formed by buckled six-membered rings through 180°.

The lattices of the two polymorphs are related by symmetry changes in the lattice, associated with changes of symmetry of the lattice particles through rotation. One example of this type of behavior is given by sodium nitrate, which at 250–275°C transforms from the trigonal modification into a regular modification [16]. This phase change takes place as a result on the onset of rotation of the nitrate ions about the trigonal axis. A considerable elongation of the lattice in the direction of the trigonal c-axis takes place at the same time, leaving essentially unchanged the magnitudes of the two basal axes. This dilatational change forces the trigonal lattice to transform into a lattice of hexagonal symmetry.

Different polymorphs can exist under certain conditions as a result of varying coordination schemes that become possible when the radius ratio for the particles lies close to one of the limits shown in Table 2. Examples of this behavior are the calcite and aragonite modifications of calcium carbonate [17]. In the rhombohedral structure of calcite, each calcium ion has the six oxygen atoms of two carbonate ions as neighbors, while in aragonite the nine oxygen atoms of three carbonates are nearest neighbors. It happens that the radius ratio for the calcium ion in the two forms is close to 0.73, which places it at the cusp that separates 8-coordination from 6-coordination. The thermal expansion of the lattice that accompanies heating a sample of calcite is sufficient to cause the phase transformation to aragonite.

Polymorphs can arise from the displacement of atomic layers or groups of layers in layer lattices. Examples from this class are the red tetragonal and yellow rhombic modifications of mercuric iodide, which are enantiotropically transformable at 127°C [18]. The red phase consists of a layered lattice, in which the mercury ions possess the unusual coordination number of four and a pseudo-tetrahedral configuration. This coordination number arises as a result of the very tight binding of the stable modification and the close packing of alternate layers of mercury and iodine atoms. In the high-temperature yellow form, the mercury atoms assume a coordination number of six in a pseudo-octahedral structure. The transformation arises from a shift in the relationship between the alternate layers of mercury and iodine atoms, originating from both a lattice expansion and a lateral shift. The transition is enantiotropic in nature since the phase change only requires that each mercury atom shift from the center of a pseudo-tetrahedron to the center of a structurally related pseudo-octahedron.

In other systems, new polymorphs can originate in the migration of atoms within a lattice in which the lattice particles have a relatively high mobility because the lattice can be regarded as one with unoccupied positions. This can also take the form of a change from a rigid lattice to one with mobile lattice particles. For instance, in the α and

γ forms of aluminum oxide, the substance is able to crystallize in an open lattice with unoccupied positions (γ-alumina) [19] or in a compact lattice (α-alumina) [20]. In the most stable α-form, the oxygen atoms in the lattice form a hexagonal close packing, the gaps of which are filled by a regular arrangement of aluminum atoms. The structure of the γ-phase is much more open, being a spinel lattice in which one-third of the spaces available for the metal atoms are unoccupied. The arrangement of the oxygen atoms in the γ-form approximates that of cubic close packing, possessing both octahedral and tetrahedral sites for the aluminum atoms.

Finally, new polymorphic structures can arise from different ways of assembling closely related coordination polyhedra. Three possibilities are known that lead to polymorphism of this type, which are most commonly found in oxidic lattices. First, the coordination polyhedron of one lattice particle is the same, while that of the other is only slightly changed (e.g., silicon dioxide). Second, the coordination polyhedron of one lattice particle is the same, and that of the other is considerably changed (e.g., titanium dioxide). Finally, the coordination polyhedra are both deformed to varying degrees, with the basic lattice type being the same but the type of distortion different (e.g., zirconium dioxide).

V. POLYMORPHISM AMONG MOLECULAR LATTICES

The survey given above cannot possibly represent a systematization of polymorphic substances, but the examples chosen from the range of inorganic structural possibilities serve to provide an insight into the underlying causes of polymorphism. The systems become more complex when one considers the structural causes of polymorphism among molecular lattices, where it becomes essential to develop the structural considerations by a closer examination of the states of bonding of the atoms in the different lattices.

For substances that form molecular lattices, different modifications can arise in two main distinguishable ways. Should the molecule exist as a rigid group of atoms with definite symmetry, these can be stacked differently to occupy the points of different lattices. This type of polymorphism is then attributable to packing phenomena and will fall into the categories of packing equivalent or nonequivalent spheres (or ellipsoids). On the other hand, if the molecule is not rigidly constructed and can exist in multiple conformational states, then each of these conformationally distinct modifications may be able to crystallize in its own lattice structure.

A. Polymorphism Resulting from the Packing of Conformationally Similar Molecules

The general concept of packing polymorphism will be illustrated through a discussion of representative examples where polymorphs have originated through different assembly modes of conformationally equivalent molecules. One theme that will be encountered throughout this section is that differing packing modes are readily available to compounds for which hydrogen bonding is important, and for which there exist multiple possibilities to achieve the necessary intramolecular and intermolecular interactions.

1. Maltol

Maltol (3-hydroxy-2-methyl-4H-pyran-4-on) is a flavoring agent used to impart the "freshly baked" odor to breads and cakes. The compound is conformationally rigid, and the most commonly encountered crystal form has a reported melting point of 161–162°C [21]. Two polymorphs have been reported, one of which was isolated from ethanol/water and the other from ethanol equilibrated with heptane, and the crystallographic data for these forms are summarized in Table 3 [22]. The two polymorphs exhibit different packing arrangements through variations in the intermolecular hydrogen bonding patterns.

In Form 1, there are two crystallographically unique molecules arranged alternately in the chain, with successive pairs related by a cell translation along the c-axis. The hydrogen bonds link the molecules to form a near planar chain. In Form 2, there are three crystallographically unique molecules. One of these molecules forms hydrogen bonded dimeric pairs mutually linked across a center of symmetry, while the other two molecules are hydrogen bonded to each other in a dimeric arrangement.

Table 3 Crystallographic Data for the Polymorphs of Maltol

	Form 1	Form 2
Crystal class	orthorhombic	monoclinic
Space group	Pca2$_1$	P2$_1$/c
Crystal habit	blocks	rods
Unit cell lengths	a = 7.134 Å	a = 7.166 Å
	b = 12.152 Å	b = 35.909 Å
	c = 13.304 Å	c = 6.982 Å
Unit cell angles	α = 90°	α = 90°
	β = 90°	β = 109.79°
	γ = 90°	γ = 90°
Molecules in unit cell	8	12
Cell volume	1153.4 Å^3	1690.5 Å^3
Density	1.453 g/mL	1.486 g/mL

Source: Ref. 22.

The maltol system represents a classic example of how different hydrogen bonding schemes can result in different polymorphs.

2. Ethyl Maltol

The crystal structures of three polymorphs of ethyl maltol (2-ethyl-3-hydroxy-4-pyranone) have been reported, with differing modes of hydrogen bonding giving rise to the three crystal forms [23]. Form 1 contains nearly planar chains of molecules, Form 2 contains three-dimensional or spiral chains, and Form 3 contains hydrogen bonded dimers. The crystallographic data reported for these three forms are summarized in Table 4.

In Forms 1 and 2, the hydrogen bonds link the molecules to form infinite chains. For Form 1, the result is a near-planar chain in which adjacent molecules are related by a twofold screw axis. Form 2 contains a three-dimensional chain in which adjacent molecules spiral along a threefold screw axis. In Form 3, pairs of molecules form hydrogen bonded dimers mutually linked across a center of symmetry. Representative views of the three structures are found in Fig. 4.

It is interesting to note that Forms 2 and 3 were obtained as unreacted components in reactions involving silicon trichloride or antimony

Table 4 Crystallographic Data for the Polymorphs of Ethyl Maltol

	Form 1	Form 2	Form 3
Crystal class	monoclinic	hexagonal	triclinic
Space group	$P2_1/n$	$\bar{R}3$	$\bar{P}1$
Crystal habit	blocks	plates	plates
Unit cell length	$a = 4.867$ Å	$a = 20.828$ Å	$a = 7.144$ Å
	$b = 6.895$ Å	$b = 20.828$ Å	$b = 7.641$ Å
	$c = 20.200$ Å	$c = 7.832$ Å	$c = 7.700$ Å
Unit cell angles	$\alpha = 90°$	$\alpha = 90°$	$\alpha = 62.13°$
	$\beta = 95.23°$	$\beta = 90°$	$\beta = 68.28°$
	$\gamma = 90°$	$\gamma = 120°$	$\gamma = 76.39°$
Molecules in unit cell	4	18	2
Cell volume	675.0 Å^3	2942 Å^3	344.2 Å^3
Density	1.379 g/mL	1.423 g/mL	1.352 g/mL

Source: Ref. 23.

trichloride. The effect on polymorphism of the other substances present in solution during the growth of crystals is not known.

3. Sulfamerazine

Sulfamerazine (4-amino-*N*-(4-methyl-2-pyrimidinyl)benzene-sulfon-amide) is an important antibacterial agent, for which the predominant polymorph is reported to exhibit a melting point of 234–238°C [21]. Owing to a variety of possible hydrogen bonding arrangements and ring-stacking modes, sulfonamides exhibit an array of polymorphic forms. The existence of at least two orthorhombic forms of sulfamerazine has been demonstrated by various physical methods, and the physical properties of these have been studied [24,25]. The crystallographic data obtained on Forms I and II are collected in Table 5.

In both forms, the repeating motif is a molecular dimer, which is pseudocentrosymmetric in Form I and centrosymmetric in Form II. The dimer is assembled through the formation of two *N*(amide)-*H*-*N*(pyrimidinyl) hydrogen bonds. The extent of hydrogen bonding is equivalent in the two structures, which accounts for the relatively low energy required to effect the Form-II-to-Form-I solid–solid phase transition. As

Form 1

Form 2

Form 3

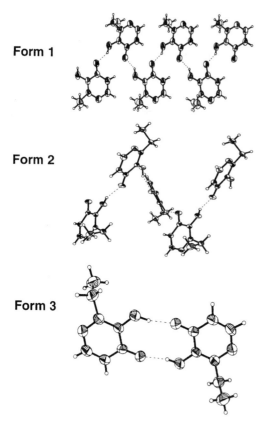

Fig. 4 Hydrogen bonding schemes associated with the three polymorphs of ethyl maltol. (The figure is adapted from data presented in Ref. 23.)

would be anticipated, the phase transition is accompanied by a decrease in the crystal density.

4. Iodoanilinium Picrate

Crystals of 2-iodoanilinium picrate can be isolated in three polymorphic forms, each of which exhibits a characteristically different color [26]. The structures of yellow Form I and green Form II have been obtained, but the structure of the red Form III could not be solved owing to its instability. Crystals of Form II were found to change into Form

Table 5 Crystallographic Data for the Polymorphs of Sulfamerazine

	Form I [25]	Form II [24]
Crystal class	orthorhombic	orthorhombic
Space group	Pn2$_1$a	Pbca
Crystal habit	rhombic prisms	tabular
Unit cell lengths	$a = 14.474$ Å	$a = 9.145$ Å
	$b = 21.953$ Å	$b = 11.704$ Å
	$c = 8.203$ Å	$c = 22.884$ Å
Unit cell angles	$\alpha = 90°$	$\alpha = 90°$
	$\beta = 90°$	$\beta = 90°$
	$\gamma = 90°$	$\gamma = 90°$
Molecules in unit cell	8	8
Cell volume	2606.5 Å^3	2449.3 Å^3
Density	1.347 g/mL	1.43 g/mL

Source: Refs. 24, 25.

I in a day. The crystallographic data obtained on the two characterized structures are collected in Table 6.

In Forms I and II, the iodoanilinium cations and picrate anions form nearly planar ion pairs, with the iodoanilinium ion being hydrogen bonded to the picrate ion by a N—H····O hydrogen bond. As illustrated in Fig. 5, the pairs related by translational symmetry in Form I are stacked to form segregated columns, in which the same ion types are stacked with each other. In addition, all the nitro groups lie essentially in the plane of the benzene ring. This structure may be contrasted with that of Form II, where the anions and cations are alternatively stacked along the *a*-axis to form continuous columns. In Form II, the oxygen atoms of the nitro groups at the ortho positions deviate from the plane of the benzene ring.

The iodoanilinium picrate system was found to exhibit thermochromism for both Form I and Form II. Form I crystals changed from yellow to red upon heating to 60°C, while the Form II crystal changed from green to red at temperatures above 60°C. The thermochromism is due to the polymorphic transformation to the red form (Form III), which is only stable at elevated temperatures. The changes in stacking mode alter the nature of the charge-transfer transitions be-

Table 6 Crystallographic Data for the Polymorphs of Iodoanilinium Picrate

	Form I	Form II
Crystal class	monoclinic	triclinic
Space group	$P2_1/n$	$P\bar{1}$
Crystal habit	prismatic	plates
Unit cell lengths	$a = 14.492$ Å	$a = 7.323$ Å
	$b = 4.3733$ Å	$b = 8.245$ Å
	$c = 24.153$ Å	$c = 13.656$ Å
Unit cell angles	$\alpha = 90°$	$\alpha = 75.71°$
	$\beta = 98.81°$	$\beta = 74.05°$
	$\gamma = 90°$	$\gamma = 75.64°$
Molecules in unit cell	4	2
Cell volume	1512.6 Å^3	753.7 Å^3
Density	1.968 g/mL	1.974 g/mL

Source: Ref. 26.

tween 2-iodoaniline and picric acid that are responsible for the observed color. In the yellow Form I, only the local excitation band at 400 nm is observed. However in the green Form II, the local excitation band (400 nm) is superimposed with the charge-transfer band observed at 500–700 nm. Upon passing into the red Form III, the charge-transfer band shifts toward higher energies (wavelength of 500 nm), and its combination with the local excitation band (400 nm) results in the observed change in color.

5. Lapachol

Lapachol (2-hydroxy-3-(3-methyl-2-butenyl)-1,4-naphthalenedione) is a natural pigment derived from the heartwood of certain tropical plants and is known to be an active antineoplastic agent. Two polymorphs of this compound have been obtained and characterized, one of which (Form I) was obtained by slow evaporation of an ethanol–ether mixture, and the other (Form II) being isolated by slow evaporation of an ethanol–water mixture [27]. The crystal packings of both forms show that the molecules form dimers through O—H⋯O hydrogen bonds

Form I

Form II

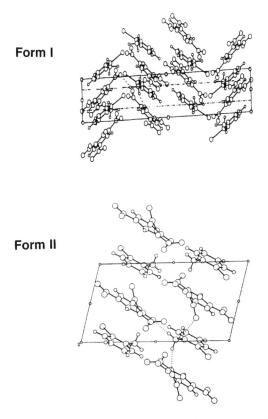

Fig. 5 The two types of hydrogen bonding and how these affect the molecular packing in the two polymorphs of iodoanilinium picrate. The projection of Form I is down the *a*-axis, while the projection of Form II is down the *b*-axis. (The figure is adapted from data presented in Ref. 26.)

around centers of symmetry. A summary of the crystallographic data is found in Table 7.

In both forms, the molecules were found to be composed of two nearly planar parts (the aromatic ring system and the allylic side chain) that are almost perpendicular to each other. The crystal packing in both forms is determined primarily by the π-interactions between aromatic rings, which leads to stacking forces and van der Waals forces. Only one intermolecular hydrogen bond was found in each structure, forming

Table 7 Crystallographic Data for the Polymorphs of Lapachol

	Form I	Form II
Crystal class	triclinic	monoclinic
Space group	P$\bar{1}$	P2$_1$/c
Crystal habit	plates	needles
Unit cell lengths	a = 5.960 Å	a = 6.035 Å
	b = 9.569 Å	b = 9.427 Å
	c = 10.670 Å	c = 20.918 Å
Unit cell angles	α = 96.82°	α = 90°
	β = 98.32°	β = 98.27°
	γ = 90.32°	γ = 90°
Molecules in unit cell	2	4
Cell volume	598.2 Å^3	1177.7 Å^3
Density	1.345 g/mL	1.366 g/mL

Source: Ref. 27.

dimers in both cases around centers of symmetry. The closer packing of the molecules in Form II is reflected in the higher crystal density observed for this polymorph.

6. Nitrofurantoin

Nitrofurantoin (*N*-(5-nitro-2-furfurylidene)-1-aminohydantoin) is widely used as a urinary tract antibacterial agent, but it has poor water solubility, and consequently its dissolution is the rate-determining step in its bioavailability. Complicating the situation is that the compound can exist in various crystal forms, among which are two anhydrate forms [28]. The structures of the two anhydrates have been reported, and the crystallographic data obtained for these are collected in Table 8. The α-form was crystallized from a hot acetic acid–water solution, while the β-form was obtained from acetone.

    In both polymorphs, the nitrofurantoin molecules pack in layers that are held together by van der Waals forces. The molecules associate to form centrosymmetric dimers through two identical intermolecular N—H····O hydrogen bonds. These dimers are themselves linked by a weaker C—H····O hydrogen bond in both anhydrous crystal structures. There is a considerable amount of intralayer hydrogen bonding dictat-

Table 8 Crystallographic Data for the Polymorphs of
Nitrofurantoin

	α-Form	β-Form
Crystal class	triclinic	monoclinic
Space group	P1̄	P2$_1$/n
Crystal habit	tabular	plates
Unit cell lengths	a = 6.774 Å	a = 7.840 Å
	b = 7.795 Å	b = 6.486 Å
	c = 9.803 Å	c = 18.911 Å
Unit cell angles	α = 106.68°	α = 90°
	β = 104.09°	β = 93.17°
	γ = 92.29°	γ = 90°
Molecules in unit cell	2	4
Cell volume	477.6 Å^3	960.2 Å^3
Density	1.656 g/mL	1.648 g/mL

Source: Ref. 28.

ing the packing mode in the β-form, which is less extensive for the α-
form. This accounts for the very slight difference (0.5%) in crystal
densities between the two forms.

7. FK664

FK664 ((*E*)-6-(3,4-dimethoxyphenyl)-1-ethyl-4-mesitylimino-3-methyl-
3,4-dihydro-2(1*H*)-pyrimidinone) is an orally effective noncatechol
and nonglycoside cardiotonic agent and has been found to exist in two
polymorphs and a hemihydrate. The crystal structures of the two anhydrate phases have been reported, and the crystallographic data obtained
on these are found in Table 9 [29]. Although the two polymorphs are
characterized by the same space group, and even contain the same number of molecules in their respective unit cells, the two forms still yield
very different x-ray powder diffraction patterns.

As is visible in Fig. 6, the packing of FK664 molecules in the
two anhydrate crystal forms is quite different. There are no notable
hydrogen bonds holding the molecules together; instead the two forms
are held together by different arrays of van der Waals forces. In crystals
of Form B, some close molecular contact results in π–π interactions

Table 9 Crystallographic Data for the Polymorphs of FK664

	Form A	Form B
Crystal class	monoclinic	monoclinic
Space group	$P2_1/c$	$P2_1/c$
Crystal habit	not reported	not reported
Unit cell lengths	$a = 13.504$ Å	$a = 8.067$ Å
	$b = 6.733$ Å	$b = 15.128$ Å
	$c = 24.910$ Å	$c = 18.657$ Å
Unit cell angles	$\alpha = 90°$	$\alpha = 90°$
	$\beta = 96.55°$	$\beta = 102.34°$
	$\gamma = 90°$	$\gamma = 90°$
Molecules in unit cell	4	4
Cell volume	2250.2 Å^3	2226.4 Å^3
Density	1.203 g/mL	1.216 g/mL

Source: Ref. 29.

between two dimethozyphenyl groups related with the inversion center in the crystal lattice. This interaction seems to contribute to stabilizing the crystal structure of Form B relative to Form A, and it may account for the slightly higher density observed for Form B. It was noted in this work that grinding Form A produced an amorphous material that easily transformed into Form B.

8. 4-Stryrlcoumarin

4-Styrylcoumarin has been shown to crystallize out of chloroform and hexane in two different polymorphs [30]. The crystal structures of the two forms differed in the nature of the molecular packing, and these differences were sufficient to permit different pathways of photochemistry in the solid state. A summary of the crystallographic data obtained on the two forms is located in Table 10.

When excited by ultraviolet radiation, Form I undergoes a stereospecific photodimerization to produce a head-to-tail dimer across the pyrone double bond. However, an examination of the structure reveals that considerable molecular motion must take place before the photoreaction can take place. In fact, the reaction might not proceed at all if

Form A

Form B

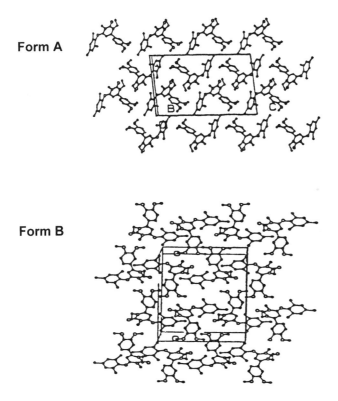

Fig. 6 Views of the molecular packing along the short axis for the two polymorphs of (*E*)-6-(3,4-dimethoxyphenyl)-1-ethyl-4-mesitylimino-3-methyl-3,4-dihydro-2(1*H*)-pyrimidinone. The projection of Form A is along the *b*-axis, while the projection of Form B is along the *a*-axis. (The figure is adapted from data presented in Ref. 29.)

it were not for the cavities that exist in the structure. The molecular motions during the photodimerization process do not favor the juxtaposition of the correct groups, and consequently the photodimer yield is relatively poor.

Form II reacts to produce a photodimer product across the styrene double bond. The reactivity of this form follows directly from topochemical considerations, since the pyrone double bonds of the centro-

Table 10 Crystallographic Data for the Polymorphs
of 4-Styrylcoumarin

	Form I	Form II
Crystal class	triclinic	monoclinic
Space group	P$\bar{1}$	P2$_1$/c
Crystal habit	prisms	needles
Unit cell lengths	a = 11.082 Å	a = 13.418 Å
	b = 11.215 Å	b = 5.720 Å
	c = 12.127 Å	c = 17.840 Å
Until cell angles	α = 103.35°	α = 90°
	β = 116.41°	β = 110.79°
	γ = 98.62°	γ = 90°
Molecules in unit cell	4	4
Cell volume	1265.9 Å^3	1280.1 Å^3
Density	1.299 g/mL	1.290 g/mL

Source: Ref. 30.

symmetrically related pairs are well positioned in the structure for overlap. The reaction proceeds rapidly and is 100% complete in about 10
hours of irradiation.

B. Polymorphism Resulting from the Packing of Conformationally Different Molecules

Organic molecules (which of course account for most pharmaceutical
agents) can exist in a variety of conformations that have similar energies. In many cases, the molecule adopts the conformation of lowest
energy when it packs into a crystal. The packing-type polymorphism
discussed in the previous section usually represents different modes by
which molecules in their most favorable conformational state have been
assembled into alternate solid structures. In some cases, however, a
molecule for which different conformational states are possible yields a
different polymorph when its different conformations are packed. This
conformational polymorphism is therefore defined as the existence of
different conformers of the same molecule in different polymorphic
modifications.

Bernstein has provided a comprehensive explanation for the polymorphism that can arise from the packing of molecules in different conformational states [31]. Although it is ordinarily taken that the various conformational states available to a molecule are essentially isoenergetic, computational chemists have long known that this is not the case. One can define conformations as those various shapes of molecules that arise by rotations about single bonds, and that correspond to potential energy minima. Since the crystal structure of a molecule will yield information about the preferred conformation of that molecule in a given polymorph, it follows that the various arrangements of atoms or conformation cannot differ too significantly in free energy content from the equilibrium structure of the molecule. It is important to emphasize that both the crystal structure and the molecular conformation represent potential energy minima that are not necessarily unique, and that there may exist a number of possible molecular geometries of very nearly the same energy for both the molecular conformation and the crystal structure. It is this proximity of molecular conformational energies that makes possible the existence of different molecular conformations in a single crystal structure, or conformational polymorphism when they are in different crystal structures.

The differences in lattice energy among different crystal forms of an organic compound can be expected to be in the range of 1–2 kcal/mol, especially when van der Waals interactions dominate the structure. From estimates of the magnitudes of intermolecular interactions, this is within the range of energy required to bring about changes in molecular torsional parameters about single bonds. However, the energy required to perturb bond lengths and bond angles significantly far exceeds this range. Therefore for those molecules that possess torsional degrees of freedom, various polymorphs can exhibit significantly different molecular conformations. This represents the rationale for crystal forces playing a role in determining the conformation of a molecule and yielding the phenomenon of conformational polymorphism.

1. p-(N-chlorobenzylidene)-p-chloroaniline

Bernstein has described methodology suitable for the study of how crystal forces can influence molecular conformation, and he has eluci-

dated the basic tenets of conformational polymorphism. [32]. A combination of quantitative analysis of the molecular packing in different space groups along with *ab initio* molecular orbital calculations was used. Choosing the model system of *p*-(*N*-chlorobenzylidene)-*p*-chloroaniline, lattice energy calculations involving minimization of the energy of the triclinic and orthorhombic crystal forms of this molecule were performed. The results of these calculations were able to explain the stability of the lattice structure in which the unstable planar conformation of the molecule is obtained. Three different potential functions were evaluated, and all potential functions yielded lower energies for the triclinic lattice (in agreement with experimental observation).

The crystallographic data for the two forms are summarized in Table 11, while the crystal packing and molecular conformations are illustrated in Fig. 7. The molecule was found to adopt a planar conformation in the triclinic phase, but in the orthorhombic phase the phenyl rings were rotated by 24.8° about the N-phenyl and CH-phenyl bonds. Both structures may be described as being composed of blocks of molecules between two neighboring parallel planes. For the triclinic polymorph the blocks are stacked along the *b*-axis, while in the orthorhom-

Table 11 Crystallographic Data for the Polymorphs of *p*-(*N*-chlorobenzylidene)-*p*-chloroaniline

	Form I	Form II
Crystal class	orthorhombic	triclinic
Space group	Pccn	$P\bar{1}$
Crystal habit	prisms	needles
Unit cell lengths	$a = 24.503$ Å	$a = 5.986$ Å
	$b = 6.334$ Å	$b = 3.933$ Å
	$c = 7.326$ Å	$c = 12.342$ Å
Unit cell angles	$\alpha = 90°$	$\alpha = 87.38°$
	$\beta = 90°$	$\beta = 78.40°$
	$\gamma = 90°$	$\gamma = 89.53°$
Molecules in unit cell	4	4
Cell volume	284.2 Å^3	284.3 Å^3
Density	not reported	not reported

Source: Ref. 32.

**Triclinic
Form**

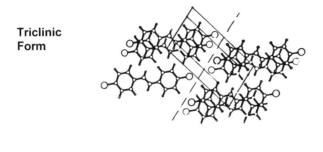

**Orthorhombic
Form**

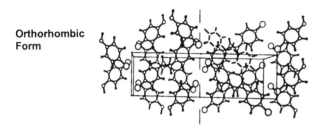

Fig. 7 Views of the molecular packing for the triclinic and orthorhombic polymorphs of *p*-(*N*-chlorobenzylidene)-*p*-chloroaniline. (The figure is adapted from data presented in Ref. 32.)

bic structure the stacking is due to a glide along the *c*-axis rather than a simple translation. Within the blocks, the stacks are related by a translation along the *a*-axis in the triclinic structure and a translation along the *b*-axis in the orthorhombic form. In the triclinic form, disorder was noted about a center of symmetry, while in the orthorhombic form the disorder was about a twofold crystallographic axis.

Subsequently, additional studies of the effect of crystal forces on molecular conformations were reported for various methyl analogues of *N*-(*p*-chlorobenzylidene)-*p*-chloroaniline [33]. The *ab initio* calculations were able to predict the relative stability of the various polymorphs, and by using a number of potential functions the authors were able to demonstrate the ruggedness of the method. Taken together, the two works demonstrate that use of partial atomic energies for the analysis of changes in molecular environments in crystals is valid, even when the energy differences involved are small in magnitude.

2. Iminodiacetic Acid

Iminodiacetic acid (N-(carboxymethyl)glycine) has been found to crystallize in three polymorphs, for which the structures are composed of molecules having different conformational states. The crystal structure of Form I was reported by Boman et al. [34], while the structures of Forms II and III were reported by Bernstein [35]. A summary of the crystallographic data obtained for the three polymorphic forms is given in Table 12. The molecular conformations, as measured by comparison of torsion angles, were found to differ significantly among the three forms. These differences were related to the packing scheme, as determined predominantly by the hydrogen bonding network.

The differences in molecular conformation are evident in Fig. 8, where ORTEP drawings of the iminodiacetic acid unit in the three structures are shown. Significant differences in torsional angles around the N—C bond serve to distinguish the three conformations. The other

Table 12 Crystallographic Data for the Polymorphs of Iminodiacetic Acid

	Form I [34]	Form II [35]	Form III [35]
Crystal class	monoclinic	orthorhombic	monoclinic
Space group	$P2_1/c$	$Pbc2_1$	$P2_1/n$
Crystal habit	short prismatic	thin plates	needles
Unit cell lengths	$a = 6.341$ Å	$a = 5.267$ Å	$a = 5.258$ Å
	$b = 9.136$ Å	$b = 14.140$ Å	$b = 12.206$ Å
	$c = 9.378$ Å	$c = 14.933$ Å	$c = 8.709$ Å
Unit cell angles	$\alpha = 90°$	$\alpha = 90°$	$\alpha = 90°$
	$\beta = 92.72°$	$\beta = 90°$	$\beta = 90°$
	$\gamma = 90°$	$\gamma = 90°$	$\gamma = 100.14°$
Molecules in unit cell	4	8	4
Cell volume	542.7 Å^3	1112.2 Å^3	550.2 Å^3
Density	1.629 g/mL	1.590 g/mL	1.607 g/mL

Source: Refs. 34, 35.

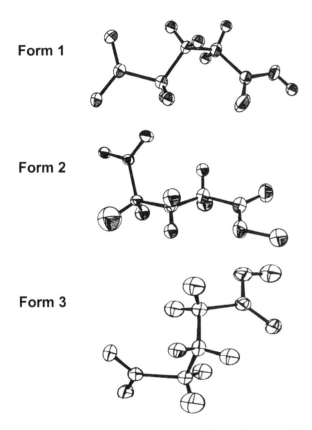

Fig. 8 Views of the molecular conformations associated with the three poly-morphs of iminodiacetic acid. (The figure is adapted from data presented in Ref. 34 and 35.)

major conformational differences relate to the carboxyl groups and the nitrogen atom.

All hydrogen atoms present on the carboxylate groups participate in a single hydrogen bond, but the role of the oxygen atoms differs significantly among the three forms. Differences in the crystallographic symmetry elements relating hydrogen bonded molecules were also re-ported. The variations noted in hydrogen bonding are seen to be a direct consequence of the packing of molecules having different conforma-tional states.

3. p-*methyl*-N-(p-*methylbenzylidine*)aniline

Bar and Bernstein continued their work on conformational polymorphism, using lattice energy minimization techniques to characterize the influence of crystal forces (as well as orientational and positional disorder) on the molecular conformation of p-methyl-*N*-(p-methylbenzylidine)aniline [36]. This compound has been obtained in three polymorphic forms, in which the title molecule was found to adopt different conformations in each form. A summary of the reported crystallographic data is found in Table 13.

Form I was found to crystallize in the monoclinic space group $P2_1/c$, with four molecules in the unit cell. This form exhibited a positional disorder, and it was found that there were two orientations of the bridge atoms with nonequivalent occupancy. The observed disorder led to fairly large displacements of the bridge atoms, but significantly smaller displacements for the ring atoms and methyl groups. Form II was found to crystallize in a nondisordered chiral monoclinic space group ($P2_1$) and hence would exhibit optical activity within this crystalline state. Finally, Form III also crystallizes in the monoclinic space group $P2_1/c$ but contains only two molecules per unit cell. Although

Table 13 Crystallographic Data for the Polymorphs of p-methyl-*N*-(p-methylbenzylidine)aniline

	Form 1	Form 2	Form 3
Crystal class	monoclinic	monoclinic	monoclinic
Space group	$P2_1/c$	$P2_1$	$P2_1/c$
Crystal habit	not reported	not reported	not reported
Unit cell lengths	$a = $ 6.089 Å	$a = $ 6.891 Å	$a = $ 9.878 Å
	$b = $ 7.751 Å	$b = $ 7.153 Å	$b = $ 4.884 Å
	$c = $ 26.766 Å	$c = $ 12.600 Å	$c = $ 12.018 Å
Unit cell angles	$\alpha = $ 90°	$\alpha = $ 90°	$\alpha = $ 90°
	$\beta = $ 103.16°	$\beta = $ 102.70°	$\beta = $ 90.48°
	$\gamma = $ 90°	$\gamma = $ 90°	$\gamma = $ 90°
Molecules in unit cell	4	2	2
Cell volume	1230.07 Å^3	605.88 Å^3	579.78 Å^3
Density	not reported	not reported	not reported

Source: Ref. 36.

disorder was observed at the bridge atoms, this phenomenon was different in character from that noted for Form I.

Lattice energy calculations were performed on the three polymorphs in order to account for the relative stability of Forms I and III, where the unstable planar conformation is prevalent. The minimizations were carried out with three different potential functions, and it was found that Form II was characterized by the highest energy. The energies of the other two forms were found to be somewhat comparable and were considerably less than the energy of Form II. The minimized lattice energies were analyzed in terms of partial atomic energy contributions to the total energy, and it was shown that the relative contribution of various groups to the total energy was the same for the three polymorphs. The stability of Forms I and III with respect to Form II was attributed to the relatively favorable environment of the bridge atoms, demonstrating the important role of disorder in the stabilization of the energetically less favorable planar conformation.

4. Probucol

Probucol (4,4'-[(1-methylethylidene)bis(thio)]-bis-[2,6-bis(1,1-dimethylethyl)phenol]) is reported in the *Merck Index* to exist as either white crystals (having a melting point of 124.5–126°C) or yellow crystals (having a melting point of 125.5–126.5°C) [21]. A third form having an onset of melting of 116°C has also been reported. Form II will eventually transform into Form I upon standing, and the conversion has also been observed upon grinding. The structures of Forms I and II have been reported [37], and a summary of the crystallographic data is provided in Table 14.

The conformations of the probucol molecule in each form are quite different, as is illustrated in Fig. 9. In Form II, the C—S—C—S—C chain is extended, and the molecular symmetry approximates C_{2v}. This symmetry is lost in Form I, where the torsional angles around the two C—S bonds deviate significantly from 180°. Steric crowding of the phenolic groups by the *t*-butyl groups was evident from deviations from trigonal geometry at two phenolic carbons in both forms. Using a computational model, the authors found that the energy of Form II was 26.4 kJ/mol higher than the energy of Form

Table 14 Crystallographic Data for the Polymorphs of Probucol

	Form I	Form II
Crystal class	monoclinic	monoclinic
Space group	$P2_1/c$	$P2_1/n$
Crystal habit	not reported	not reported
Unit cell lengths	$a = 16.972$ Å	$a = 11.226$ Å
	$b = 10.534$ Å	$b = 15.981$ Å
	$c = 19.03$ Å	$c = 18.800$ Å
Unit cell angles	$\alpha = 90°$	$\alpha = 90°$
	$\beta = 113.66°$	$\beta = 104.04°$
	$\gamma = 90°$	$\gamma = 90°$
Molecules in unit cell	4	4
Cell volume	3116.0 Å^3	3272.0 Å^3
Density	1.102 g/mL	1.049 g/mL

Source: Ref. 37.

I, indicating that the less symmetrical conformer is more stable. The Form I conformer packs in the polymorph whose density is 5% higher than that of Form II, indicating the existence of more efficient packing.

The authors examined space-filling models of the two conformers and concluded that the phenolic groups were poorly exposed on the molecular surface. The inaccessibility of the phenolic groups was taken to explain both the low solubility of the drug in water and the absence of hydrogen bonding in either of the two crystal forms.

5. *Spironolactone*

Spironolactone (17-hydroxy-7α-mercapto-3-oxo-17α-pregn-4-ene-21-carboxylic acid), a diuretic steroidal aldosterone agonist, has been obtained into anhydrous polymorphs as well as four solvated crystalline forms [38]. The compound shows variable and incomplete oral absorption due to its poor water solubility and dissolution rate, factors that are compounded by the existence of polymorphism and solvate formation. The two anhydrate forms (Forms I and II) both crystallize in orthorhombic space groups, and the crystallographic data for these are collected in Table 15. The spironolactone system exhibits the classic

Form I

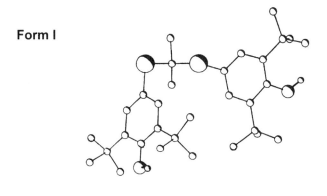

Form II

Fig. 9 Molecular conformations adopted by probucol in its two polymorphic forms. (The figure is adapted from data presented in Ref. 37.)

behavior where the metastable Form I crystallizes more rapidly than does the thermodynamically more stable Form II.

Form III is a 2:1 acetonitrile solvate, Form IV is a 2:1 ethanol solvate, Form V is a 4:1 ethyl acetate solvate, and Form VI is a 1:2 methanol solvate. Interestingly, a definite formula could only be obtained for Form III that agreed with the results of the crystallographic study. The solvent content of the other solvates was in conflict with the determined space group and may be indicative of disorder in the solid.

Examination of the intermolecular distances in the various forms indicates that the predominant cohesive forces are van der Waals interactions, and all crystals are characterized by close molecular packing.

Table 15 Crystallographic Data for the Polymorphs of
Spironolactone

	Form I	Form II
Crystal class	orthorhombic	orthorhombic
Space group	$P2_12_12_1$	$P2_12_12_1$
Crystal habit	needle-like	prisms
Unit cell lengths	$a = 9.979$ Å	$a = 10.584$ Å
	$b = 35.573$ Å	$b = 18.996$ Å
	$c = 6.225$ Å	$c = 11.005$ Å
Unit cell angles	$\alpha = 90°$	$\alpha = 90°$
	$\beta = 90°$	$\beta = 90°$
	$\gamma = 90°$	$\gamma = 90°$
Molecules in unit cell	4	4
Cell volume	2209.8 Å^3	2212.6 Å^3
Density	not reported	not reported

Source: Ref. 38.

The steroid nuclei are almost planar in all forms, and they are perpendicular to the E-ring and to the 7-α-acetylthio side chain. Comparison of torsion angles demonstrates the existence of conformational polymorphism. Differences in carbon–carbon bond angles are noted in the flexible A-rings of all forms, with this ring being in a "half-chair" conformation in Form I and a "sofa" conformation in Form II. The conformation of the rigid B-ring and C-ring is relatively fixed in the various forms, while the conformation of the flexible D-ring and E-ring varies significantly in the various forms.

6. *Lomeridine Dihydrochloride*

Lomeridine dihydrochloride (1-[bis(4-fluorophenyl)methyl]-4-(2,3,4-trimethoxygenzyl)-piperazine dihydrochloride) has been shown to be capable of existing in two polymorphic forms, for which the structures have been reported [39]. Form I was obtained from acetonitrile solution, while Form II was crystallized from methanol, ethanol, n-propanol, i-propanol, n-butanol, or methyl ethyl ketone. Crystallographic data reported for the two forms are found in Table 16.

Table 16 Crystallographic Data for the Polymorphs of Lomeridine
Dihydrochoride

	Form I	Form II
Crystal class	triclinic	trigonal
Space group	P$\bar{1}$	R$\bar{3}$
Crystal habit	not reported	not reported
Unit cell lengths	a = 19.487 Å	a = 20.060 Å
	b = 11.930 Å	b = 20.060 Å
	c = 6.920 Å	c = 20.060 Å
Unit cell angles	α = 118.27°	α = 90°
	β = 90.99°	β = 90°
	γ = 97.35°	γ = 117.30°
Molecules in unit cell	2	6
Cell volume	1399.6 Å^3	4504 Å^3
Density	1.285 g/mL	1.198 g/mL

Source: Ref. 39.

Although the molecular conformations in the two forms are similar in many respects, torsion about one bond is sufficient to yield the conformational polymorphism. The two conformations of the lomeridine unit are illustrated in Fig. 10, where it is evident that the sole conformational difference consists in the rotation about a single C—N bond.

However, the packing implications of this slight conformational difference are profound. Form I exhibits head-to-head and tail-to-tail packing toward the diagonal axis of the *ab*-plane. On Form II, a three fold screw axis passes by one of the *p*-fluorobenzenes, which stacks continuously, and the other *p*-fluorobenzene lies along the axis. It was deduced that the difference in packing stability between the two polymorphs is caused mainly by the stability of the packing of the trimethoxy-benzene units.

7. 5-Methyl-2-[(2-nitrophenyl)amino]-thiophenecarbonitrile
5-Methyl-2-[(2-nitrophenyl)amino]-thiophenecarbonitrile represents a system that exhibits both conformational and color polymorphism [40].

Form I **Form II**

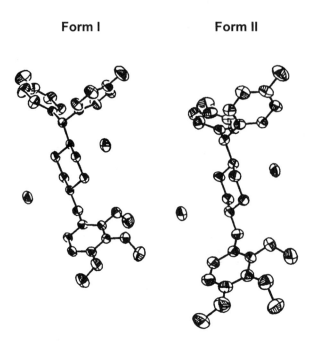

Fig. 10 Views of the molecular conformations adopted by lomeridine dihydrochloride in its two polymorphic forms. (The figure is adapted from data presented in Ref. 39.)

Although the compound is known to crystallize in five crystal forms, of greatest interest are the red, orange, and yellow polymorphs. These crystal forms are stable at room temperature for many years, with the yellow form being most stable at room temperature. Crystallographic data for this polymorphic system are collected in Table 17.

The red, orange, and yellow forms differ in the angle between the plane of the phenyl and that of the thiophene ring, with values of 106° (yellow form), 54° (orange form), and 46° (red form) being reported. It was hypothesized that the different colors were due to the different degrees of overlap of the double bonds in these two rings and accompanying perturbation of the molecular orbitals in each case. This hypothesis is consistent with calculations of the overlap and absorption spectra of 5-methyl-2-[(2-nitrophenyl)amino]-3-thiophenecarbonitrile.

Table 17 Crystallographic Data for the Polymorphs of 5-methyl-2-[(2-nitrophenyl)amino]-thiophenecarbonitrile

	Yellow form	Orange form	Red form
Crystal class	monoclinic	monoclinic	triclinic
Space group	P2$_1$/n	P2$_1$/c	P$\bar{1}$
Crystal habit	blocks	needles	blocks
Unit cell lengths	$a = 8.5001$ Å	$a = 3.9453$ Å	$a = 7.4918$ Å
	$b = 16.413$ Å	$b = 18.685$ Å	$b = 7.7902$ Å
	$c = 8.5371$ Å	$c = 16.3948$ Å	$c = 11.9110$ Å
Unit cell angles	$\alpha = 90°$	$\alpha = 90°$	$\alpha = 75.494°$
	$\beta = 91.767°$	$\beta = 93.830°$	$\beta = 77.806°$
	$\gamma = 90°$	$\gamma = 90°$	$\gamma = 63.617°$
Molecules in unit cell	not reported	not reported	not reported
Cell volume	not reported	not reported	not reported
Density	1.447 g/mL	1.428 g/mL	1.438 g/mL

Source: Ref. 40.

8. 8-(2-Methoxycarbonylamino-6-methylbenzyloxy)-2-methyl-3-(2-propynyl)-imidazo-[1,2-a]-pyridine

8-(2-Methoxycarbonylamino-6-methylbenzyloxy)-2-methyl-3-(2-propynyl)-imidazo-[1,2-a]-pyridine has been investigated as an imidazopyridine derivative and an orally effective antiulcer agent potentially useful as an inhibitor of H$^+$ - K$^+$ ATPase. The title compound has been obtained in two polymorphic forms, and these have been characterized as to their crystal structures to clarify the differences in conformational features and crystal packing [41]. Crystallographic data for both polymorphs are found in Table 18.

The conformational differences between the two forms were found to be fairly subtle, but they were sufficient to yield different crystal structures. The main differences noted were variations in hydrogen bonding between Forms A and B, and several torsional angles. Consideration of the structures indicated that differences in steric repulsion between the amide group and the benzene ring in Form B might be slightly higher than the corresponding interaction in Form A.

However, when the molecules consisting of different conformational states are packed into the crystalline form, they are assembled

Table 18   Crystallographic Data for the Polymorphs of 8-(2-Methoxycarbonylamino-6-methylbenzyloxy)-2-methyl-3-(2-propynyl)-imidazo-[1,2-a]-pyridine

	Form I	Form II
Crystal class	monoclinic	monoclinic
Space group	C2/c	$P2_1/c$
Crystal habit	needles	needles
Unit cell lengths	$a = 42.936$ Å	$a = 4.367$ Å
	$b = 4.356$ Å	$b = 38.214$ Å
	$c = 21.536$ Å	$c = 11.253$ Å
Unit cell angles	$\alpha = 90°$	$\alpha = 90°$
	$\beta = 109.92°$	$\beta = 95.47°$
	$\gamma = 90°$	$\gamma = 90°$
Molecules in unit cell	8	4
Cell volume	3786.7 Å^3	1869.4 Å^3
Density	1.275 g/mL	1.292 g/mL

Source: Ref. 41.

in significantly different patterns. As shown in Figure 11, one observes a partial overlap of imidazopyridine planes along the shortest axis, which suggests the existence of a stacking interaction. Since the area of the superimposed planes in Form A is slightly larger than that in Form B, the stacking forces differ. The authors concluded that the imidazopyridine planes in the Form A crystal would be stacked more rigidly than in Form B. Since the stacking interaction is the main intermolecular force, the packing forces are dominated by the stacking in the direction of the shortest axis, and by van der Waals forces in the other two crystal directions.

VI.   SUMMARY

It is apparent that the molecules are capable of adopting multiple solid-state forms upon crystallization and that the rationales for the existence of different polymorphic states are wide and varied. Understanding the origins of such phenomena becomes the first step in prediction, and

Form A

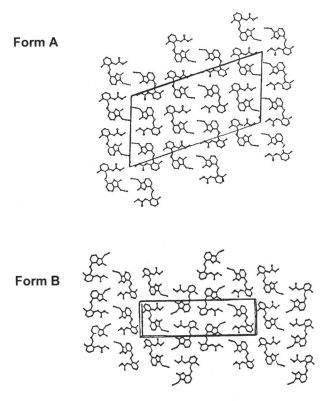

Form B

Fig. 11 Views of the molecular packing along the short axis for the two polymorphs of 8-(2-methoxycarbonylamino-6-methylbenzyloxy)-2-methyl-3-(2-propynyl)-imidazo-[1,2-*a*]-pyridine. The projection of Form A is along the *b*-axis, while the projection of Form B is along the *a* -axis. (The figure is adapted from data presented in Ref. 41.)

prediction in turn should yield mechanisms for the control of polymorphs. Approaches to identifying the most stable polymorphs are important and must be used in order to guide the selection of the best form for development.

In recent years, approaches to the calculation of the crystal structure of various polymorphs have been introduced and discussed. Such methods were introduced in Chapter 1 of this book. If crystal structures can be calculated with certainty it will, of course, be possible to predict the polymorphism of a compound and use this information to guide experimental studies. This remains a difficult goal to achieve, and it requires the development of better force fields. However, Desiraju has recently presented a balanced view of the possibility of calculating crystal structures [43]. In some cases it is possible to calculate them, especially if the space group is known. The calculation is more difficult for other cases, particularly those involving salts and when several molecules are present in an asymmetric unit. However, it will be interesting to observe the developments in this area as force fields improve and the speed of computation increases.

Compounds under pharmaceutical development will continue to adopt structures differing in their free energy, and it will probably be true that the most thermodynamically favored form will only appear after some time has elapsed. However, if the research teams have done their job properly, the full details of the polymorphism will be known long before the drug reaches Phase 3, and then we will encounter no surprises.

REFERENCES

1. R. J. Haüy, Essai d'une théorie sur la structure des cristaux, Paris, 1784, in *Traité de Cristallographie*, Paris, 1801.
2. A. L. Seeber, Ann. Phys., *76*, 229, 349 (1824).
3. W. Barlow, *Nature, 29*, 186, 205, 404 (1883).
4. W. Friedrich, P. Knipping, and M. Laue, *Math.-Phys. Klasse, 303* (1912).
5. W. L. Bragg, *Proc. Roy. Soc. London, A89*, 248 (1913).
6. A. Bravais, *Etudes Crystallographiques*, Gautier-Villars, Paris, 1866.
7. G. Burns and A. M. Glazer, *Space Groups for Solid State Scientists*, Academic Press, New York, 1978.
8. A. F. Wells, *Structural Inorganic Chemistory*, 5th ed., Clarendon Press, Oxford, 1984.
9. K. A. Gschneider, Crystallography of the rare-earth metals, in *The Rare Earths* F. H. Spedding and A. H. Daane, eds), John Wiley, New York, 1961.
10. F. A. Cotton and G. Wilkinson, *Advanced Inorganic Chemistry*, 5th ed., John Wiley, New York, 1988, pp. 654–655.

11. R. Hultgren, N. S. Gingrich, and B. E. Warren, *J. Chem. Phys., 3*, 351 (1935).
12. B. E. Warren and J. T. Burwell, *J. Chem. Phys., 3*, 6 (1935); J. T. Burwell, *Z. Krist., 97*, 123 (1937).
13. W. Hückel, *Structural Chemistry of Inoganic Compounds*, Vol. 2, Elsevier, Amsterdam, 1951, pp. 687–688.
14. G. Wagner and L. Lippert, *Z. Physik. Chem., B31*, 263 (1936).
15. W. L. Bragg, *Proc. Roy. Soc. London, A89*, 468 (1913).
16. F. C. Kracek, E. Posnjak, and S. B. Hendricks, *J. Am. Chem. Soc., 53*, 3339 (1931).
17. L. Bragg and G. F. Claringbull, *Crystal Structures of Minerals*, Cornell Univ. Press, Ithaca, New York, 1965, pp. 127–134.
18. J. M. Bijvoet, A. Claassen, and A. Karssen, *Proc. K. Acad. Wetensch. Amsterdam, 29*, 529 (1926).
19. G. Hägg and G. Söderholm, *Z. Physik. Chem., B29*, 88 (1935); G. Hägg, *Z. Krist., 91*, 114 (1935).
20. V. Schmaeling, *Z. Krist., 67*, 1 (1928); J. Topping, *Proc. Roy. Soc. London, A122*, 251 (1928).
21. The Merck Index, 12th ed. Merck 1996.
22. J. Burgess, J. Fawcett, D. R. Russell, R. C. Hider, M. B. Hossain, C. R. Stoner, and D. van der Helm, *Acta Cryst., C52*, 2917 (1996).
23. S. D. Brown, J. Burgess, J. Fawcett, S. A. Parsons, D. R. Russell, and E. Waltham, *Acta Cryst., C51*, 1335 (1995).
24. K. R. Acharya, K. N. Juchele, and G. Kartha, *J. Cryst. Spect. Res., 12*, 369 (1982).
25. M. R. Caira and R. Mohamed, *Acta Cryst., B48*, 492 (1992).
26. M. Tanaka, H. Matsui, J.-I. Mizoguchi, and S. Kashino, *Bull. Chem. Soc. Jp., 67*, 1572 (1994).
27. I. K. Larsen and L. A. Andersen, *Acta Cryst., C48*, 2009 (1992).
28. E. W. Pienaar, M. R. Caira, and A. P. Lötter, *J. Cryst. Spect. Res., 23*, 785 (1993).
29. A. Miyamae, S. Kitamura, T. Tada, S. Koda, and T. Yasuda, *J. Pharm. Sci., 80*, 995 (1991).
30. J. N. Moorthy and K. Venkatesan, *Bull. Chem. Soc. Jp., 67*, 1 (1994).
31. J. Bernstein, Conformational polymorphism, Chap. 13 in *Organic Solid State Chemistry* (G. R. Desiraju, ed.), Elsevier, Amsterdam, 1987.
32. J. Bernstein and A. T. Hagler, *J. Am. Chem. Soc., 100*, 673 (1978).
33. A. T. Hagler and J. Bernstein, *J. Am. Chem. Soc., 100*, 6349 (1978).
34. C.-E. Boman, H. Herbertsson, and A. Oskarsson, *Acta Cryst., B30*, 378 (1974).

35. J. Bernstein, *Acta Cryst.*, *B35*, 360 (1979).

36. I. Bar and J. Bernstein, *J. Phys. Chem.*, *86*, 3223 (1982).

37. J. J. Gerber, M. R. Caira, and A. P. Lötter, *J. Cryst. Spect. Res.*, *23*, 863 (1993).

38. V. Agafonov, B. Legendre, N. Rodier, D. Wouessidejewe, and J.-M. Cense, *J. Pharm. Sci.*, *80*, 181 (1991).

39. Y. Hiramatsu, H. Suzuke, A. Kuchiki, H. Nakagawa, and S. Fujii, *J. Pharm. Sci.*, *85*, 761 (1996).

40. G. A. Stephenson, T. B. Borchardt, S. R. Byrn, J. Bowyer, C. A. Bunnell, S. V. Snorek, and L. Yu, *J. Pharm. Sci.*, *84*, 1385 (1995).

41. A. Miyamae, S. Koda, S. Kitamura, Y. Okamoto, and Y. Morimoto, *J. Pharm. Sci.*, *79*, 189 (1990).

42. N. Feeder and W. Jones, *Acta Cryst.*, *C50*, 816 (1994).

43. G. R. Desiraju, *Science*, *278*, 404 (1997).

4

Structural Aspects of Hydrates and Solvates

Kenneth R. Morris

Purdue University
West Lafayette, Indiana

I. PHARMACEUTICAL IMPORTANCE OF CRYSTALLINE HYDRATES

The potential pharmaceutical impact of changes in hydration state of crystalline drug substances and excipients exists throughout the development process. The behavior of pharmaceutical hydrates has become the object of increasing attention over the last decade, primarily due (directly or indirectly) to the potential impact of hydrates on the development process and dosage form performance. Substances may hydrate/dehydrate in response to changes in environmental conditions, processing, or over time if in a metastable thermodynamic state [1].

It may not be practical or possible to maintain the same hydrate isolated at the discovery bench scale synthesis during scale-up activities for a hydrated compound. The choice of counterions to produce a more soluble salt form may also be dictated by the extent and type of hydration observed for a given salt and/or by the moisture level that may be safely accommodated by the dosage form [2].

The physicochemical stability of the compound may raise issues during preformulation. Some hydrated compounds may convert to an amorphous form upon dehydration and some may become chemically labile. This is true of cephradine dihydrate that dehydrates to become amorphous and undergoes subsequent oxidation. Other compounds may convert from a lower to a higher state of hydration yielding forms with lower solubility. In any case, the resulting "new" forms would represent unique entities that, depending on the dosage form, might have to be maintained throughout the manufacturing process and in the clinic and would impact on the regulatory status of the compound. Most often this demands that the form (usually crystalline) be identified and characterized with respect to handling conditions during the early pre-IND stage of the development process.

As dosage form development proceeds, changes in hydration state can result in variable potencies depending on handling conditions during weighing steps, the kinetics of the hydration/dehydration process, and the environmental conditions during processing. Differences in powder flow can result from changes in crystal form and/or morphology that may accompany the hydration/dehydration process. This can affect content uniformity in solid processing either in the mixing process or during transfer to other processing equipment such as tablet presses. Aqueous granulation, particle size reduction, film coating, and tablet compression all provide opportunities to "trap" a compound in a metastable form that may "relax" to a more stable form at some unpredicted point in the life of a dosage form. Alternately, a kinetically favored but thermodynamically unstable form may be converted during these processes to a more stable and less soluble form.

During and after manufacturing, moisture from the environment or that sealed in the package may redistribute throughout the dosage form and change the hydration state(s). These changes can, in turn, visit the negative consequences discussed above for the bulk drug on the dosage form. These can be manifest as changes in tablet/capsule dissolution rates (and perhaps bioavailability), changes in lyophile reconstitution times, tablet capping, chemical instability, discoloration, and more. Of course, the potential for changes in hydration state also exists for many pharmaceutical excipients (such as lactose or magnesium stearate).

Such problems are typically magnified as both synthetic and dosage form production is scaled up. This may be caused by solvent limitations, heat transfer differences in production equipment, changes in raw materials and/or raw material suppliers, changes in processing times, and time and control constraints on product storage, to name a few.

The arguments just provided detail the potential issues around hydrates in the development process. The other consideration is the frequency with which hydrates are encountered in real life. Focusing on active drug substances, it is estimated that approximately one-third of the pharmaceutical actives are capable of forming crystalline hydrates [3]. A search of the Cambridge Structural Database (CSD) shows that approximately 11% of all the reported crystal structures contain molecular water [4]. This represents over 16,000 compounds. If organometallic compounds are excluded, this number drops to approximately 6,000 (3.8%), and the breakdown of these according to hydration number is shown in Fig. 1. This shows the expected trend in which monohydrates are most frequently encountered, and where the frequency decreases almost exponentially as the hydration number increases. The hemihydrate stoichiometry occurs approximately as frequently as the trihydrate, which should serve as a caution to explore fully the occurrence of fractional hydration. That is, an apparent stoichiometry of 0.6 water molecules could be a partially dehydrated monohydrate, or it

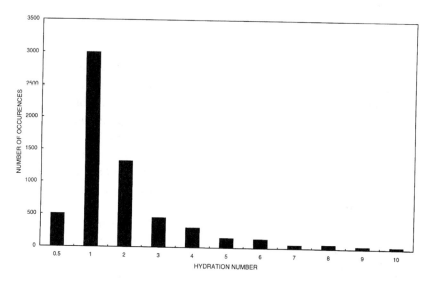

Fig. 1   Occurrence of various crystalline hydrate stoichiometries.

could be a hemihydrate with additional sorption due to defects or amorphous material.

The symmetry of these hydrate crystals follows fairly closely with that reported for organic structures overall [5]. Table 1 shows the breakdown for space groups, organized by crystal system, accounting for the top approximately 90% of the structures. $P2_1/c$ (number 14) is the most common space group here as with the general population of organic molecules contained in the CSD. It has been reported for inorganic species that hydrated structures are generally of lower symmetry than are their anhydrous counterparts [6]. This is attributed to the fact that the highest symmetry associated with the water molecule is C_{2v} and most inorganic structures are of higher symmetry. This is not obviously the case for organic structures. Regardless of the solvation state, organic molecules generally exhibit lower symmetry than do inorganic compounds, so the impact of the symmetry constraints imposed by water does not appear to be the controlling element. Further comparisons would be required to explore the phenomena fully.

Table 1   Space Groups for the Top 90% of Organic Crystalline Hydrates in the Cambridge Structural Database

Space group	Crystal system	Percent occurrence
P_{-1}	Triclinic	15.5
P_1	Triclinic	2.6
$P_{21/c}$	Monoclinic	23.2
P_{21}	Monoclinic	13.4
$C_{2/c1}$	Monoclinic	5.8
C_2	Monoclinic	2.8
P_{212121}	Orthorhombic	17.8
P_{bca}	Orthorhombic	2.3
P_{21212}	Orthorhombic	1.8
P_{na21}	Orthorhombic	1.8
P_{nma}	Orthorhombic	1.3
Unknown		1.2

II. HYDRATE THERMODYNAMICS

The equilibrium thermodynamics of stoichiometric hydrates has been
described by several authors. The overview presented here is intended
both to review the basic thermodynamics of crystalline hydrate
formation/stability and to highlight the intrinsic differences between
polymorphic systems and hydrate systems (a discussion of the kinetics
of dehydration/hydration will be given in Section IV). The following
description is a hybrid based on the work of Grant and Higuchi [7]
and that of Carstensen [8].

A. Classical Higuchi/Grant Treatment

The equilibrium between a hydrate and an anhydrous crystal (or between levels of hydration) may be described by the following relationship.

$$A(solid) + mH_2O \leftrightarrows A \cdot mH_2O(solid) \tag{1}$$

where

$$K_h = \frac{a[A \cdot mH_2O(solid)]}{a[A(solid)]a[H_2O]^m}$$

Here a represents the activity of the hydrate ($a[A \cdot mH_2O(solid)]$) and
anhydrate ($a[A(solid)]$), respectively. When the water activity
($a[H_2O]$) is greater than the ratio

$$\frac{a[A \cdot mH_2O(solid)]}{a[A(solid)k_h]^{1/m}} \tag{2}$$

then the hydrate species is the stable form. The anhydrate species will
be stable if the water activity is less than the ratio in Eq. (2). If the
pure solids are taken as the standard states for the hydrate and anhy-
drous materials (i.e., as the states with unit activity), then $K_h = a[H_2O]^{-m}$ (and $m = 1$ for a monohydrate). So, clearly, the stability of
a hydrate relative to the anhydrate (or lower hydrate) depends upon
the activity of water in the vapor phase, or the partial vapor pressure
or relative humidity (the ratio of the vapor pressure of water to the
saturation vapor pressure at that temperature P/P_0). This straightforward thermodynamic description of hydrate equilibria is the key to un-

derstanding not only the stability of hydrated forms but the inherent differences between hydrates and polymorphs.

Just as the state of hydration depends upon the water vapor activity, so will the water activity (relative humidity, RH) in a closed system depend upon the state of hydration of the solid phase. These microenvironmental RH values can be of significance for the redistribution of moisture within a dosage form and/or package. An excellent illustration of this was given by Carstensen for sodium phosphate [9]. Figure 2 shows the relation between water vapor pressure (P) and the number of moles of water for the compound. Here it is seen that as water is added to a closed system, the compound takes it up until it no longer has any capacity in a given form (i.e., all of the solid is converted to a given hydrate). During this time, however, the RH of the system is constant. As more water is added, the RH rises until the critical value is reached that is sufficient to initiate the formation of the next higher hydration state. This cycle repeats as long as there are more states to be attained. Ultimately, the RH drifts up if the compound deliquesces. One would not, therefore, expect to maintain a constant RH with differences in water content in a system unless it contains hydratable components to buffer the changes. The type of behavior shown here is most common with inorganic compounds, but the principle is the same for

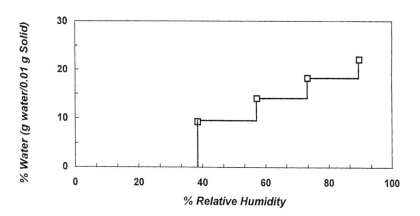

Fig. 2 Vapor pressure–hydration state diagram of sodium phosphate (reproduced with permission).

organic molecules even though the number of stable hydrates that form may be less.

B. Similarities and Differences Between Polymorphs and Hydrates

Hydrates and polymorphs are typically discussed together (as in this volume), and there are good reasons for this. In the scope of characterization of pharmaceuticals, many of the behaviors of polymorphic systems are at least apparently shared by compounds that can exist in various crystalline states of hydration. For the purposes of this chapter, such systems (including the anhydrous) will be referred to simply as hydrates. Members of both polymorphic and hydrate systems have different crystal structures and exhibit different x-ray powder diffraction patterns (XRPD), thermograms (DSC or TGA), infrared spectra, dissolution rates, hygroscopicity, etc. Interconversion between polymorphs or hydrates may occur as a function of temperature and/or pressure or be solution mediated. The potential for interconversion during processing, stability testing, and storage is, therefore, present for both polymorphs and hydrates. Given this long list of similar behaviors, it is generally proper that polymorphs and hydrates be addressed in the same general area of the pharmaceutical development process (for both technical and regulatory concerns).

The differences between polymorphs and hydrates are significant. The basis for all these differences is that polymorphs are different crystal structures of the same molecules(s) while hydrates are crystals of the drug molecule with different numbers of water molecules. As discussed above, the hydration state (and therefore the structure) of a crystalline hydrate is a function of the water vapor pressure (water activity) above the solid. Polymorphs, however, are typically only affected by changes in water vapor pressure if water sorption allows molecular motion, which in turn allows a reorganization into a different polymorph (i.e., a solution mediated transformation). This distinction is particularly important in defining the relative free energy of hydrates. A simple (only one molecule) anhydrous crystalline form is a one component system, and the free energy is, practically, specified by temperature and pressure. A crystalline hydrate is a two-component system and is specified by temperature, pressure, and water activity. In both cases it is assumed that the activity of the pure solid is unity.

Consider the thermodynamic stability of a phase that crystallizes from water. If the phase is anhydrous (assuming no specific interaction between the molecule and the water), when the phase is removed from the solvent it is usually stable at that temperature (i.e., the free energy of the phase is independent of the solvent of origin). If the phase is hydrated, when it is removed from the solvent the situation changes completely. All that can be known is that the phase was thermodynamically stable at a water activity of approximately unity. Although it is a rule of thumb that the higher the hydration state that forms at a temperature the more thermodynamically stable, Grant et al. [10] have reported the opposite behavior. Once removed from the water, the activity of water needed to maintain the form (the critical activity) had to be determined by other methods. These may include water vapor sorption data or a titration of the amount of water in a cosolvent system [11]. A typical constant pressure G–T diagram (free energy vs. temperature, recalling that $\delta G/\delta T = -S$) for a polymorphic system is really analogous to a logarithm solubility vs. reciprocal temperature plot (ln X vs. $1/T$) for a system of hydrates (Figs. 3a and b). Alternatively, a G–T plot for a hydrate system at constant water activity is analogous. The relationship between free energy and the ideal solubility of a solid is seen from the following equation.

$$\Delta G = -RT \ln X \tag{3}$$

or

$$\ln X = \frac{\Delta S_f}{R} - \frac{\Delta H_f}{R} \cdot \frac{1}{T}$$

where ΔS_f and ΔH_f are the entropy and enthalpy of fusion at the melting point, respectively, and R is the gas constant. A plot of ln X against $1/T$ should yield a straight line with a negative slope equal to $-\Delta H_f/R$ and an intercept of $\Delta S_f/R$ for each phase. This conclusion assumes that the enthalpy of solution is equal to the enthalpy of melting at the melting point. A more general expression may be derived, but the reciprocal dependence of solubility and temperature is preserved. Just as with a

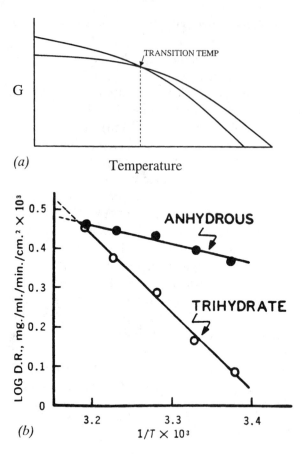

(a) Temperature

(b)

Fig. 3 (a) G–T diagram, two polymorphs, temperature plot. (b) Log dissolution vs. reciprocal temperature plot for ampicillin anhydrate and trihydrate (reproduced with permission).

G–T plot, the intersection of curves generated for two different crystalline phases represents a point of equal free energy and a transition temperature.

There are many implications of the relatively more complex structure of hydrates. As water is included or lost from the crystal structure, there must be a change in the volume of the unit cell (corrected for Z, the number of molecules per unit cell) at least as large as the volume of the water molecule (15–40 Å^3) [12]. Although there is no study known to the author comparing the relative volume change between polymorphic pairs vs. hydrate pairs, it must be assumed that the trend would be that the volume change is larger for hydrates, which have to accommodate the additional volume occupied by the water molecules.

The obvious problem for pharmaceutical development is that the water activity can vary throughout the lifetime of the compound, and it is for this reason that knowledge of the water sorption behavior of active substances and excipients is so critical.

C. Hydrogen Bonding in Hydrates

The ability of water to form hydrogen bonds and hydrogen bonding networks gives it unique behavior with respect to colligative properties such as boiling and melting points. Similarly, hydrogen bonding between water molecules and drug molecules in the solid state dictates its role in the structure of all classes of crystalline hydrates (i.e., the ability of water to form cocrystals with the drug molecule). Water will, of course, be hydrogen bonded whenever physically possible. This may take the form of hydrogen bonding to other water molecules, with functional groups on other molecules, or to anions. Hydrogen bonding to other water molecules is common both in the crystal lattice and in interstitial cavities or channels. Hydrogen bonding to other moieties and anions in crystalline hydrates is primarily within the lattice. In addition, the lone pair electrons of the water oxygen may be associated with metallic cations present in many salts. This interaction is largely electrostatic in nature for the metal cations common to pharmaceutics (Na, Ca, K, Mg). These main-group metal ions lack the d-orbitals of suitable energy that are necessary to form coordinate covalent or coordination bonds that ions of the transition series form with oxygen. It is often stated that Mg(II) has a coordination number of 4, but this is a result of packing (or geometric) restrictions arising when fitting water molecules around the cation in response to the electrostatic attraction. Since these "bonds" are mostly electrostatic in nature, they are not properly described by a molecular orbital but are best defined by classical electrostatics [13]. These "bonds" are often stronger than hydrogen bonds but exist with less directional dependence. A typical water hydrogen bond is on the order of 4.5 kcal/mol, whereas a sodium–oxygen lone pair electrostatic interaction can be four to five times stronger. These bonds also exert their influence through hydrogen bonds in the form of cooperative effects. The specific characteristics of the hydrogen bond are presented here in the formalism of Falk and Knop [6].

The ubiquitous hydrogen bonding of water is largely a result of its being both a hydrogen bond donor and acceptor. It may participate in as many as four hydrogen bonds, one from each hydrogen and one for each lone pair on the oxygen. Classification schemes based solely on the type of coordination of the water oxygen have been proposed [6]. As each bond is formed, it makes the other sites more attractive as partners for additional bonds. Hydrogen bond acceptors must be electronegative and include one of the following: oxygen atoms from other water molecules, oxygen and nitrogen atoms from other functional groups, and chlorine atoms. Hydrogen bond donors include protons on nitrogen, oxygen, and sulfur, of the types usually found on water, alcohols, amines, and the like.

Free water (vapor) has an OH bond length of 0.957 Å and an HOH angle of 104.52°. As soon as the molecule starts interacting with other molecules through hydrogen bonds, coordination, or other electrostatic "bonds", the molecule is distorted from its free conformation. The formation of hydrogen bonds weakens the OH bond, usually resulting in an increase in its length. This increase can be up to 0.01 Å for an exceptionally strong hydrogen bond, but it is more typically on the order of 0.01 to 0.02 Å for organic hydrates with hydrogen bond lengths of 2.7 to 2.9 Å (O—O distance).

The limits of length of a hydrogen bond are defined at the lower end by the van der Waals radii of the two atoms and at the upper end arbitrarily by the length of the weakest hydrogen bond observed. This can be seen more quantitatively by expressing the hydrogen bond distances shown in Fig. 4 in terms of the van der Waals radii.

$$R(H—Y) < r(H) + r(Y) - 0.20 \text{ Å} \tag{4}$$

and since $r(H) = 1.2$ Å,

$$R(H—Y) < r(Y) - 1.00 \text{ Å}$$

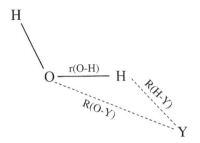

Fig. 4 Formalism used in the discussion of hydrogen bond strength and length. (From Ref. 6.)

The factor of 0.2 represents the combined experimental and statistical uncertainty. The compounds studied by Falk and Knop were inorganic and small organic hydrates [6]. Of the 129 compounds studied, only one failed this criterion. Often the hydrogen bond lengths are given as the O—Y (e.g., oxygen–oxygen or oxygen–chlorine) distances, because in x-ray diffraction studies it is often difficult or impossible to locate accurately the hydrogen atoms due to their inherently low scattering and their relatively high mobility. Because of the large cross sections, hydrogen atoms are often located by neutron diffraction studies. Under these circumstances, crystallographers will report the O—Y distance they know to be reliable, and geometric constraints may also be applied to such data.

$$R(O\!-\!Y) \leq R(H\!-\!Y) + r(OH) \tag{5}$$

substituting above, then setting $r(OH) = 0.98$ Å,

$$R(O\!-\!Y) \leq r(Y) + 1.98 \text{ Å}$$

As the electrostatic bond strength of the donor to the water oxygen (X) increases, the length of the H—Y bond decreases. This cooperative effect is also seen as the number of hydrogen bonds per water molecule increases. Hydrogen bonds prefer to be linear but may adopt a range of angles at the expense of the strength of the bond [6,15].

All of these aspects of water hydrogen bonding are evident in the infrared spectra of crystalline hydrates. When a molecule absorbs infrared radiation, this energy is used to excite the molecule into higher vibrational energy level states. The occupancy of these higher states manifests itself in greater degrees of molecular vibration. The frequencies at which the molecule absorbs are a function of the mass of the bonded atoms, the geometry of the molecule, and the force constant (strength) of the bond. This relationship can be described by analogy to classical mechanics through Hooke's law, which states that the frequency of motion (v) for a harmonic oscillator is inversely proportional to the square root of the reduced mass of the system (μ). The force constant (f, in units of dyn/cm) is the proportionality constant. Thus

$$v = \frac{1}{(2\pi c)} \left(\frac{f}{\mu}\right)^{1/2} \tag{6}$$

where c is the velocity of light in units of cm/s. As illustrated in Fig. 5, water in crystalline hydrates has nine potential degrees of freedom: two stretching modes (symmetric and asymmetric), one bending mode, three vibrational modes (hindered rotation), and three hindered transla-

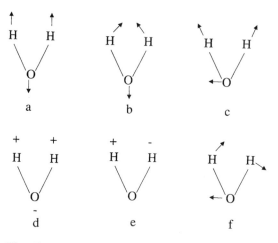

Fig. 5 Vibrational modes of water. Shown are the internal modes: (a) symmetric stretch, (b) bending, (c) asymmetric stretch, and the librational modes (d) wag, (e) twist, and (f) rock. Not shown are the three hindered translational modes.

tional modes. These vibrational modes and their characteristic absorption frequencies are presented in Table 2 and contrasted with those of water vapor.

The dominant feature in the infrared spectrum of a hydrate is band system associated with the O–H stretching frequencies between 4000 and 3000 cm^{-1}. These peaks are unusually intense due to the effect of hydrogen bonding on the changes in dipole moment that are associated with the wave functions describing the molecular motion. If an O–H stretching band is not observed, then no water is associated with the compound. When present, the contributing OH groups must be properly assigned to distinguish water from alcoholic, phenolic, hydroxide, or other interfering absorptions. This is accomplished by first assigning energy values to the known structure of the molecule from tables. If resolved, the presence of an H_2O bending frequency around 1600 cm^{-1} is proof that the sample contains water. Comparing this to the IR spectrum of the anhydrous material or the dehydrated sample shows which peaks in the region are due only to water. The IR spectra of ampicillin before and after dehydration exhibits such behavior.

Table 2 also shows that the OH stretching frequency of water occurs at lower wave numbers (longer wavelength and lower energy) in the crystal than in the vapor. This is due to the reduction of the force constant (f) by interaction of the water with neighboring groups, in particular hydrogen bonding and lone pair interactions involving the water oxygen. The weakening of the bond results in a slight elongation of the bond length, in the range of 0.01 to 0.02 Å [15]. This shift in

Table 2 Vibrational Modes of Water in Various Phases

Vibrational mode	Frequency in vapor phase (cm^{-1})	Frequency in solid phase (cm^{-1})
Stretching	3755.8 (symmetric) 3656.7 (asymmetric)	2850–3625
Bending	1594.6	1498–1732
Rotation/libration		355–1080
Translation		200–490

the OH stretching frequency of water can be used to evaluate the interaction energy between water and the other molecules. Specifically, the higher is the degree of water hydrogen bonding, the lower the frequency will be of the OH stretch. In fact, good correlation between OH stretching frequency and the length of hydrogen bonds is available for inorganics and very small organic crystals [14]. While repulsive lattice energy tends to increase this frequency, it typically yields only a very small shift. In large molecular crystals, however, the energetics become more complicated and the correlation is not as good. Adventitious adsorbed water tends to have broad peaks in the lower part of the frequency range. The broadening is due to the vibrational coupling between water molecules, and the lower frequencies are due to the multiple hydrogen bonds. This "dispersion of stretching frequencies" is analogous to the broadening of DSC peaks due to the multiple energetic environment that adventitious water can experience. If the water occupies only one type of crystal site, the DSC and IR peaks should be sharp relative to those of adventitious water. This is seen in the ampicillin example of the classification section. The O–H stretching peaks from water in the crystal lattice will occur at various frequencies depending upon the strength of the hydrogen bonding.

In addition to shifts in frequency and peak shape, peaks may become split owing to the interaction of the two water hydrogens if they participate in different hydrogen bonds. Therefore in some crystalline hydrates there may be two or more peaks associated with the OH stretching mode.

Metal cations affect the infrared absorption behavior in several ways. First, if the water oxygen is bound at the inner hydration sphere of the cation, polarization of the electron density causes stronger hydrogen bonds to be formed. This effect will lower the OH stretching frequency in a manner proportionate to the degree of bond strengthening, a decrease that can be up to 640 cm^{-1} in inorganic compounds [6,15]. Second, cation–water interaction can increase the bending frequencies (which are observed in the 1600 cm^{-1} region) by as much as 50 cm^{-1}. Finally, the bonds formed between the cation and the atoms in its inner coordination sphere will be observed as low-frequency modes below 400 cm^{-1}.

III. CLASSIFICATION OF HYDRATES

The combination of the vibrational mode information, the hydrogen bonding characteristics, and the thermodynamic relationships serves to form a clear picture as to why water can and does participate in hydrate formation with drug molecules. The possible structures that may result from such interactions are quite diverse. For practical purposes, an identification of types or classes of possible resulting structures is useful. Water is small enough to fill many commonly occurring periodic "voids" formed when larger molecules are packed, and it interacts through hydrogen bonds to overcome some of the entropy of mixing. The ability of the water molecules to self-associate, combined with the small size, allows them to fill larger periodic spaces conforming to different shapes. The characteristics give water a chameleon-like quality (also seen in protein hydration), which gives rise to "motifs" of water arrangements in crystal structures.

Crystalline hydrates have been classified by either structure or energetics [6]. The idea of the structural classification scheme presented here is to divide the hydrates into three classes that are discernible by the commonly available analytical techniques. The classification of crystalline hydrates of pharmaceutical interest by their structural characteristics is the most common, intuitive, and useful approach. A good classification system should direct the preformulation/formulation scientist to the characteristics of the particular class that will help in identifying a new sample, in selecting the proper form of the substance, and in estimating boundary conditions for safe handling (Table 3).

Table 3 Classification of Crystalline Hydrates

Class	Description
1	Isolated lattice sites
2	Lattice channels
−a	Expanded channels (non-stoichiometric)
−b	Lattice planes
−c	Dehydrated hydrates
3	Metal-ion coordinated water

An example of each class and the analytical manifestations of its crystalline structure will be presented in the succeeding sections. For this section, each class will be examined with respect to a specific example for which single crystal structure, DSC/TGA, XRPD, and IR data are available. Starting with the packing diagrams from the solved structure, the data from the other methods will be compared to that expected based on the information presented in the discussions above.

A. Class 1: Isolated Site Hydrates

These hydrate species represent the structures with water molecules isolated from direct contact with other water molecules by intervening drug molecules. To illustrate some of the characteristics of this hydrate class, consider the example of cephradine dihydrate [17], whose structure is shown in Fig. 6. This compound has the chemical formula $C_{16}H_{18}N_3O_4S$ (molecular weight 385.45) and is identified by the Cambridge reference code SQ22022. It crystallizes in the $P2_1$ space group and has the unit cell dimensions of a = 10.72 Å, b = 7.31 Å, and c = 11.87 Å. Finally, the cell volume is 908 Å^3, and its density is 1.41 g/cm^3.

The characterization of the crystal structure of cephradine dihydrate was particularly important because it becomes amorphous and very unstable upon dehydration [17]. Figure 7 shows the packing diagram of cephradine dihydrate, while Fig. 8 shows the measured and calculated XRPD patterns. Each unit cell contains two molecules of

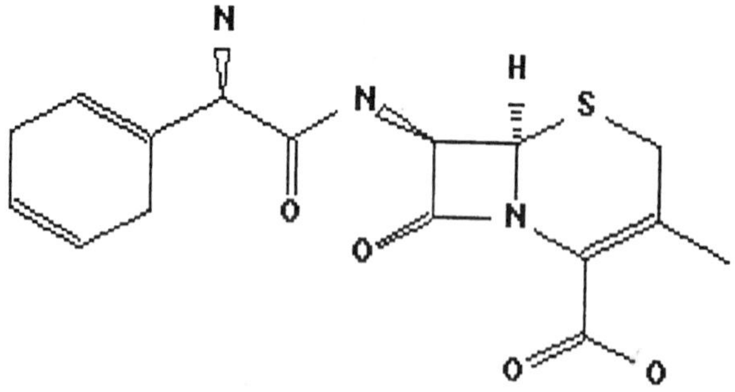

Fig. 6 Structure of cephradine.

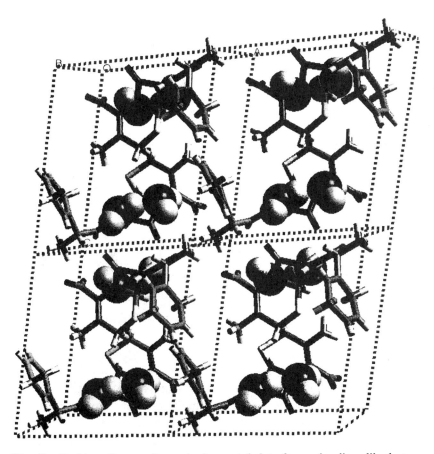

Fig. 7 Packing diagram from single crystal data for cephradine dihydrate. The van der Waals radii are included for the water hydrogens and oxygen. The pairs of water molecules reside in isolated lattice sites.

cephradine and four molecules of water. The diagram shows isolated pairs of water molecules arranged at intervals in the lattice. The two water molecules hydrogen bond with each other, and with the carbonyl, carboxyl, and amide groups on the two cephradine molecules.

No axis can be drawn between separate water pairs without passing through an intervening portion of the cephradine molecule. This means that a pair of water molecules on the surface may be easily

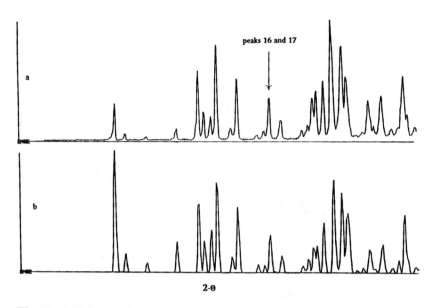

peaks 16 and 17

2-θ

Fig. 8   (a) Measured and (b) calculated x-ray powder diffraction patterns for cephradine dihydrate.

lost, but the creation of the hole does not leave other water molecules accessible. Similarly, no network of hydrogen bonds exists solely involving water molecules on any axis through the crystal. The diagram has been constructed to show the water molecules with their van der Waals dimensions while the drug is in a stick representation. Areas that may appear void between pairs are actually filled by the electron clouds of the neighboring atoms. Since each pair is located in the same environment, it is expected that the interaction energy between each water molecule and its neighbor be similar (low energy dispersion).

This structure should yield sharp DSC endotherms, a narrow TGA weight loss range, and sharp O–H stretching frequencies in the infrared spectrum (Figs. 9a and b and 10). The DSC thermogram shows two incompletely resolved, but sharp, endotherms at approximately 100°C, and the TGA thermogram shows the anticipated sharp weight loss over a similar range. In addition, the onset of dehydration observed in the DSC curve is quite sharp. The diffuse reflectance infrared (DRIFT)

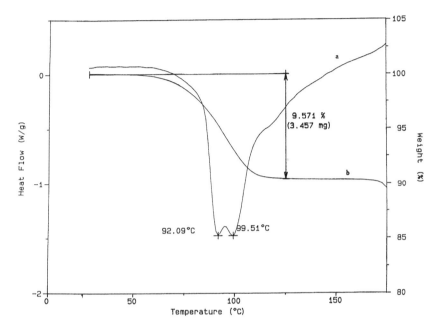

Fig. 9 (a) Differential scanning calorimetry and (b) thermogravimetric analysis thermograms for cephradine dihydrate.

spectrum of cephradine dihydrate shows sharp O–H stretches at approximately 3520 and 3425 cm^{-1} that are absent in the anhydrous material [17]. The thermal and spectroscopic data are seen to be consistent with the known single-crystal structure. Were the single crystal structure not available, the general features of the water association could have been deduced using the rationale previously developed.

B. Class 2: Channel Hydrates

Hydrates in this class contain water in lattice channels, where the water molecules included lie next to other water molecules of adjoining unit cells along an axis of the lattice, forming "channels" through the crystal. The empty channels are actually a conceptual construct, since a corresponding low-density structure with empty channels would not be

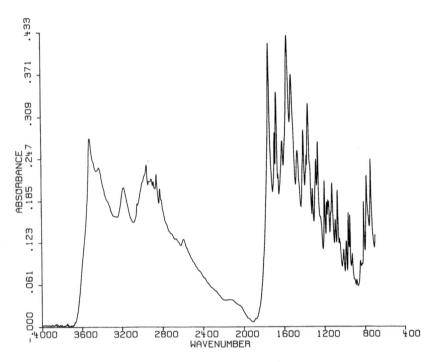

Fig. 10 Diffuse reflection infrared absorption (DRIFT) spectrum of cephradine dihydrate.

expected to be physically stable without some associated change in lattice parameters.

To illustrate some of the characteristics of the second hydrate class, we will examine the instance of ampicillin trihydrate [18,19], whose structure is shown in Fig. 11. This compound has the chemical formula $C_{16}H_{19}N_3O_4S$ (molecular weight 349.41) and is identified by the Cambridge reference code AMPCIH. It crystallizes in the $P2_12_12_1$ space group and has the unit cell dimensions of a = 15.49 Å, b = 18.89 Å, and c = 6.66 Å. Finally, the cell volume is 1949.4 Å^3, and its density is 1.37 g/cm^3.

Figure 12 shows the packing diagram, which consists of eight unit cells, and Figs. 13a and b show the measured and calculated XRPD patterns. As is obvious from the diagram, the water molecules line up

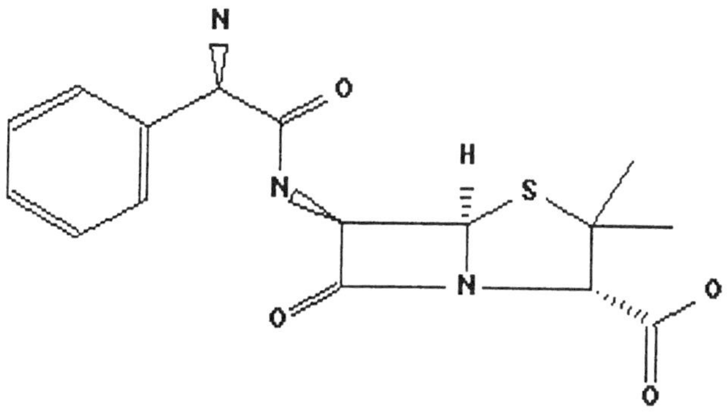

Fig. 11 Structure of ampicillin.

along the c screw axis. A channel of 3.5 Å would be formed if the water molecules were to leave without changing the structure. The water molecules occupy specific sites in the lattice (ordered water), as has been noted for cephradine. The water molecules are hydrogen bonded to four (or more) other water molecules and at least two ampicillin molecules through the carboxylate ion, the carbonyl on the β-lactam, the amino group, or through the ammonium group. Ampicillin trihydrate crystallizes as the zwitterion [19].

Examination of the DRIFT spectrum of ampicillin (Fig. 14) shows a sharp O–H stretching frequency at 3334 cm^{-1}, and an O–H bend at 1650 cm^{-1}, which are absent in the spectrum of the anhydrous form. The sharp O–H stretch indicates that the interaction energies between the OH groups on the water and the drug fall into the expected relatively narrow range. In addition, the band frequency is lower than that for cephradine, which suggests that the hydrogen bonding may be stronger in ampicillin. This is supported by the comparison of the DSC curves, which show dehydration at about 100°C for cephradine and 120°C for ampicillin.

The DSC/TGA data (Figs. 15a and b) also show one of the interesting characteristics of channel hydrate dehydration. Notice the early onset of dehydration for ampicillin compared to cephradine (isolated sites). Ampicillin loses water continuously to 125°C, while cephradine endotherms are very narrow even though the dehydration temperature of ampicillin is higher. This is due to dehydration beginning at the

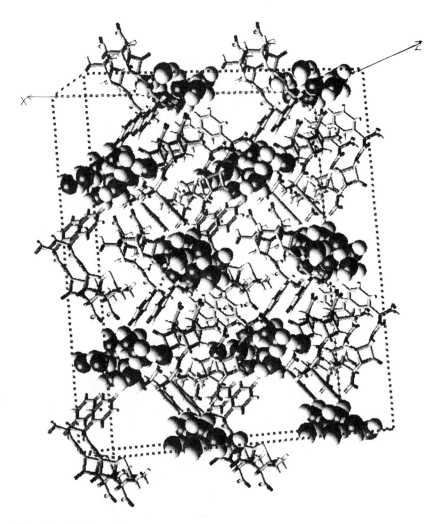

Fig. 12 Packing diagram for ampicillin trihydrate deduced from the single-crystal structural data. The van der Waals radii are included for the water hydrogens and oxygen, and the "channels" are along the screw axes.

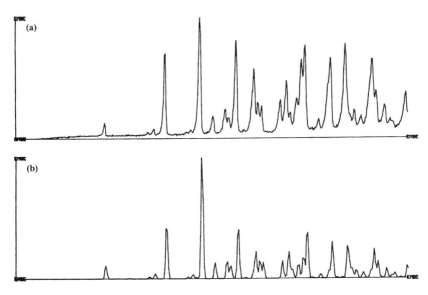

Fig. 13 (a) Measured and (b) calculated x-ray powder diffraction patterns for ampicillin trihydrate.

"ends" of the crystal and continuing toward the center along the channels. As the temperature increases so does the probability of losing the first water molecules on the surface at the channel ends. The loss of these water molecules leaves a channel for the next and sets up a thermodynamic gradient in the same direction. The drug molecules need not reorganize to lose many water molecules from the crystal. Using light microscopy, Byrn has observed and documented this phenomenon for ampicillin and theophylline [20]. As the crystal is heated on the microscope hot stage, the dehydration appears as a progressive darkening (increasing opacity) from the ends of the crystal toward the center along the c crystallographic axis. At some point in the dehydration, the crystal may either change its structure or become amorphous. Ampicillin would be recognizable as a channel hydrate based solely on microscopic/hot stage observations.

1. Expanded Channels

Another characteristic of some channel hydrates is that they may take up additional moisture in the channels when exposed to high humidity.

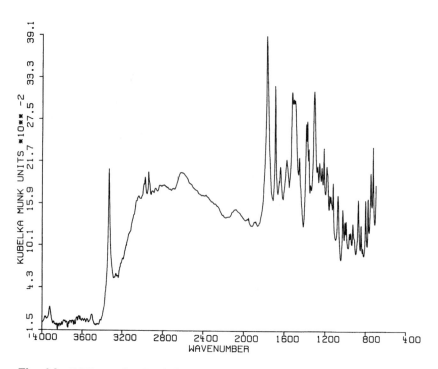

Fig. 14 Diffuse reflection infrared absorption (DRIFT) spectrum of ampicillin trihydrate.

As the hydration or dehydration proceeds, the crystal lattice may expand or contract by as much as 0.8 Å [21]. This lattice expansion must effect changes in the dimensions of the unit cell. This is manifested in the XRPD pattern as slight shifts in some or all of the scattering peaks. If the dimension increases, the associated peak is shifted to smaller 2-θ values, which correspond to larger d-spacings, and visa versa. This was shown qualitatively for chromylin sodium by Cox et al. [21]. As the chromylin absorbed more moisture, the lattice expanded to accommodate the additional water. The d-spacings and unit cell dimensions increased with water content, as determined by analysis of the film, but the structure was not solved for each level of hydration. Their subsequent model was used to determine the number of water molecules necessary to account for the observed expansion. As hydration contin-

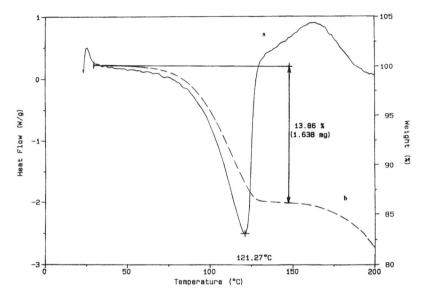

Fig. 15 (a) Differential scanning calorimetry and (b) thermogravimetric analysis thermograms for ampicillin trihydrate.

ues, the crystal expands until the changes are too large to maintain the same crystal structure. At this point, another crystal structure can result (with its own XRPD pattern), or it may revert to an amorphous material. This behavior is often observed with channel hydrates and warrants further division into subclasses having the characteristics of being non-stoichiometric or of continuously variable hydration.

Another example for which the single crystal structure was not known was (S)-4-[[[1-(4-fluorophenyl)-3-(1-methylethyl)-1H-indol-2-yl]-ethynyl]-hydroxyphosphinyl]-3]-hydroxy-butanoic acid, disodium salt (also known as SQ33600, whose structure is shown in Fig. 16); physical investigations provided considerable insight into the structure [22]. SQ33600 was found to take up moisture variably, lose it on exposure to low humidity (and/or heating), and appeared to undergo lattice expansion. Figure 17 shows the XRPD patterns for material stored at different relative humidities. It is clear that the largest changes take place in the low angle diffractions (larger d-spacings) as the degree of exposure to relative humidity increases. The same is true for several

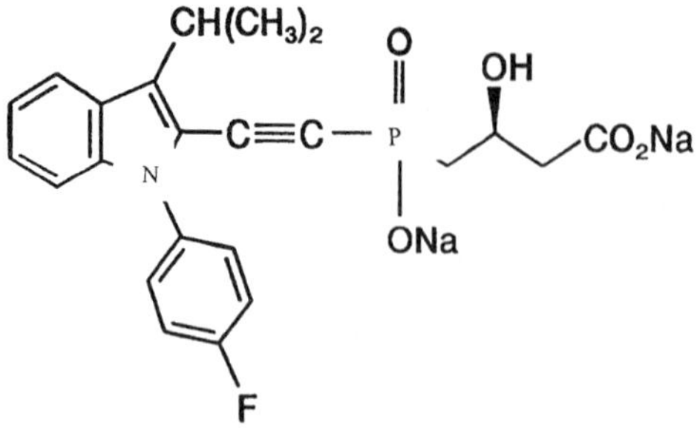

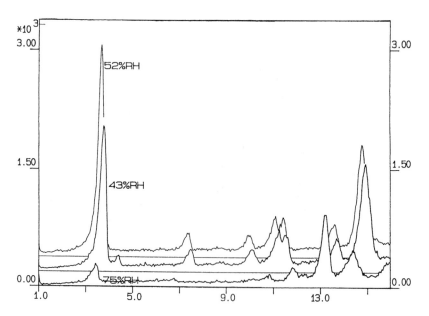

Fig. 16 Structure of (S)-4-[[[1-(4-fluorophenyl)-3-(1-methylethyl)-1H-indol-2-yl]-ethynyl]-hydroxyphosphinyl]-3]-hydroxybutanoic acid, disodium salt, SQ33600.

Fig. 17 X-ray powder diffraction patterns for SQ33600 at various relative humidity values.

other diffractions in material stored at 43% and 52% RH, while at 75% RH the structure has changed into something quite different (retaining only the low angle diffraction peaks in common). In fact, the material will deliquesce at a relative humidity of 75%.

2. Planar Hydrates

This subclass of crystalline hydrates has its water localized in a two-dimensional order, or plane. Figure 18 shows the packing diagram for sodium ibuprofen, where the waters of hydration are associated with the sodium ions localized in the a–c plane of the lattice. A similar structure has been reported for nedocromil zinc [23]. In both cases the water is ion associated, but there is no obvious reason that such structures require this to be the case. In the case of nedocromil zinc, the long axis of the crystal is perpendicular to the hydration plane, and under crossed polarizing microscopic optics was observed to dehydrate

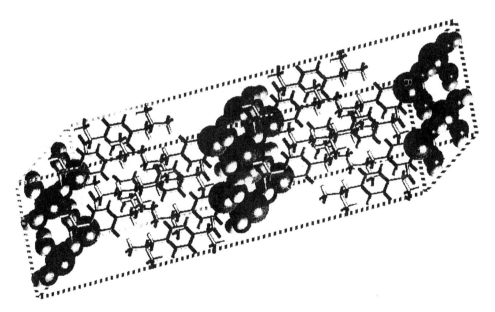

Fig. 18 Packing diagram for sodium ibuprofen with waters and sodium shown as van der Waals spheres. (From Jack Z. Gougoutas, unpublished data.)

primarily along the planar axes. This finding is consistent with the behavior cited earlier for the dehydration of channel hydrates.

3. Dehydrated Hydrates

Dehydrated hydrates may in principle belong to any of the classes just discussed, but the cases with which the author is familiar (findings not yet published) have all been either channel hydrates or ''clathrate'' type structures where water is the guest instead of the host in a cavity and in a nonstoichiometric amount. This subclass deals with crystals that dehydrate even at relatively high partial pressures of water. Therefore, the hydrate that forms in solution dehydrates almost immediately on removal from the mother liquor. When dehydration leaves an intact anhydrous structure that is very similar to the hydrated structure but with lower density, it is classified as a dehydrated hydrate. If there already exists an anhydrous crystalline form of the molecule, the dehydrated hydrate is classified as a polymorph.

Ulrich Griesser (University of Innsbruck) and Jack Gougoutas (Bristol-Myers Squibb) have raised the important question whether the new form is anything but another polymorph. If the water is truly a part of the crystalline structure and if in the hydrated state the waters are found, then removal of the water should change the cell parameters by some, possibly small, amount. The structure of the dehydrated form should similarly have voids large enough to accommodate some fraction of the water of hydration (hence the lower density). However, if the dehydrated form is well behaved this may not be an important distinction in the development of the compound (beyond a basic understanding of the origin of the form) unless later changes in final crystallization solvents produce solvates that do not give up their solvent quite as readily.

Another source of confusion can arise if their exists a pair of polymorphs (such as the anhydrous form and the dehydrated form) that have physicochemical properties that appear to complicate the classification of the system as enantiotropic or monotropic. For example, if density data are at odds with the relative stabilities and melting point data, then the relationship between the forms is not easily determined.

C. Class 3: Ion Associated Hydrates

Hydrates of this type contain metal ion coordinated water, and the major concern with these is the effect of the metal–water interaction on the structure of crystalline hydrates. The metal–water interaction can be quite strong relative to the other "bonding" in a molecular crystal, so that dehydration takes place only at very high temperatures [13]. Drugs with solubility, dissolution, or handling problems are most often recrystallized as Na(I), K(I), Ca(II), or Mg(II) salts and are often hygroscopic to some degree [16].

     The characteristics of this hydrate class can be illustrated through a consideration of the tetra-decahydrate and tetra-dihydrate species formed by calteridol calcium (24). This compound, whose structure is shown in Fig. 19, has the chemical formula $Ca(H_2O)_2[C_{17}H_{29}N_4 O_7Ca]_2$ (molecular weight 923.12) and is identified by the Cambridge reference code SQ33248. Although both the tetra-decahydrate and the tetra-dihydrate phases crystallize in the $C_{2/c}$ space group, the unit cell dimensions of these differ. For the tetra-decahydrate phase a = 33.625 Å, b = 9.517 Å, c = 20.949 Å, and β = 125.356, while for the tetra-

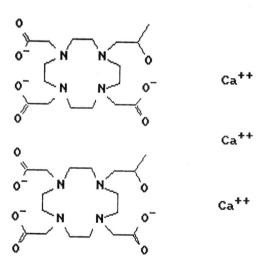

Fig. 19   Structure of calteridol calcium.

dihydrate phase a = 33.573 Å, b = 9.373 Å, c = 20.207 Å, and β = 131.620. The cell volume of the tetra-decahydrate was found to be 5467.2 Å³, and the cell volume of the tetra-dihydrate phase was 4753.6 Å³. Finally, the density of the tetra-decahydrate phase was determined to be 1.43 g/cm³, while the density of the tetra-dihydrate phase was 1.44 g/cm³. For reasons that will become clear in the following discussion, the tetra-decahydrate phase will be referred to as the reactant material, while the tetra-dihydrate phase will be referred to as the product material.

Calteridol calcium is used as a chelating excipient in a parenteral formulation and was chosen as an example for two reasons. It contains water associated with a metal cation and thus represents a Class 3 hydrate. In addition, water is also contained in channels, as with the second example. What is unique about the calteridol system is that a single crystal structure has been solved for the tetra-decahydrate (reactant) and the tetra-dihydrate (product) using the same crystal. This procedure permits an interesting look not only into the structural differences but also into the energetic differences of the water environment through thermal analysis. A similar study was performed on the dihydrate and trihydrate phases of di-sodium adenosine 5'-triphosphate [25].

Figures 20a and 20b show the packing diagrams of the initial crystal (or reactant) and the final (or product), respectively. In the packing diagrams, the water oxygens and the central calcium of the calteridol moiety are shown by their van der Waals spheres. Water molecules in the lattice (but not associated with the calcium) are represented by a sphere of one-half of their van der Waals radii, and the calteridol backbone is in stick representation. Figure 20a shows that for each calteridol, there are four water molecules associated with each central calcium, and ten lattice water molecules in each channel that run perpendicular to the plane of the figure (the b-axis). One would anticipate that the channel water would be lost more easily than would be the water associated with the cation. One reason for this is that the energy of association of water with calcium through the lone pair electrons on oxygen is greater than that of hydrogen bonding in water, and the other reason is that the loss of water in channels occurs more easily than would be expected by its interaction energy alone. Figure 20b shows the single crystal structure of the crystal after dehydration at room tem-

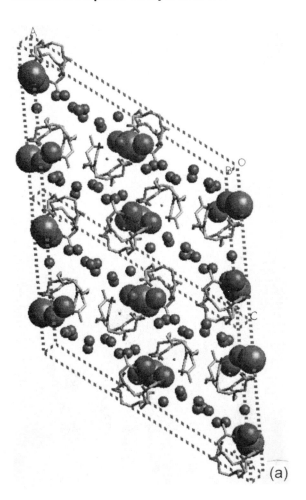

Fig. 20　Packing diagram from single crystal data for calteridol calcium (a) reactant and (b) product. The van der Waals radii are included for the water oxygens. The radii for the oxygens directly associated with the calcium are full van der Waals radii, while the lattice and channel water oxygens are shown as one-half of the van der Waals radii for illustration purposes.

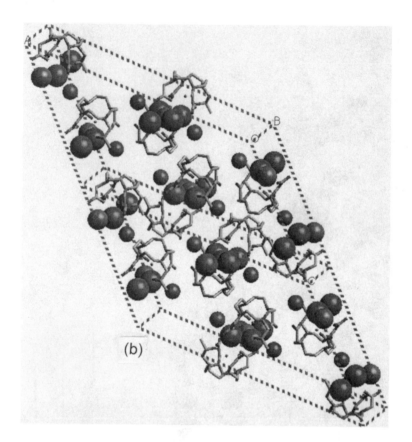

(b)

Fig. 20 Continued

perature for 6 hours. Most of the channel water (eight moles) has been lost, leaving the four calcium-associated water molecules and two water molecules in the "secondary sphere of hydration" of the calcium.

Figures 21a and b, 22a and b, and 23a and b contain the DSC, TGA, and XRPD data for the two hydrate phases of calteridol calcium. The DSC thermogram exhibits a large broad endotherm beginning at ambient temperature and peaking at approximately 75°C, which is associated with a weight loss of 13.56% (based on dry weight) or eight moles of water (calculated 13.96%). Further heating shows a smaller

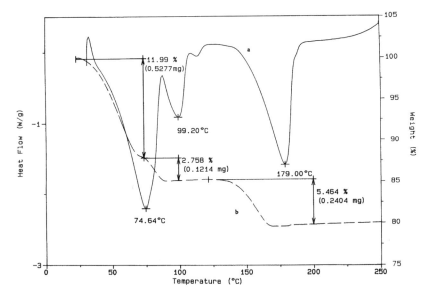

Fig. 21 (a) Differential scanning calorimetry and (b) thermogravimetric analysis thermograms for calteridol calcium tetra-decahydrate.

endotherm and weight loss corresponding to two moles of water. As the heating continues, a sharper endotherm is seen at approximately 179°C, and a weight loss corresponding to the four moles of water directly associated with the calcium ion. The molar heat of dehydration is between two and three times larger for the loss of the calcium-bound water than for the channel or lattice water.

The calculated XRPD patterns show distinctly different peak positions and intensities for the reactant and product. This is because the loss of water results in an entirely new crystal structure, unlike the situation noted for cromolyn sodium [21], in which the lattice "contracts" upon water loss without changing the crystal form. XRPD patterns collected on samples stored at 31% and 70% relative humidity show similar behavior (Figs. 22a and 23a). Notice that several diffraction peaks not present in the calculated powder pattern of the reactant are present in the measured powder pattern of material stored at 70% RH (Fig. 22). The same situation is noted for the calculated XRPD

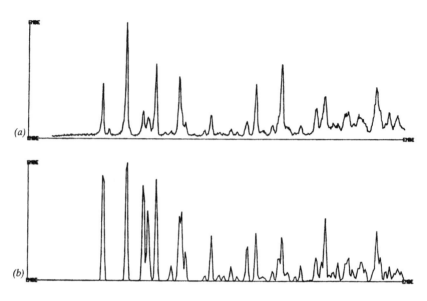

Fig. 22 (a) Measured and (b) calculated x-ray powder diffraction patterns for the calteridol calcium reactant (tetra-decahydrate).

pattern of the product pattern when compared to the measured powder pattern. This is because neither the reactant nor the product are ''pure'' in the sense that each contains a small amount of the other.

The thermal, Karl Fisher, and powder XRD data, coupled with the knowledge that the compound is a calcium salt, have suggested two deductions. First, the compound contains water in several crystalline environments or thermodynamic states, probably being lattice and cation associated. Second, the low-temperature water is probably contained in channels.

The characterization of pharmaceutical hydrates must be sufficient to provide confidence that the behavior of the material is predictable and reproducible. This requires the application of considerable molecular level intuition along with the available data. Data from all the techniques discussed are not always available, so in the absence of a complete data set the gaps must be filled using the types of energetic and structural principles described in the earlier chapter [1]. When the available data are consistent with what is expected from these relation-

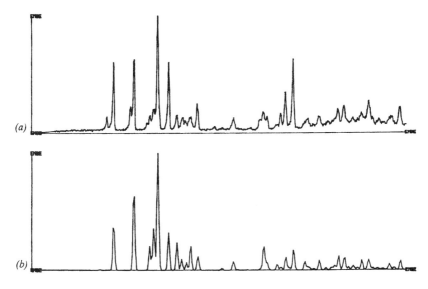

Fig. 23 (a) Measured and (b) calculated x-ray powder diffraction patterns for the calteridol calcium product (tetra-dihydrate).

ships, the behavior of crystalline hydrates should be at least qualitatively predictable.

IV. DEHYDRATION/HYDRATION KINETICS

Many models have been developed to account for the dehydration kinetics of crystalline hydrates [26]. These all assume a certain geometry and rely on some consistency of the system as the process proceeds. Often these models are indistinguishable for a given system due to experimental variation and because many structures (and, therefore, mechanisms) change during dehydration. The change in structure may also contribute to the hysteresis observed upon rehydration of a dehydrated structure. These models will not be discussed here, and the reader is referred to the literature [20] for a thorough review. The practical consequences of dehydration kinetics may be found in the activities associated with determining boundary conditions for allowable exposure of bulk drug substances during development and processing; proper packaging; allowable temperature ranges for shipping, storage, and labeling of the final product; and the initial selection of a form for development.

A. Dehydration and Hydrate Class

Some general ideas of dehydration rates by class that were implicitly introduced in the classification section can be made for hydrates at room temperature and above their critical relative humidity percentage. For isolated site hydrates, rates would be very low even when heated almost to the dehydration temperature. At that temperature, the kinetics would be expected to be very rapid unless the structure collapsed and presented a barrier to loss of the last water (although this should be minimal for this class). This is because the waters are all in similar energetic environments (hydrogen bonded to drug molecules), so that when sufficient energy is supplied to free one water, there is enough to liberate them all. This is obviously a generalization, since even in an homogeneous environment there will be a relatively narrow distribution of energies.

The story should be different for channel hydrates, where the hydrogen bonding network is dominated by water–water interactions. First, one needs to revisit the concept of sigma cooperativity [27]. The idea that all the hydrogen bonds of a given water molecule may get stronger as subsequent hydrogen bonds are formed is generally accepted. The opposite situation has not been described but must apply. So if one considers a water molecule involved in 3–4 hydrogen bonds, the remaining bonds must become weaker as a bond breaks due to "inverse cooperativity." Consider the dehydration of channel hydrates. At a given temperature, the energy available for dehydration will be sufficient for some fraction of the water molecules to leave from the end on the channel. However, because the hydrogen bonding network connects through the other water molecules in the channel, and unlike the drug molecules, the inverse cooperativity now diminishes the energy with which the next water molecules are associated with the lattice. This increases the number of molecules within the range where the energy is sufficient to dislodge them from the structure. This explains both the continuous dehydration observed with onset at relatively low temperatures for channel hydrates and the relatively discrete loss for isolated site and ion associated hydrates not subject to this cascade effect.

B. Impact of Particle Size and Morphology

In general, the smaller the particle size, the more rapidly dehydration occurs. This can be justified on the basis of surface area and/or, in some cases, the existence of crystal defects. The latter is particularly true for substances that have been subjected to high-energy particle size reduction processes. The impact of morphology may be more subtle and is certainly not independent of size considerations. The balance between the two mechanisms has been a topic of discussion, but very little quantitative work has been done to elucidate the relative contribution to the dehydration process.

This topic is complicated by the reality that particle size analysis is a very subjective measurement prone to several types of sampling errors. This makes it difficult to know how to approach the problem of obtaining two morphologies of a crystalline hydrate form with the same size. This could mean the same longest dimension, the same surface area, the same aspect ratio, etc. At very small particle sizes (less than 10–20 μm), the crystallite size may not be the primary particle size, and the rate may be determined by the size of the agglomerated material. Another complication has to do with the dehydration rate measurements. As discussed earlier, the stability of the hydrated form cannot be specified without knowledge of the partial pressure (or RH) of water at each temperature of interest. Most dehydration experiments are done against very low values of relative humidity and may not be very realistic or revealing. The literature does contain some interesting studies of dehydration at various relative humidity values that clearly show the expected trends [28]. The use of moisture sorption/desorption data is also becoming more common, although most workers still use only the equilibrium values even though the kinetic data are usually available from the same experiments. The other major complication was discussed above, which is that as mechanisms change, dehydration characteristics also change.

The discussion on dehydration characteristics of the different classes suggests that, for equivalent particle size, channel hydrate dehydration kinetics should be more sensitive to changes in morphology than would be the other classes. This hypothesis should be tested by taking an isolated site and a channel hydrate, each in two morphologies with similar surface areas. The phenomena can be seen conceptually by simulating the morphology from a single crystal structure and the Bravis–Friedel–Donnay–Harker relationship [29]. Choosing the channel hydrate ampicillin trihydrate as an example, it can be seen that the water lies on the crystallographic c-axis. As the morphology changes by increasing along the channel axis (Fig. 24), the relative density of dehydration sites decreases. This should lead to slower dehydration (at least early in the process before structural changes can occur) as the average distance a water molecule must travel increases.

A precursory study was performed on two different morphologies of carbamazapine dihydrate [30]. A photomicrograph of the two habits is shown in Figs. 25a and b, which show a brick shape and a needle shape. The surface areas were not determined due to the difficulties of such determinations on easily dehydrated channel hydrates. However, the longest dimensions are close to each other, which may be perceived as an equivalent particle size (such as would be deduced by traditional sieve analysis). This would be true unless the needles pass through lengthwise. A typical problem with needles is the clogging of screens due to their stacking. The isothermal water loss (at 40°C) was followed for both samples by TGA. The results (Fig. 26) show substantial differences in the loss rates, with the needles losing mass more rapidly than do the bricks. While this example points out the possible impact of morphology on dehydration, it is apparently the opposite of what was predicted for dehydration of channel hydrates. This is possibly the result of changes in mechanism, as carbamazapine was shown to crack upon dehydration [30].

To summarize, the primary influence on dehydration rates is most likely differences in particle size (and aggregation), with morphology representing a secondary influence. However, there may be cases where morphology contributes significantly to the process and must at least be ruled out as a factor.

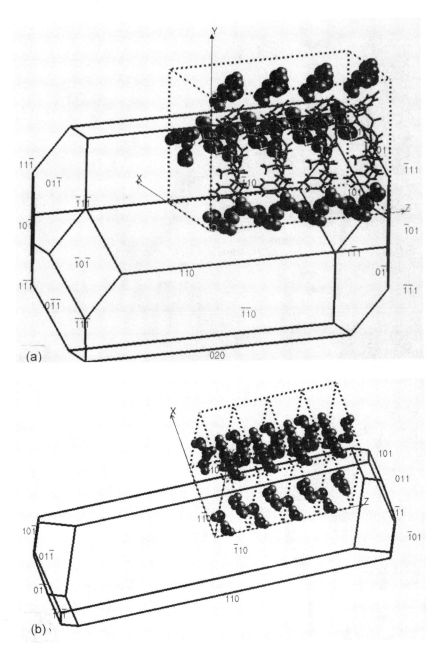

Fig. 24 (a) BFDH morphology for ampicillin, (b) computationally grown along the 110 zonal axis. (From Ref. 29.)

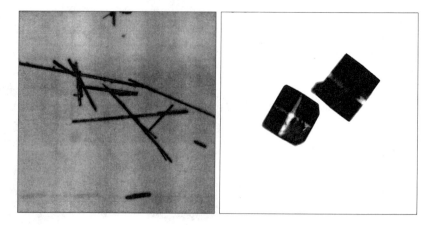

Fig. 25 Carbamazepine grown in water (left) and surfactant (right) at 52%
RH for five days.

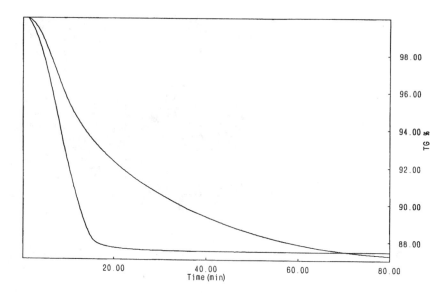

Fig. 26 Isothermal dehydration thermogravimetric analysis curves for carbamazapine dihydrate. The upper curve is for the block-like crystals grown
in surfactant, and the lower is for needle-like crystals grown in water.

V. BEHAVIOR OF HYDRATES DURING PROCESSING, HANDLING, AND STORAGE

We have so far reviewed the thermodynamic concept of phase transitions and introduced a classification system for hydrates. It remains to explore where in the dosage form development process such transitions are most likely to occur and what we can say about them in light of the preceding discussion. The following discussion will be divided into situations where processing induces transitions, and transitions taking place in the final product. When appropriate, polymorphic systems are also illustrated for contrast and completeness.

A. Processing Induced Transitions

Of particular interest in pharmaceutical development is the induction of phase transitions during processing, which can occur for several possible reasons. The crystal (or amorphous) form of the bulk drug may be a metastable form that is able to ''relax'' to the more stable form during processing. The final state may be either a more stable polymorph or a different hydration state. The processing may kinetically trap a metastable form in the dosage form. It is also possible, in principle, that a new form may appear that is only stable in the formulation matrix. All of these possibilities are explicable in terms of G–T or G–P diagrams. This, of course, assumes that the diagrams are known, which is not typically the case. However, the combination of some data and knowledge of possible relationships should help resolve such problems.

1. Relaxation

Examples of an initially metastable phase that relaxes to a more stable phase are to be found in solutions, in suspensions, and in some solid dosage forms [31]. In solutions, the metastable form will have a higher solubility than its more stable form (and a higher free energy), so a solution of the less stable form may be supersaturated with respect to the more stable form, which then crystallizes during processing. The stable form can be a solvate or an anhydrous form. Cortisone acetate suspensions represent a well-known example where a transition to a more stable polymorph takes place during processing. The triazinoindoles are an example where an anhydrous material transforms to a hydrate in suspension [32].

These examples are all solution-mediated transformations, which depend upon the solution phase to provide the mobility necessary to rearrange in the most stable form. Solid-state transformations are also possible when temperature and pressure changes can move the system across a phase boundary. These conditions often favor the metastable phase, however, unless there is a temperature reduction. Solid-state relaxation is more likely to be a slow (kinetically controlled) process, which is also possible during storage.

2. Kinetic Trapping

Processes that kinetically trap a metastable phase are not unusual. To trap a phase, one must restrict its molecular mobility once it is formed, which is easier to accomplish in a solid dosage form. Once formed, these phases may be quite stable, and some are marketed (e.g., progesterone) to take advantage of their potentially higher solubility, dissolution rate, and bioavailability. If the transformation results in a phase other than the desired one, however, concerns will arise. The two common processes in solid dosage form processing that are most likely to produce conditions conducive to kinetic trapping are tableting and wet granulation.

The heat and pressure generated during tableting has long been known to produce polymorphic and pseudopolymorphic conversions [33]. A recent example of this is found in the work of Otsuka and Matsuda, who showed that tableting can induce polymorphic transitions between the known forms of chlorpropamide [34]. Figure 27 compares the XRPD patterns for chlorpropamide tablets after repeated compression cycles with the patterns of the two polymorphs. The comparison clearly demonstrates that the compound has undergone a partial polymorphic phase transformation during the process. A similar example with a hydrate was found for SQ33600 (Fig. 16). The XRPD patterns for the surface and cross sections of regular and high-compression tablets were compared (Fig. 28), and it was obvious that the amount of change depended upon the compression force and position in the

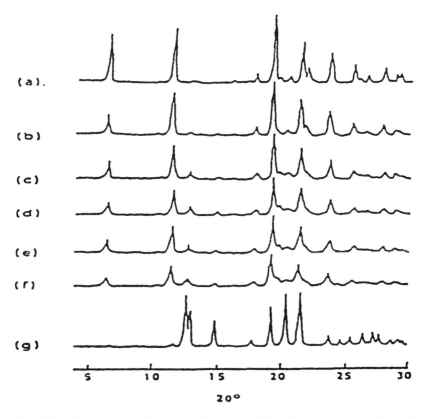

Fig. 27 Changes in chlorpropamide x-ray diffraction patterns of form A produced by repeated compression. (Reproduced with permission.)

tablet. The effect of spatial distribution may be due to the variable compression force experienced at a given depth, or it may be related to the variation in temperature with depth upon compression. It is not clear if the substance simply converted to another crystal form, or if it became amorphous due to dehydration.

Wet granulation is a particularly efficient process for kinetic trapping. In this processing methodology, one prepares a highly concentrated solution of the drug by adding hot solvent to the dry substance and mixing vigorously. The mixture is then rapidly dried, using an evaporative (cooling) process. The resulting product can exhibit char-

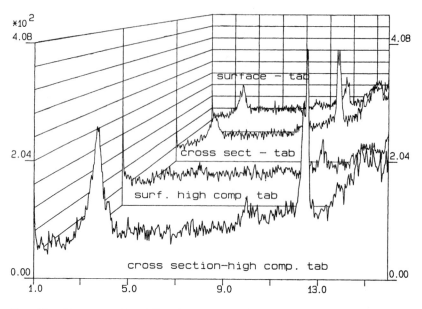

Fig. 28 X-ray powder diffraction patterns for SQ33600 tablet surface and cross section, taken from tablets formed using two different compression forces.

acteristics varying from no effect, to a mixture of forms, to a metastable crystalline form, or to a metastable amorphous form.

One example of this behavior is given by the ACE inhibitor fosinopril sodium (whose structure is provided in Fig. 29). Initially, this formulation was to be wet granulated and then compressed. Figures 30a and b show the XRPD patterns for the two known polymorphic forms of this drug. Based on the thermal data, the direction of solution mediated transformations, and thermodynamic rationale, it was determined that the forms were probably enantiotropically related, with form B being the metastable form at room temperature. During preformulation studies, it was found that the metastable B form could be produced by flash evaporation from methanol in a watch glass. This finding led to studies designed to see if one could produce form B in a laboratory granulation. Figures 31a, b, and c show the XRPD patterns for form A, form B, and the pattern generated from the lab granulation. The

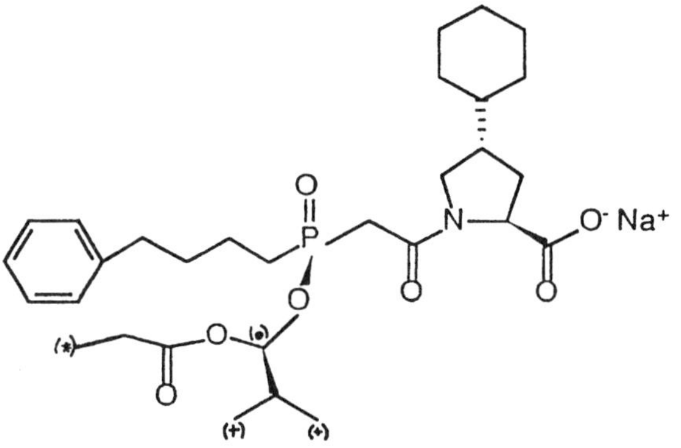

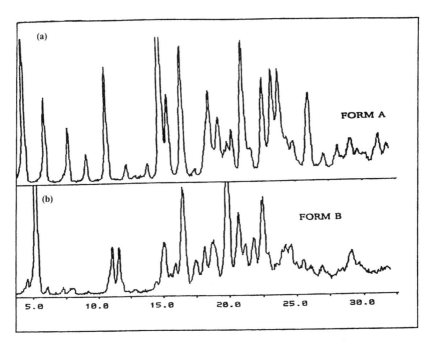

Fig. 29 Structure of fosinopril sodium.

Fig. 30 X-ray powder diffraction patterns for fosinopril sodium, (a) form A and (b) form B.

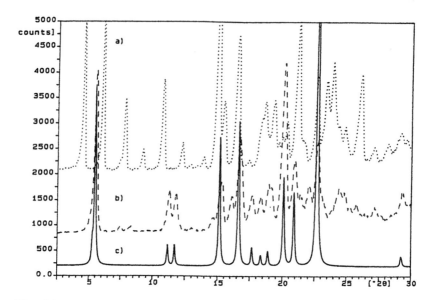

Fig. 31 X-ray powder diffraction patterns for fosinopril sodium, (a) form A, (b) form B, and (c) the material generated from a laboratory granulation.

data clearly show that the manufacturing process converted the initial form A material into the metastable form B in the dosage form. The XRPD method was found not to be sufficiently sensitive to show any form B in pilot lab tablets. Solid state ^{31}P NMR spectra (Fig. 32) clearly show the form B resonance at 55.0 ppm (form A resonating at 52.8 ppm). The process was adjusted to prevent the formation of form B.

The second example involves the channel hydrates of SQ33600. Examination of aqueous granulations of the compound with excipients showed that the crystal structure had disappeared, indicating that the drug substance had converted to the amorphous form. This is evident in the XRPD pattern comparing the dry blend to the granulate and to the tablet (Fig. 33). Apparently the tableting process also has the effect of changing the crystallinity of the compound, which further complicates the interpretation. The crystal lattice of SQ33600 is much less stable than that of fosinopril sodium, which caused significant problems in its development.

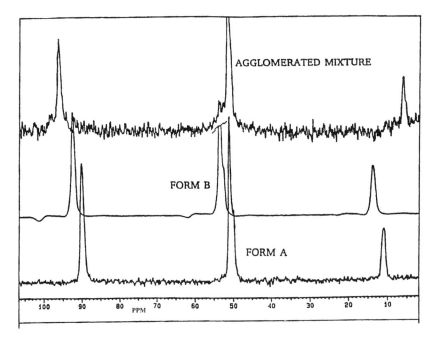

Fig. 32 Solid-state ^{31}P NMR spectra of fosinopril sodium.

B. Transitions in the Final Product

Once a final dosage form has been produced, there remains the possibility of in situ phase transitions. These can happen either because a metastable form was produced or because the ambient conditions have forced the transition. These changes can be hydration or dehydration, alterations in the polymorphic state, or amorphous-to-crystalline phase transitions.

Accelerated stability studies are potentially problematic for at least three reasons. Stress testing conditions may exceed the temperature of a polymorphic phase transition or dehydration. The use of accelerated conditions may make a relaxation of a metastable phase more rapid due to the increased molecular mobility. Finally, the relative humidity of the station may be in the range sufficient to cause a "pseudopolymorphic" transition due to dehydration or hydration. It is sobering

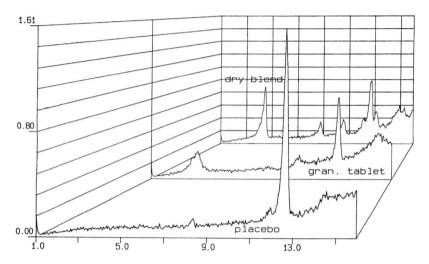

Fig. 33 X-ray powder diffraction patterns of SQ33600 dry blend (upper), granulated tablet (middle), and placebo (lower).

to think that any transition that may be observed during stability studies is also possible during storage and handling. The impact of all of these transitions is unknown until the physicochemical behavior of the related phases is determined. More and more it is becoming expected that all phase transitions that might occur in a product will be explored during the course of dosage form development.

It is easy to see from the G–T diagram of an enantiotropic polymorphic system that if the storage condition of the stable phase of a drug exceeds the transition temperature, the order of stability changes, and a new phase will form if the kinetics permit. A more common problem is exceeding the glass transition temperature of amorphous compounds (or coatings), which may in turn induce crystallization and/ or other mobility related problems. The general G–T diagram (of the type shown in Fig. 3a) also shows that a metastable form at a given temperature can convert to the stable form if the kinetics are favorable. As the temperature is increased, the conversion kinetics become more favorable, even though the free energy gradient decreases. This is an unpredictable balance, but prudent workers would typically opt for maintaining a lower temperature if their formulation included a metastable crystal form.

Using fosinopril sodium as the example, one can see evidence of such a relaxation transition. This was not observed in the final dosage form, but it probably would have occurred if the metastable form B had been chosen for development instead of the stable form A. In one study, a sample of form B was characterized by solid-state NMR, and then again after three months of storage at ambient laboratory conditions. As is evident from the data presented in Figs. 34a and b, the

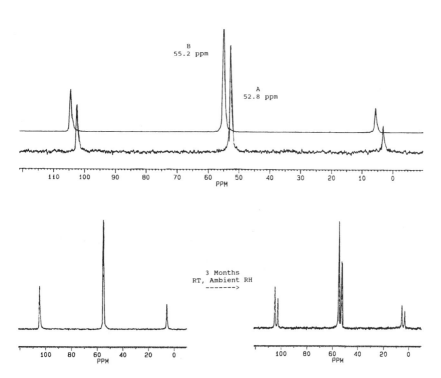

Fig. 34 Solid-state ^{31}P NMR spectra of fosinopril sodium, form A and form B (upper traces), showing the chemical shifts for the phosphorous nucleus in each form. The lower spectra show the partial interconversion from form B to form A after storage at room temperature and ambient humidity for three months.

sample had partially converted to form A in the solid state. This is a case where sufficient driving force existed for the transition (negative free energy change) but was somewhat retarded by the kinetics.

An example of a transition induced by a change in the relative humidity at a fixed temperature is again provided by SQ33600. Recall from the hydrate classification section that SQ33600 that has been equilibrated at different relative humidity values yielded XRPD patterns consistent with channel hydrate behavior. It was found that after wet granulation, the crystal form was substantially diminished and appeared to have become amorphous (see Fig. 33). When these granules were stored for five months in HDPE bottles at room temperature or at 40°C and 75% RH, the powder patterns showed that the amorphous phase had recrystallized to the forms expected under the environmental conditions (see Fig. 35). This raised concerns that the same effect could happen in tablets exposed to elevated relative humidities. An experiment was conducted in which tablets were exposed to 52% and 75% RH for four days. The resulting XRPD patterns confirmed the concern (Fig. 36). The drug had recrystallized into the hydrated forms expected from the work on the pure compound. Great care had to be taken to

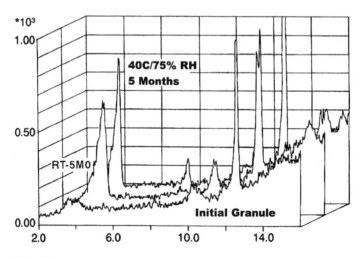

Fig. 35   X-ray powder diffraction patterns of SQ33600 granules after storage at 75% relative humidity.

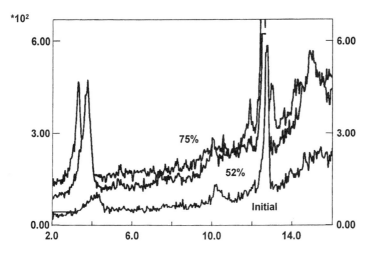

Fig. 36 X-ray powder diffraction patterns of SQ33600 tablets after their exposure for four days at 75 and 52% RH.

maintain a given hydration state and associated crystal form. This was true for clinical supplies and would obviously be true for stability studies. Fortunately, the drug solubility was sufficiently high so as to obviate any potential effects of crystal form on the dissolution rate.

C. Kinetics of Transformation

Only a few generalities can be advanced regarding the kinetics of phase transformation. Constant temperature solution mediated transformations typically proceed much faster than do those in the solid state. The rate of a solution mediated transformation is proportional to the solubility of the species involved, and this is particularly true of the relaxation transformations previously mentioned. Some transformations that proceed rapidly and are apparently occurring in the solid state may be taking place in the absorbed water layer or in the amorphous fractions. Hydrate transformations can be thought of as reactions that are not necessarily occurring in the solid state, since the water may be in the vapor phase. These may even be considered as solution mediated, which would explain the relative rapidity with which many hydration/dehydration reactions occur.

If a G–T analysis shows only a very small driving force for a solid-state polymorphic transformation, even favorable kinetics may be insufficient to allow the transition to occur rapidly. In general, the existence of favorable kinetics has more effect than does a relatively large driving force, so increasing temperature and/or pressure are the main factors over which one has any control. Obviously, relative humidity is the main factor in hydrate transformation, and here a large driving force in the partial pressure of water may enhance the kinetics by inducing mobility in the system.

VI. SUMMARY

Based on their structural characteristics, crystalline hydrates were broken into three main classes. These were (1) isolated lattice site water types, (2) channel hydrates, and (3) ion associated water types. Class 2 hydrates were further subdivided into expanded channel (nonstoichiometric) types, planar hydrates, and dehydrated hydrates. The classification of the forms together with a suitable phase diagram provides a rationale for anticipating the direction and likelihood of a transition, including transitions that may be solution mediated.

Phase transitions possible during the development process were also broadly classified. These were typified into process induced transitions (relaxation of metastable forms, and kinetic trapping of metastable forms) and transitions of the final product (accelerated stability induced changes, and relaxation of metastable forms during storage). Examples of each case were reviewed, and some specific steps in the development process were identified as being potentially problematic. Wet granulation, tableting, particle size reduction, and stress testing stability conditions were all singled out as areas of possible concern.

It is generally desirable to select the most thermodynamically stable crystal form for formulation development. This form will usually have the least liability for phase transitions and will often be the most chemically stable form. Sometimes, however, a metastable form is selected either for performance issues or because the stable form is not obtainable directly (or for historical regulatory reasons). Under these conditions, it is imperative that the solid state system be thoroughly

understood so as to avoid problems that might not have been anticipated. It is rapidly becoming expected that pharmaceutical scientists
should characterize and understand their dosage form ''systems'' at
the time when regulatory documents are filed.

Acknowledgments

The work presented here represents the efforts of many scientists. The
data and insight contributions of Dr. Jack Z. Gougoutas (Bristol-Myers
Squibb Institute for Medical Research) pervade this chapter. In addition, the contributions of Dr. Ann Newman, Dr. David Bugay, and Dr.
Beverly Bowman (also of Bristol-Myers Squibb) make up a significant
fraction of this work. Finally, I must credit the many helpful discussions
I have had with Dr. David Grant (University of Minnesota), Dr. Jens
Carstensen (University of Wisconsin), Dr. Steven Byrn (Purdue University), and Dr. Ulrich Griesser (University of Innsbruck).

REFERENCES

1. K. R. Morris, and N. Rodriguez-Hornado, ''Hydrates,'' in *Encyclopedia
   of Pharmaceutical Technology* (J. Swarbrick, and J. Boylan, eds.), Vol.
   7, Marcel Dekker, New York, 1993.
2. K. R. Morris, M. G. Fakes, A. B. Thakur, A. W. Newman, A. K. Singh,
   J. J. Venit, C. J. Spagnuolo, and A. T. M. Serajuddin, *Int. J. Pharm.*,
   105, 209–217 (1994).
3. H. P. Stahl, Heinrich, ''The Problems of Drug Interactions with Excipients,'' in *Towards Better Safety of Drugs and Pharmaceutical Products*
   (D. D. Braimar, ed.), Elsevier/North-Holland Biomedical Press, 1980,
   pp. 265–280.
4. Cambridge Structural Database, V 2.3.7, 1996.
5. F. Leusen, personal communication, Molecular Simulations Inc., Cambridge, England.
6. M. Falk, and O. Knop, *Water, A Comprehensive Treatise*, Vol. 2 (F.
   Franks, ed.), Plenum Press, New York, 1973, pp. 55–113.
7. D. J. W. Grant, and T. Higuchi, *Solubility Behavior of Organic Compounds*, John Wiley, New York, 1990.
8. J. T. Carstensen, ''Two-Component Systems,'' in *Solid Pharmaceutics:*

Mechanical Properties and Rate Phenomena, Academic Press, New York, 1990, pp. 102–133.

9. J. T. Carstensen, "Two-Component Systems," in *Solid Pharmaceutics: Mechanical Properties and Rate Phenomena*, Academic Press, New York, 1990, pp. 102–133.

10. H. Zhu, R. K. Khankari, B. E. Padden, E. J. Munson, W. B. Gleason, and D. J. W. Grant, *J. Pharm. Sci.*, *85*, 1026–1934 (1996).

11. J. Zhu, and D. J. W. Grant, *Int. J. Pharm.*, *139*, 33–43 (1996).

12. J. Z. Gougoutas, personal communication.

13. I. Dzidic, and P. Kebarle, *J. Phys. Chem.*, *74*, 1466–1474 (1970).

14. W. Mikenda, *J. Mol. Struct.*, *147*, 1–15 (1986).

15. H. D. Lutz, "Bonding and Structure of Water Molecules in Solid Hydrates," in *Structure and Bonding 69* (Leeberg, ed), Springer-Verlag, Berlin, Heidelberg, 1988, pp. 97–125.

16. S. M. Berge, L. D. Bighley, and D. C. Monkhouse, *J. Pharm. Sci.*, *66*, 1–18 (1977).

17. K. Florey, "Cephradine," in *Analytical Profiles of Drug Substances*, Vol. 2 (K. Florey, ed.), Academic Press, New York, 1973, pp. 1–62.

18. H. G. Brittain, D. E. Bugay, S. J. Bogdanowich, and J. DeVincentis, *Drug Dev. Ind. Pharm.*, *14*, 2029–2046 (1988).

19. E. Ivashkiv, "Ampicillin," in *Analytical Profiles of Drug Substances*, Vol. 2 (K. Florey, ed.), Academic Press, New York, 1973, pp. 1–62.

20. S. R. Byrn, *Solid-State Chemistry of Drugs*, Academic Press, New York, 1982.

21. J. S G. Cox, G. D. Woodard, and W. C. McCrone, *J. Pharm. Sci.*, *6*, 1458–1465 (1971).

22. K. R. Morris, A. W. Newman, D. E. Bugay, S. A. Ranadive, A. K. Singh, M. Szyper, S. A. Varia, H. G. Brittain, and A. T. M. Serajuddin, *Int. J. Pharm.*, *108*, 195–206 (1994).

23. J. Zhu, B. E. Padden, E. J. Munson, and D. J. W. Grant, *J. Pharm. Sci.*, *86*, 418–428 (1997).

24. J. Z. Gougoutas, unpublished data.

25. Y. Sugawara, N. Kamiya, H. Iwasaki, T. Ito, and Y. Satow, *J. Am. Chem. Soc.*, *113*, 5440–5445 (1991).

26. D. Giron, *Acta Pharm. Jugosl.*, *40*, 95–147 (1990).

27. L. Pauling, *The Nature of the Chemical Bond*, 3d ed., Cornell Univ. Press, New York, 1960, Chapter 12.

28. A. K. Galwey, G. M. Laverty, N. A. Baranov, and B. Okhotnikov, *Phil. Trans. Royal. Soc. London*, *A347*, 157–184 (1994).

29. Cerius[2] Version 3.5 (Morphology predictor), Molecular Simulations Inc., San Diego, CA (1997).
30. D. Murphy, Ph.D. dissertation, Univ. of Michigan, 1997.
31. J. Haleblian, and W. C. McCrone, *J. Pharm. Sci.*, *58*, 911–929 (1969).
32. J. Haleblian, *J. Pharm. Sci.*, *64*, 1269–1288 (1975).
33. M. G. Wall, *Pharm. Manuf.* (February), 33–42 (1986).
34. M. Otsuka, and Y. Matsuda, *Drug Dev. Ind. Pharm.*, *19*, 2241–2269 (1993).

5

Generation of Polymorphs, Hydrates, Solvates, and Amorphous Solids

J. Keith Guillory

The University of Iowa
Iowa City, Iowa

I. METHODS EMPLOYED TO OBTAIN UNIQUE POLYMORPHIC FORMS

Organic medicinal agents that can exist in two or more solid phases often can provide some distinct advantages in particular applications. The metastable solid may be preferred in those instances where absorption of the drug is dissolution rate dependent. The stable phase may be less susceptible to chemical decomposition and may be the only form that can be used in suspension formulations. Often a metastable polymorph can be used in capsules or for tableting, and the thermodynamically stable form for suspensions. Factors related to processing, such as powder flow characteristics, compressibility, filterability, or hygroscopicity, may dictate the use of one polymorph in preference to another. In other cases, a particular form may be selected because of the high reproducibility associated with its isolation in the synthetic procedure.

It is essential to ascertain whether the crystalline material that results from a synthetic procedure is thermodynamically stable before conducting pivotal trials, since a more stable form may be obtained subsequently, and it may be impossible to produce the metastable form in future syntheses. Conversion from one polymorph to another can occur during processing or upon storage. An additional incentive for isolating and identifying polymorphs that provides certain advantages is the availability of subsidiary patents for desirable polymorphic forms, or for retaining a competitive edge through unpublished knowledge. In 1990 Byrn and Pfeiffer found more than 350 patents on crystal forms granted on the basis of an advantage in terms of stability, formulation, solubility, bioavailability, ease of purification, preparation or synthesis, hygroscopicity, recovery, or prevention of precipitation [1].

One question that is likely to arise during the registration process is "What assurance can be provided that no other crystalline forms of this compound exist?" It is incumbent on the manufacturer of a new drug substance to show that due diligence has been employed to isolate and characterize the various solid-state forms of a new chemical entity. This may seem to be a daunting task, particularly in light of the widely quoted statement by Walter C. McCrone [2] that "Those who study polymorphism are rapidly reaching the conclusion that all compounds, organic and inorganic, can crystallize in different crystal forms or polymorphs. In fact, the more diligently any system is studied the larger the number of polymorphs discovered." On the other hand, one can take comfort from the fact that some important pharmaceuticals have been in use for many years and have, at least until now, exhibited only one stable form. Indeed, it seems to this author that there must be particular bonding arrangements of some molecules that are so favorable energetically as to make alternate arrangements unstable or nonisolatable.

In the future, computer programs using force-field optimization should be perfected to the point where it will be possible to predict, with confidence, that a particular crystalline packing arrangement is the most stable that is likely to be found. These programs also may make it possible to predict how many alternate arrangements having somewhat higher energy can potentially be isolated [3,4]. Until that time, the developmental scientist is handicapped in attempting to predict how many solid forms of a drug are likely to be found. The situation is further complicated by the phenomenon of "disappearing polymorphs" [5], or metastable crystal forms that seem to disappear in favor of more stable ones.

Some polymorphs can be detected, but not isolated. Hot stage microscopy has been used extensively to study polymorphic transformations. The microscopist can detect numerous polymorphic transformations, but the individual polymorphs often prove to be so unstable that they cannot be isolated by the usual methods. An excellent example of this is the work of Grießer and Burger on etofylline [6]. These authors identified five polymorphic forms by thermomicroscopy, but only stable Modification I could be obtained by recrystallization, even when seed crystals from the hot stage were used. Similarly, Kuhnert-Brandstätter, Burger, and Völlenklee [7] described six polymorphic forms of piracetam, only three of which could be obtained by solvent crystallization. All the others were found only by crystallization from the melt. What, then, is a careful investigator to do?

In this chapter, the various methods used to isolate polymorphs, hydrates, and solvates will be described. As Bernstein [8] has observed, "The conditions under which different polymorphs are obtained exclusively or together also can provide very useful information about the relative stability of different phases and the methods and techniques that might be necessary to obtain similar structures of different chemical systems." In this context, it is hoped that the following information will prove useful in devising a "screening" protocol for the preparation of the various solid state forms of pharmaceuticals. While one cannot be absolutely certain that no additional forms will be identified in the future, this approach should provide some assurance that "due diligence" has been exercised to isolate and identify crystalline forms that are likely to arise during the normal course of drug development and storage.

A. Sublimation

On heating, approximately two-thirds of all organic compounds are converted partially from the solid to the gaseous state and back to solid, i.e., they sublime [9]. While strictly speaking the term sublimation refers only to the phase change from solid to vapor without the intervention of the liquid phase, it is often found that crystals are formed on cooler surfaces in close proximity to the melt of organic compounds when no crystals were formed at temperatures below the melting point. The most comprehensive information concerning sublimation temperatures of compounds of pharmaceutical interest can be found in tables in the textbook of Kuhnert-Brandstätter [9]. While the information in these tables is designed primarily for the microscopic examination of compounds, it is also possible to utilize it to determine which compounds might be susceptible to the application of techniques (such as vacuum sublimation) that can be carried out on larger scales and at lower temperatures.

The sublimation temperature and the distance of the collecting surface from the material undergoing sublimation have a great influence on the form and size of the crystals produced. The occurrence of polymorphic modifications depends on the temperature of sublimation. In general, it may be assumed that unstable crystals form preferentially at lower temperatures, while at higher temperatures stable forms are to be expected. Nevertheless, mixtures consisting of several modifications are frequently found together. This is the case for barbital and for estradiol benzoate. It should be obvious that the sublimation technique is applicable only to those compounds that are thermally stable.

A simple test can be used to determine if a material sublimes. A small quantity (10–20 mg) of the solid is placed in a petri dish that is covered with an inverted watch glass. The petri dish is heated gently on a hot plate and the watch glass is observed to determine if crystals are growing on it. According to McCrone [2], one of the best methods for obtaining a good sublimate is to spread the material thinly over a portion of a half-slide, cover with a large cover glass, and heat slowly using a Kofler block. When the sublimate is well formed, the cover glass is removed to a clean slide for examination. It is also possible to form good crystals by sublimation from one microscope slide to a second held above it, with the upper slide also being heated so that its temperature is only slightly below that of the lower slide. Cooling of the cover slip by placing drops of various low-boiling solvents on the top surface will cause condensation of the more unstable forms, the lower temperatures leading to the most unstable forms. On a larger scale, a glass cold finger or a commercial sublimator can be employed. Once crystals of various modifications have been obtained, they can be used as seeds for the solution phase crystallization of larger quantities.

Form I of 9,10-anthraquinone-2-carboxylic acid was obtained as needle-like crystals upon sublimation at temperatures exceeding 250°C [10]. Fokkens et al. have used sublimation to purify theophylline for vapor pressure studies [11]. Sakiyama and Imamura found that stable phases of both 1,3-dimethyluracil and malonamide could be prepared by vacuum sublimation [12].

B. Crystallization from a Single Solvent

Slow solvent evaporation is a valuable method for producing crystals. Solutions of the material being crystallized, preferably saturated or nearly so, are filtered to remove most nuclei and then left undisturbed for a reasonable period of time. The rate of evaporation is adjusted by covering the solution with aluminum foil or Parafilm® containing a few small holes. For a solvent to be useful for recrystallization purposes, the solubility of the solute should be on the order of 5–200 mg/mL at room temperature. If the solubility exceeds 200 mg/mL, the viscosity of the solution will be high, and a glassy product is likely to be obtained. A useful preliminary test can be performed on 25–50 mg of sample, adding a few (5–10) drops of solvent. If all the solid dissolves, the solvent will not be useful for recrystallization purposes. Similarly, highly viscous solvents, and those having low vapor pressures (such as glycerol or dimethylsulfoxide) are not usually conducive to efficient crystallization, filtration, and washing operations. The solvents selected for recrystallization should include any with which the compound will come into contact during synthesis, purification, and processing, as well as solvents having a range of boiling points and polarities. Examples of solvents routinely used for such work are listed in Table 1 together with their boiling points.

The process of solution mediated transformation can be considered the result of two separate events, (a) dissolution of the initial phase, and (b) nucleation/growth of the final, stable phase. If crystals do not grow as expected from a saturated solution, the interior of the vessel can be scratched with a glass rod to induce crystallization by distributing nuclei throughout the solution. Alternatively, crystallization may be promoted by adding nuclei, such as seed crystals of the same material. For example, Suzuki showed that the α-form of inosine could be obtained by crystallization from water, whereas isolation of the β-form required that seeds of the β-form be used [13].

If two polymorphs differ in their melting point by 25–50°C, for

Table 1 Solvents Often Used in the Preparation of Polymorphs

Solvent	Boiling point (°C)
Dimethylformamide	153
Acetic acid	118
Water	100
1-Propanol	97
2-Propanol	83
Acetonitrile	82
2-Butanone	80
Ethyl acetate	77
Ethanol	78
Isopropyl ether	68
Hexane	69
Methanol	65
Acetone	57
Methylene chloride	40
Diethyl ether	35

monotropic polymorphs the lower melting, more soluble, form will be difficult to crystallize. The smaller the difference between the two melting points, the more easily unstable or metastable forms can be obtained.

A commonly used crystallization method involves controlled temperature change. Slow cooling of a hot, saturated solution can be effective in producing crystals if the compound is more soluble at higher temperatures; alternatively, slow warming can be applied if the compound is less soluble at higher temperatures. Sometimes it is preferable to heat the solution to boiling, filter to remove excess solute, then quench cool using an ice bath or even a dry ice–acetone bath. High boiling solvents can be useful to produce metastable polymorphs. McCrone [2] describes the use of high boiling solvents such as benzyl alcohol or nitrobenzene for recrystallization on a hot stage. Behme et al. [14] showed that when buspirone hydrochloride is crystallized above 95°C the higher melting form is obtained; below 95°C the lower melting form is obtained. Thus the lower melting polymorph could be converted to the higher melting polymorph by recrystallizing from xylene (boiling point 137–140°C).

To understand how temperature influences the composition of crystals that form, it is useful to examine typical solubility–temperature diagrams for substances exhibiting monotropic and enantiotropic behavior [15]. In Fig. 1a, Form II, having the lower solubility, is more stable than Form I. These two noninterchangeable polymorphs are monotropic over the entire temperature range shown. For indomethacin, such a relationship exists between Forms I and II, and between Forms II and III.

In Fig. 1b, Form II is stable at temperatures below the transition temperature T_t, and Form I is stable above T_t. At the transition temperature the two forms have the same solubility, and reversible transformation between enantiotropic Forms I and II can be achieved by temperature manipulation. The relative solubility of two polymorphs is a

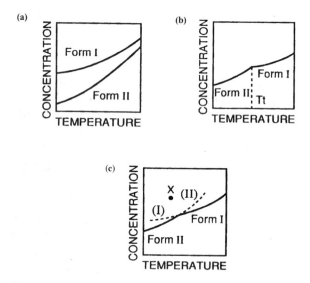

Fig. 1 Solubility curves exhibiting (a) monotropy, (b) enantiotropy, and (c) enantiotropy with metastable phases. (Reprinted with permission of the copyright holder [15].)

convenient measure of their relative free energies. The polymorph having the lower solubility is the more thermodynamically stable form, i.e., the form with the lower free energy at the temperature of the solubility measurement. At room temperature, carbamazepine Form I (m.p. 189°C) is more soluble than is Form III (m.p. 174°C), so the form with the higher melting point is more soluble. The polymorphs are enantiotropic with respect to each other [16].

There are situations in which kinetic factors can for a time override thermodynamic considerations. Figure 1c depicts the intervention of metastable phases (the broken line extensions to the two solubility curves). If a solution of composition and temperature represented by point X (supersaturated with respect to both I and II) is allowed to crystallize, it would not be unusual if the metastable Form I crystallized out first even though the temperature would suggest that Form II would be the more stable (i.e., less soluble) form. This is an extension of Ostwald's law of stages [17], which states that "when leaving an unstable state, a system does not seek out the most stable state, rather the nearest metastable state which can be reached with loss of free energy." This form then transforms to the next most soluble form through a process of dissolution and crystallization. Crystallization of Form I when Form II is more stable would be expected if Form I had the faster nucleation and/or crystal growth rate. However, if the crystals of Form I were kept in contact with the mother liquor, transformation could occur as the more soluble Form I crystals dissolve and the less soluble Form II crystals nucleate and grow. For crystals that exhibit this type of behavior, it is important to isolate the metastable crystals from the solvent by rapid filtration so that phase transformation will not occur.

In the general case, if there are any other polymorphic forms with solubilities below that of Form II, the above-described process will continue between each successive pair of forms until the system finally contains only the most stable (the least soluble) form. The implication of this hypothesis is that, by controlling supersaturation and by harvesting crystals at an appropriate time, it should be possible to isolate the different polymorphic forms. Furthermore, the theory predicts that at equilibrium the product of any crystallization experiment must be the stable form, regardless of the solvent system. It is apparent, however, from the literature that for some solutes it is the choice of solvent rather than the effects of supersaturation that determines the form that crystallizes [18].

Crystallization of mannitol as a single solute was found to be influenced by both the initial mannitol concentration and by the rate of freezing [19]. In the range of 2.5% to 15%, the δ-polymorph is favored by higher concentrations, whereas the β-polymorph is favored at lower concentrations. At constant mannitol concentration (10%), the α-polymorph is favored by a slow freezing rate, whereas the δ-polymorph is favored by a fast freezing rate.

Kaneko et al. [20] observed that both the cooling rate and the initial concentration of stearic acid in *n*-hexane solutions influenced the proportion of polymorphs A, B, C, and E that could be isolated. Garti et al. [21] reported that for stearic acid polymorphs crystallized from various organic solvents, a correlation was observed between the polymorph isolated and the extent of solvent–solute interaction.

The reason for using crystallization solvents having varying polarities is that molecules in solution often tend to form different types of hydrogen-bonded aggregates, and that these aggregate precursors are related to the crystal structures that develop in the supersaturated solution [22]. Crystal structure analysis of acetanilide shows that a hydrogen-bonded chain of molecules is aligned along the needle axis of the crystals. This pattern is characteristic of secondary amides that crystallize in a trans conformation so that the carbonyl acceptor group and the –NH hydrogen bond donor are anti to one another. The morphology of acetanilide crystals can be controlled by choosing solvents that promote or inhibit the formation of this hydrogen-bond chain. Hydrophobic solvents such as benzene and carbon tetrachloride will not participate in hydrogen-bond formation, so they will induce the formation of rapidly growing chains of hydrogen-bonded amides. Crystals grown by evaporation methods from benzene or carbon tetrachloride are long needles. Solvents that are proton donors or proton acceptors inhibit chain formation by competing with amide molecules for hydrogen-bonding sites. Thus acetone inhibits chain growth at the –NH end, and methanol inhibits chain growth at the carbonyl end of the chain. Both solvents encourage the formation of rod-like acetanilide crystals, while mixtures of benzene and acetone give hybrid crystals that are rod-shaped, with fine needles growing on the ends [23].

Some solvents favor the crystallization of a particular form or forms because they selectively adsorb to certain faces of some polymorphs, thereby either inhibiting their nucleation or retarding their growth to the advantage of others. Among the factors affecting the types of crystal formed are (a) the solvent composition or polarity, (b) the concentration or degree of supersaturation, (c) the temperature, including cooling rate and the cooling profile, (d) additives, (e) the presence of seeds, (f) pH, especially for salt crystallization, and (g) agitation [22].

Martínez-Ohárriz et al. [24] found that Form III of diflunisal is obtained from polar solvents, whereas Forms I and IV are obtained from nonpolar solvents. Likewise, Wu et al. [25] observed that when moricizine hydrochloride is recrystallized from relatively polar solvents (ethanol, acetone, and acetonitrile), Form I is obtained, whereas nonpolar solvents (methylene chloride or methylene chloride/ethyl acetate) yield Form II.

In determining what solvents to use for crystallization, one should be careful to select those likely to be encountered during formulation and processing. Typically these are water, methanol, ethanol, propanol, isopropanol, acetone, acetonitrile, ethyl acetate, and hexane. Matsuda employed 27 organic solvents to prepare two polymorphs and six solvates of piretanide [26].

According to McCrone [27], in a poor solvent the rate of transformation of a metastable to a more stable polymorph is slower. Hence a metastable form once crystallized can be isolated and dried before it is converted to a more stable phase by solution phase mediated transformation. In some systems the metastable form is extremely unstable and may be prepared only with more extreme supercooling. This is usually performed on a very small scale with high boiling liquids so that a saturated solution at a high temperature that is suddenly cooled to room temperature will achieve a high degree of supersaturation [28].

There are many examples in the literature of the use of single solvents as crystallization screens. Slow crystallization from acetone, acetonitrile, alcohols, or mixtures of solvents yields the Form A of fosinopril sodium, but rapid drying of a solution of this compound yields Form B, sometimes contaminated with a small amount of Form A [29]. A rotary evaporator can be used to maintain a solution at the appropriate temperature as solvent is being removed.

Form I of dehydroepiandrosterone was obtained by recrystallization from warm ethyl acetate, acetone, acetonitrile, or 2-propanol. Form II was obtained by rapid evaporation, using a vacuum from solutions in dioxane, tetrahydrofuran, or chloroform (which are higher boiling, less polar solvents) [30].

C. Evaporation from a Binary Mixture of Solvents

If single-solvent solutions do not yield the desired phase, mixtures of solvents can be tried. Multicomponent solvent evaporation methods depend on the difference in the solubility of the solute in various solvents. In this approach, a second solvent in which the solute is sparingly soluble is added to a saturated solution of the compound in a good solvent. Often a solvent system is selected in which the solute is more soluble in the component with the higher vapor pressure. As the solution evaporates, the volume of the solution is reduced and, because the solvents evaporate at different rates, the composition of the solvent mixture changes.

Occasionally, crystals are obtained by heating the solid in one solvent and then pouring the solution into another solvent or over cracked ice. Otsuka et al. [31] obtained phenobarbital Form B by adding dropwise a saturated solution of the compound in methanol to water at room temperature. Form E was obtained by the same technique, but by using a saturated solution of phenobarbital in dioxane.

Kitamura et al. have shown that the fraction of Form A of *L*-histidine decreases quickly when the volume fraction of ethanol in an ethanol–water solvent system increases above 0.2, and that pure Form B is obtained at a 0.4 volume fraction of ethanol [32]. The transformation rate for conversion of Form B to Form A decreases with ethanol concentration. The authors postulated that the concentration of the conformer that corresponds to Form A decreases more with ethanol concentration than that of Form B, and so the growth rate of Form A will also decrease.

An example of precipitation in the presence of a second solvent is seen in the case of indomethacin. The γ-crystal form of indomethacin can be obtained by recrystallization from ethyl ether at room temperature, but the α-form is prepared by dissolution in methanol and precipitation with water at room temperature [33]. Precipitation can also result from the addition of a less polar solvent. Form II of midodrine hydrochloride, metastable with respect to Form I, can be prepared by precipitation from a methanolic solution by means of a less polar solvent such as ethyl acetate or dichloromethane [34].

In Fig. 2, three crystalline modifications of thalidomide are illustrated. These were obtained by solvent recrystallization techniques and differ both in crystal habit and in crystal structure. Two of the forms were obtained from a single solvent, and one from a binary mixture.

D. Vapor Diffusion

In the vapor diffusion method, a solution of the solute in a good solvent is placed in a small, open container that is then stored in a larger vessel containing a small amount of a miscible, volatile nonsolvent. The larger vessel (often a desiccator) is then tightly closed. As solvent equilibrium is approached, the nonsolvent diffuses through the vapor phase into the solution, and saturation or supersaturation is achieved. The solubility of the compound in a precipitant used in a two-solvent crystallization method such as vapor diffusion should be as low as possible (much less than 1 mg/mL), and the precipitant (the solvent in which the compound is poorly soluble) should be miscible with the solvent and the saturated solution. The most frequent application of this technique is in the preparation of single crystals for crystallographic analysis. An illustration of the technique is provided in Fig. 3 [35].

E. Thermal Treatment

Frequently when using differential scanning calorimetry as an analysis technique, one can observe an endothermic peak corresponding to a phase transition, followed by a second endothermic peak corresponding to melting. Sometimes there is an exothermic peak between the two endotherms, representing a crystallization step. In these cases it is often

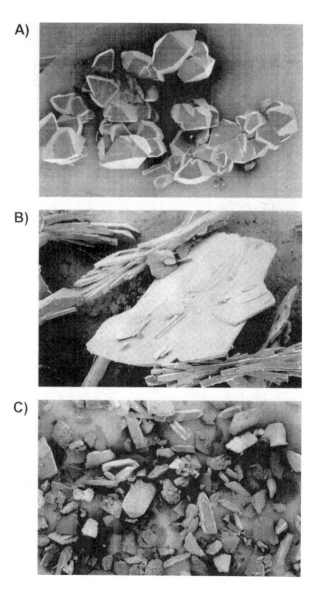

Fig. 2 Three crystalline modifications of thalidomide obtained by solvent recrystallization. (A) Form I obtained as bipyramids by slow crystallization of thalidomide in 1:1 dimethylformamide:ethanol at room temperature. (B) Form II obtained by immersing a saturated solution of thalidomide in acetonitrile in an ice bath. (C) Form III prepared as tabular crystals from a solution in boiling 1,4-dioxane, filtered, then allowed to cool to room temperature. (Photomicrographs courtesy of Dr. S. A. Botha, the University of Iowa.)

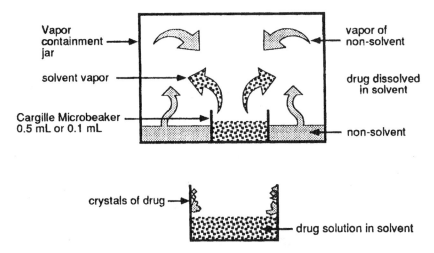

Fig. 3 Crystallization by vapor diffusion. (Reproduced with permission of the author [35] and the copyright holder, Pfizer, Inc.)

possible to prepare the higher melting polymorph by thermal treatment. Thus chlorpropamide Form A is obtained by recrystallization from ethanol solution, but Form C is obtained by heating Form A in an oven maintained at 100°C for 3 hours [36]. While the β-form of tegafur is obtained by the evaporation of a saturated methanol solution, the γ-form is obtained by heating the β-form at 130°C for one hour [37]. Form II of caffeine is prepared by recrystallization from distilled water, but Form I is prepared by heating Form II at 180°C for 10 hours [38].

F. Crystallization from the Melt

In accordance with Ostwald's rule [17], the cooling of melts of polymorphic substances often first yields the least stable modification, which subsequently rearranges into the stable modification in stages. Since the metastable form will have the lower melting point, it follows that supercooling is necessary to crystallize it from the melt. After melting, the system must be supercooled below the melting point of the metastable form, while at the same time the crystallization of the more stable form or forms must be prevented. Quench cooling a melt can sometimes result in formation of an amorphous solid that on subsequent heating undergoes a glass transition followed by crystallization [39].

On a somewhat larger scale, one can use a vacuum drying pistol and a high boiling liquid such as chlorobenzene to achieve the desired end. Form II of p-(1R,3S)-3-thioanisoyl-1,2,-2-trimethylcyclopentane carboxylic acid was obtained by recrystallization from a 50:50 v/v benzene:petroleum ether mixture. Form I then was obtained by melting Form II in the vacuum drying pistol [40]. Caffeine Form I is prepared by heating Form II at 180°C for 10 hours [38]. Yoshioka et al. [41] observed that when the amorphous solidified melt of indomethacin was stored at 40°C, it partly crystallized as the thermodynamically stable γ-form. Yet at 50°C, 60°C, and 70°C, mixtures of the α- and the γ-form were obtained. Sulfathiazole Form I is obtained by heating Form III crystals (grown from a dilute ammonium hydroxide solution at room temperature) at 170°C for 30–40 minutes [42].

G. Rapidly Changing Solution pH to Precipitate Acidic or Basic Substances

Many drug substances fall in the category of slightly soluble weak acids, or slightly soluble weak bases, whose salt forms are much more soluble in water. Upon addition of acid to an aqueous solution of a soluble salt of a weak acid, or upon addition of alkali to an aqueous solution of a soluble salt of a weak base, crystals often result. These crystals may be different from those obtained by solvent crystallization of the weak acid or weak base. Nucleation does not necessarily commence as soon as the reactants are mixed, unless the level of supersaturation is high, and the mixing stage may be followed by an appreciable time lag before the first crystals can be detected. Well-formed crystals are more likely to result in these instances than when rapid precipitation occurs.

Form I of the x-ray contrast agent iopanoic acid was prepared [43] by dissolving the acid in 0.1 N NaOH, adjusting the pH to 12.5, bubbling nitrogen into the solution, and adding 0.1 N hydrochloric acid until the pH reached 2.15. The resulting precipitate was vacuum filtered and stored *in vacuo* (380 torr) for 12 hours at 35°C. Similarly, Form III of hydrochlorothiazide was precipitated from sodium hydroxide aqueous solution by the addition of hydrochloric acid [44].

When piretanide was dissolved in 0.1 N NaOH at room temperature and acid was added in a 1:1 ratio (to pH 3.3), piretanide Form C precipitated. However, when the base:acid ratio used was 1:0.95, a mixture of amorphous piretanide and Form C precipitated [45].

H. Thermal Desolvation of Crystalline Solvates

The term "desolvated solvates" has been applied to compounds that were originally crystallized as solvates but from which the solvent has been removed (generally by vaporization induced by heat and vacuum). Frequently, these "desolvated solvates" retain the crystal structure of the original solvate form and exhibit relatively small changes in lattice parameters. For this reason, these types have been referred to as pseudopolymorphic solvates. However, in instances where the solvent serves to stabilize the lattice, the process of desolvation may produce a change in lattice parameters, resulting in the formation of either a new crystal form or an amorphous form. These solvates have been referred to as polymorphic solvates. Byrn [46] has characterized the desolvation of polymorphic solvates as occurring in four steps, (a) molecular loosening, (b) breaking of the host–solvent hydrogen bonds (or other associations), (c) solid solution formation, and (d) separation of the product phase.

The process of desolvating pseudopolymorphic solvates is simpler, involving only the two steps of (a) molecular loosening and (b) breaking of host–solvent hydrogen bonds or associations. Byrn [46] has summarized the desolvation studies performed on caffeine hydrate, theophylline hydrate, thymine hydrate, cytosine hydrate, dihydrophenylalanine hydrate, dialuric acid hydrate, cycloserine hydrate, erythromycin hydrate, fenoprofen hydrate, manganous formate dehydrate, bis(salicylaldehyde) ethylenediamine cobalt (II) chloroformate, cephatoglycine hydrates and solvates, and cephalexin solvates and hydrates. Among factors that influence the desolvation reaction are the appearance of defects, the size of tunnels in the crystal packing arrangement, and the strength of hydrogen bonding between the compound and its solvent of crystallization [46].

Rocco et al. [47] obtained Form II of zanoterone by recrystallization from ethanol and vacuum drying at 45°C. Form III was isolated by desolvating the acetonitrile solvate form at 80°C under vacuum, and this was the form chosen for use in the clinical drug product due to the high reproducibility of its isolation during manufacture. Similarly, Forms I and II of stanozolol were obtained by heating solvates of the compound to 205°C and 130°C, respectively [48].

The benzene solvate of iopanoic acid was prepared by rapidly freezing a warm benzene solution of iopanoic acid in a dry ice–acetone mixture [43]. The solid obtained was permitted to melt at room temperature, yielding crystals of the solvate suspended in benzene. When these were vacuum filtered and stored *in vacuo* (380 torr) for 12 hours at 70°C, Form II was obtained free of benzene.

Dehydration of hydrates can also lead to the formation of unique crystals. Caffeine Form II was prepared by recrystallizing caffeine from water, drying for 8 days at 30°C, and then heating for 4 hours at 80°C [38]. Chloroquine diphosphate 3:1 hydrate was converted to the anhydrous form at temperatures above 188°C [49]. Etoposide Form I (a monohydrate) was found to undergo a dehydration reaction in the temperature range of 85–115°C to yield etoposide Form 1a. This form could be melted at 198°C and transformed to etoposide Form IIa, which itself melted at 198°C and crystallized to still another polymorph, etoposide Form IIa at 206°C. Etoposide Form IIa was found to melt at 269°C and convert to its hydrated form, etoposide Form II, when exposed to the atmosphere at room temperature. This hydrate was also found to undergo a dehydration reaction at 90–120°C to yield etoposide Form IIa [50].

Differential scanning calorimetry (DSC) curves of levofloxacin hemihydrate measured under various conditions showed different thermograms. This behavior was attributed to the dehydration process that resulted in a multiple-phase transition. Dehydration at higher temperatures (above 70°C) gave a sharp endothermic peak in the DSC thermogram due to the melting of the γ-form, and at a lower temperature (50°C) it led to the observation of a sharp endothermic peak due to the melting of the α-form. In contrast, the thermal behavior of levofloxacin monohydrate was not affected by dehydration [51].

I. Growth in the Presence of Additives

The presence of impurities can have a profound effect on the growth of crystals. Some impurities can inhibit growth completely, and some may enhance growth. Still others may exert a highly selective effect, acting only on certain crystallographic faces and thus modifying the crystal habit. Some impurities can exert an influence at very low concentrations (less than 1 part per million), whereas others need to be present in fairly large amounts to have any effect [15].

Additives can be designed to bind specifically to the surfaces of particular polymorphs and so inhibit their achieving the critical size for nucleation, allowing a desired phase to grow without competition [52]. Lahav and coworkers have shown that additives at levels as low as 0.03% can inhibit nucleation and crystal growth of a stable polymorph, thus favoring the growth of a metastable polymorph [53]. They also showed that it is possible to design crystal nucleation inhibitors to control polymorphism.

Davey et al. found that Form I crystals of terephthalic acid could be obtained by crystallization only in the presence of *p*-toluic acid [54]. Form II, the more stable polymorph at ambient temperatures, was recovered from a hydrothermal recrystallization experiment.

Ikeda et al. [55] determined that indomethacin can exist in three different crystal forms, denoted α-, β-, and γ-, with the α-form possessing a higher solubility than the γ-form. On recrystallization, crystals of the α-form were the first to be deposited, but these converted gradually to the less soluble γ-form. However, in the presence of hydroxypropyl methylcellulose, conversion from the α-form to the γ-form was inhibited, leading to an increase in the solubility of indomethacin.

While the α-form of glycine normally is obtained by recrystallization from water, 3% of racemic hexafluorovaline leads to the precipitation of the γ-polymorph as trigonal pyramids [56]. This additive was designed to be strongly adsorbed at the four {011} crystal faces of the α-form and to bind at only one pole of the polar crystal, thus leaving the crystal free to grow at the opposite pole. Since it is bound at the slow growing NH_3^+ end of the polar axis, it does not interfere with the fast growing CO_2^- end.

J. Grinding

Polymorphic transformations have been observed to occur on grinding of certain materials, such as sulfathiazole, barbital, phenylbutazone, cephalexin, chloramphenicol palmitate, indomethacin, and chlorpropamide. Byrn [46] has stated that polymorphic transformations in the solid state require the three steps of (a) molecular loosening (nucleation by separation from the lattice), (b) solid solution formation, and (c) separation of the product (crystallization of the new phase). Depending on the material and the conditions employed, grinding can result in conversion to an amorphous substance. With the exercise of care, different polymorphic forms can be obtained. Otsuka et al. [57] showed that metastable Forms B and C of chloramphenicol palmitate were transformed into stable Form A upon grinding at room temperature. Indomethacin was transformed into a noncrystalline solid during grinding at 4°C, and into metastable Form A by grinding at 30°C. Caffeine Form II is converted into Form I with grinding, and a 95% phase conversion was obtained following 60 hours of grinding time [38].

II. METHODS EMPLOYED TO OBTAIN HYDRATE FORMS

Pharmaceutical solids may come into contact with water during processing steps, such as crystallization, lyophilization, wet granulation, aqueous film-coating, or spray-drying. Moreover, they may be exposed to water during storage in an atmosphere containing water vapor, or in a dosage form consisting of materials that contain water (e.g., excipients) and are capable of transferring it to other ingredients. Water may be adsorbed onto the solid surface and/or may be absorbed in the bulk solid structure. When water is incorporated into the crystal lattice of the compound in stoichiometric proportions, the molecular adduct or adducts formed are referred to as hydrates [58]. More than 90 hydrates are described in various USP monographs. Hydrates can be prepared by recrystallization from water or from mixed aqueous solvents. They can also result, in some instances, from exposure of crystal solvates (such as methanolates or ethanolates) to an atmosphere containing water vapor.

Crystalline substances often form with water molecules located at specific sites in the crystal lattice, which are held in coordination complexes around lattice cations. This type of water is denoted as water of crystallization and is common for inorganic compounds. For example, nickel sulfate forms a well-defined hexahydrate, where the waters of hydration are bound directly to the Ni(II) ion. Extraneous inclusion of water molecules can occur if a coprecipitated cation carries solvation molecules with it. Water also can be incorporated into random pockets as a result of physical entrapment of the mother liquor. Well-defined multiple hydrate species can also form with organic molecules. For example, raffinose forms a pentahydrate.

Although most hydrates exhibit a whole-number-ratio stoichiometry, an unusual case is the metastable hydrate of caffeine, which contains only 0.8 moles of water per mole of caffeine. Only in a saturated water vapor atmosphere will additional amounts of water be adsorbed at the surface of the 4/5-hydrate to yield a 5/6 hydrate [59].

In some instances, a compound of a given hydration state may crystallize in more than one form, so that the hydrates themselves exhibit polymorphism. One such example is nitrofurantoin, which forms two monohydrates that have distinctly different temperatures and enthalpies of dehydration. The monohydrates have quite different packing arrangements, with Form I possessing a layer structure and Form II exhibiting a herringbone motif. The included water molecules play a major role in stabilizing the crystal structures. Whereas water molecules are contained in isolated cavities in Form II, in Form I they are located in continuous channels, and this apparently facilitates the escape of water when these crystals are heated [60].

Another example of hydrate polymorphism is amiloride hydrochloride [61], which can be obtained in two polymorphic dihydrate forms. These forms are indistinguishable by techniques other than x-ray powder diffraction.

It is interesting that scopolamine hydrobromide has been reported to exist as the anhydrous form, a "hemihydrate," a sesquihydrate, and a trihydrate [62], while the unit cell parameters and the molecular geometry of these are all the same as those of the hemihydrate. This finding suggests that the "hemihydrate" is actually a partially desolvated sesquihydrate.

Ouabaine is another example of a compound that exhibits many different hydration levels, the most hydrated form being stable at the lowest temperature. Thus the nonahydrate phase of ouabaine is obtained from water at $0-15°C$, the octahydrate phase at $15-28°C$, and the dihydrate phase at $28-90°C$. In addition, ouabaine phases corresponding to $4.5\ H_2O$, $4\ H_2O$, and $3\ H_2O$ may be obtained from mixtures of water with other solvents. The anhydrous phase of ouabaine anhydrate is crystallized from ethanol at high temperatures [63].

Typically, hydrates are obtained by recrystallization from water. For example, trazodone hydrochloride tetrahydrate was prepared by dissolving the anhydrate in hot distilled water, allowing the solution to remain at room temperature overnight, and storing the collected crystals at 75% relative humidity and 25°C until they reached constant weight [64].

Hydrates can sometimes be obtained by simply suspending the anhydrous material in water, whereupon a form of Ostwald ripening occurs. For instance, aqueous suspensions of anhydrous metronidazole benzoate are metastable, and storage at temperatures lower then 38°C leads to monohydrate formation accompanied by crystal growth [65]. Sorbitol provides another example of this behavior, where slow cooling of a saturated aqueous solution yields long thin needles of sorbitol hydrate [66]. When suspended in water, anhydrous carbamazepine is transformed to carbamazepine dihydrate [67]. In other instances, hydrates can be obtained from mixed solvent systems. Acemetacin monohydrate can be obtained by slow evaporation from a mixture of acetone and water at room temperature [68].

Simply exposing an anhydrous powder to high relative humidity can often lead to formation of a hydrate. On exposure to a relative humidity of 100%, dexmedetomidine hydrochloride is converted to a monohydrate [69]. Droloxifene citrate is an example of a compound that is not very hygroscopic and yet forms a hydrate. Only after storage of the anhydrous form at 85% relative humidity does some sorption of water occur. The monohydrate phase can be formed by exposing the anhydrous form to 98% relative humidity for ten days at 24°C [70].

III. METHODS EMPLOYED TO OBTAIN SOLVATE FORMS

Often, when solvents are employed in the purification of new drug substances by recrystallization, it is observed that the isolated crystals include solvent molecules, either entrapped within empty spaces in the lattice or interacting via hydrogen bonding or van der Waals force with molecules constituting the crystal lattice. Solvent molecules also can be found in close association with metal ions, completing the coordination sphere of the metal atom. Coordinated solvent molecules are considered as part of the crystallized molecule. A crystal with large empty channels or cavities is not stable because of packing demands. The size and chemical environment of the cavity or channel determine what kind of solvent molecule can be included in the structure and what kind of interaction occurs between solvent and structure.

Depending on the nature of molecular packing arrangements, it may happen that the inclusion of solvent is necessary to build a stable crystal structure. van Geerestein et al. [71] found during numerous crystallization attempts of 11β-[4-(dimethylamino)phenyl]-17β-hydroxy-17α-(I-propynyl) estra-4,9-diene-3-one] that crystals were only obtainable in the presence of *n*-butyl acetate or *n*-propyl acetate. The crystal structure of the compound crystallized from *n*-butyl acetate/methylcyclohexane was solved, and one solvent molecule was found in the crystal structure that showed no strong interactions with the rest of the structure. Apparently, this solvent molecule was necessary to fill empty space resulting after the molecular packing. Solvates in which the solvent fills empty space are generally nonstoichiometric, such as the nonstoichiometric solvates formed by droloxifene citrate with acetonitrile, 2-propanol, ethanol, 1-propanol, and 1-butanol. Typically such solvates exhibit the same x-ray diffraction pattern as does the nonsolvated compound.

When solvent molecules increase the strength of the crystal lattice, they can affect the stability of the compound to solid-state decomposition. It has been observed that the four solvated and one nonsolvated structures of prenisolone *tert*-butyl acetate affect the flexibility of the steroid nucleus and the structure-dependent degradation of the compound when exposed to air and light [72].

van der Sluis and Kroon found 1,247 different compounds with cocrystallized solvents in the Cambridge Crystallographic Database [73]. Out of 46,460 total structures, they found 9,464 solvate structures, and 95% of these contained one of the 15 solvents given in Table 2.

The most commonly encountered solvates among pharmaceuticals are those of 1:1 stoichiometry, but occasionally mixed solvate species are encountered. For structures containing more than one solvent type, one generally finds nonpolar solvents crystallizing together on the one hand and polar solvents on the other. For example, the most common solvents found cocrystallizing with water are (in order of im-

Table 2 Distribution of the 15 Most Abundant Solvents in the Cambridge Crystallographic Database, as the Percentage of Solvate Structures

Solvent	Occurrence (%)
Water	61.4
Methylene dichloride	5.9
Benzene	4.7
Methanol	4.1
Acetone	2.8
Chloroform	2.8
Ethanol	2.6
Tetrahydrofuran	2.3
Toluene	2.2
Acetonitrile	1.9
N,N-dimethylformamide	0.9
Diethyl ether	0.9
Pyridine	0.7
Dimethyl sulfoxide	0.5
Dioxane	0.5

Source: From Ref. 73. Reproduced with permission of the copyright owner.

portance) ethanol, methanol, and acetone. An interesting example of a structure containing a polar and a nonpolar solvent is the sodium salt of the antibiotic K-41, *p*-bromobenzoate monohydrate *n*-hexane solvate [74], which is crystallized from *n*-hexane saturated with water. Perhaps the best known mixed solvate is doxycycline hyclate: (doxycycline · HCl)$_2$C$_2$H$_6$O · H$_2$O. Triamterene also forms a mixed solvate, containing one *N,N*-dimethylformamide molecule and one water molecule within the crystal lattice [75].

The techniques used to obtain solvates are generally similar to the solvent methods used to obtain polymorphs, i.e. crystallization from a single solvent, from mixed solvents, or by vapor diffusion. Sometimes, it is possible to exchange one solvent within the crystal structure for another. When one recrystallizes a hydrate from dry methanol, in most cases one is left with either a methanol solvate or an anhydrous, unsolvated form of the compound.

A large number of solvates have been reported, especially for steroids and antibiotics. It has been observed that cortisone acetate and dexamethasone acetate can be crystallized as 10 different solvates. Dirithromycin, a semisynthetic macrolide antibiotic, crystallizes in two anhydrous polymorphic forms and in at least nine stoichiometric solvate forms. Six of the known solvates are isomorphic, having nearly identical x-ray powder diffraction patterns [76]. In addition to the anhydrate and dihydrate, erythromycin also forms solvates with acetone, chloroform, ethanol, *n*-butanol, and *i*-propanol [77].

It may be instructive to consider some examples of solvate formation. The compound 5-methoxysulphadiazine forms 1:1 host–guest solvates with dioxane, chloroform, and tetrahydrofuran [78]. These were prepared by heating to boiling a solution of the sulfonamide in the appropriate solvent, followed by slow cooling to obtain large crystals. Spironolactone forms 1:1 solvates with methanol, ethanol, ethyl acetate, and benzene. It also forms a 2:1 spironolactone–acetonitrile solvate [79,80]. The spironolactone solvates were prepared by crystallization in a refrigerator from solutions that were nearly saturated at room temperature.

Another steroid that forms solvates is stanozolol [81]. Solvates having 1:1 stoichiometry were prepared by recrystallization from methanol, ethanol, and 2-propanol, by heating the compound in the appropriate solvent to 60–70°C and then cooling to 0°C in an ice bath to induce crystallization. The compound also forms a monohydrate and two polymorphs. The polymorphs were prepared by heating the solvates to either 130°C (Form II) or 205°C (Form I).

Mefloquine hydrochloride is an interesting case of a compound that forms stoichiometric 1:1 solvates on cooling hot (50°C) saturated acetone solutions (Form B, acetone solvate 1:1), hot (50°C) saturated isopropanol (Form I, isopropanol solvate 1:1), and a nonstoichiometric ethanol solvate (2.12% ethanol) from hot (50°C) saturated ethanol, Form E, whose x-ray powder pattern does not change following heating to 80°C, in spite of a decrease in the ethanol level to 0.12%. Mefloquine hydrochloride can also be obtained in a nonsolvated form from hot (70°C) saturated acetonitrile (Form A) and as two hemihydrates from water (Forms D and C) prepared at room temperature and at 30°C [82].

IV. METHODS EMPLOYED TO OBTAIN AMORPHOUS MATERIALS

Solids can exist in crystalline or amorphous form. Crystalline materials have defined structures, stoichiometric compositions, and melting points and are characterized by their chemical, thermal, electrical, optical, and mechanical properties [83]. By contrast, amorphous materials have no clearly defined molecular structure and no long-range order, so their structure can be viewed as being similar to that of a frozen liquid but without the thermal fluctuations observed in the liquid phase. As a result, amorphous materials exhibit the classical diffuse ''halo'' x-ray powder diffraction pattern rather than the sharp peaks observed in the pattern of a crystalline substance. When the halo is broad, it is often difficult to distinguish between a material that is truly amorphous (e.g., a true glass) and one that is merely microcrystalline. This situation exists because when microcrystallites have diameters less than about 50 Å in diameter, a similar ''halo'' effect is observed.

While crystalline solids offer the advantages of chemical and thermodynamic stability, amorphous solids are occasionally preferred because they undergo dissolution at a faster rate. Rapid dissolution is desirable in the case of solids, which must be dissolved prior to parenteral administration. Faster dissolution is also important for poorly soluble compounds administered orally, since there is often a correlation between dissolution rate and bioavailability. In fact, there are instances in which only the amorphous form has adequate bioavailability.

Amorphous solids can be precipitated from solution or obtained from melts of compounds by carrying out the solidification in such a way as to avoid the thermodynamically preferred crystallization process. They also can be prepared by disrupting an existing crystal structure. Excess free energy and entropy are incorporated into solids as they are converted into the amorphous state, since solidification occurs without permitting the molecules to reach their lowest energy states.

A. Solidification of the Melt

Amorphous solids are often created by rapidly cooling a liquid so that crystallization nuclei can neither be created nor grow sufficiently, whereupon the liquid then remains in the fluid state well below the normal freezing point. In principle, a liquid should freeze (crystallize) when cooled to a temperature below its freezing point. However, if the rate of cooling is high relative to the rate of crystallization, then the liquid state can persist well below the normal freezing point. As cooling continues there is a rise in the rate of increase of the viscosity of the supercooled liquid per unit drop in temperature. The initially mobile fluid turns into a syrup, then into a viscoelastic state, and finally into a brittle glass. A glass is, therefore, a supercooled liquid, and is characterized by an extremely high viscosity (typically of the order of 1014 Pa · s). Mechanically, if not structurally, glasses can be regarded as solids.

The characteristic temperature below which melted solids must be cooled to form a glass is the glass transition temperature T_g. The glass transition is a dynamic event that occurs at a temperature below which coordinated molecular motion becomes so slow that a liquid can be considered to take on the properties of a solid. While the exact value of this transition temperature depends on the heating rate, the glass transition temperature is generally found to be about two-thirds that of the melting temperature T_m. Glass transition temperatures reported for pharmaceuticals also follow this general rule, as can be seen in the listing of ten pharmaceuticals that form glasses (Table 3). It is often found that the presence of impurities that facilitate glass formation increases the ratio T_g/T_m either by raising T_g or by lowering T_m. Hence one might wonder if some of the high values in the last column of Table 3 are due to partial decomposition of the drug substance upon melting. Of course, this is an important concern when employing the melt solidification procedure for the preparation of amorphous materials.

There are many examples given in the monograph *Thermomicroscopy in the Analysis of Pharmaceuticals* [9] of other compounds that solidify on the microscope hot stage to form glasses. However, Table 4 contains examples from the literature in which solidification from the melt (either by slow cooling to room temperature or by quench cooling with liquid nitrogen) has been employed as the specific method for obtaining amorphous material.

B. Reduction of Particle Size

Reduction of the particle size of crystalline materials to the microcrystalline level can yield a material incapable of exhibiting an x-ray pow-

Table 3 Pharmaceuticals Forming Glasses above Room Temperature

Compound	$T_g(K)$	$T_m(K)$	T_g/T_m
Cholecalciferol	296	352	0.84
Sulfisoxazole	306	460	0.67
Stilbestrol	308	439	0.70
Phenobarbital	321	443	0.72
Quinidine	326	445	0.73
Salicin	333	466	0.71
Sulfathiazole	334	471	0.71
Sulfadimethoxine	339	465	0.73
Dehydrocholic acid	348	502	0.69
17-β-Estradiol	354	445	0.80

Source: Ref. 84.

Table 4 Amorphous Pharmaceuticals Obtained by Solidification from the Melt

Compound	Method used	Reference
Phenylbutazone	Solidification from the melt	[85]
Indomethacin	Quench cooling using liquid nitrogen or slow cooling from the melt over 30 min	[86,87]
Felodipine	Cooling of the melt in liquid nitrogen or at ambient temperature	[88,89]
Nifedipine	Melting at 180°C followed by immersion in liquid nitrogen	[90]
Benperidol	Melt in an oven at 277°C then cool to room temperature	[91]
Acetaminophen	Solidification of the melt at −5°C/min	[92]
Sulfapyridine	Melting any crystalline form and slowly cooling the melt	[93]
Lovostatin	Melting under nitrogen, rapid cooling to 20°C below the glass transition point	[94]

der diffraction pattern. Dialer and Kuessner [95] found that when sucrose was milled in a vibratory ball mill, the ordered crystal was transformed into a glass-like structure. The increase in surface energy of milled sucrose, as measured by heat of solution, could not be accounted for by an increase in surface area alone. Hence milling disrupts the crystal lattice and imparts the excess free energy and entropy associated with amorphous substances.

Particle size reduction can be achieved using a variety of methods. Sometimes it is helpful to carry out the particle size reduction at reduced temperatures, such as at 4°C or at liquid nitrogen temperature, −196°C. In other instances, grinding with an excipient has been employed as a means of obtaining amorphous materials. Cyclodextrins and microcrystalline cellulose have been used for this purpose. It is also possible that the use of polymeric excipients may inhibit crystal growth when the amorphous solid is dissolved in water. Table 5 contains a list of compounds that have been obtained in amorphous, or partly amorphous, form by milling.

Table 5 Amorphous Pharmaceuticals Obtained by Milling

Compound	Method used	Reference
Cimetidine	Milling	[96]
FR76505	Grinding in a ball mill	[97]
Cephalexin	Grinding in an agate centrifugal ball mill for 4 hours	[98]
Indomethacin	Grinding for 4 hours at 4°C in a centrifugal ball mill; grinding the γ-form at 4°C	[57,99]
(E)-6-(3,4-Dimethoxyphenyl)-1-ethyl-4-mesitylimino-3-methyl-3,4-dihydro-2(1H)-pyrimidinone	Grinding in a stainless steel shaker ball mill for 60 minutes	[100]
9,3″-Diacetyl-midecamycin	Mixed grinding with polyvinylpyrrolidone or polyvinylpyrrolidone + hydroxypropylmethylcellulose for 9 hours	[101]
Chloramphenicol stearate	Milling in a Pulverisette 5 grinder (Fritsch) (agate mortar and balls) with colloidal silica or microcrystalline cellulose	[102,103]
Calcium gluceptate	Milling in a Pulverisette 2 grinder (Fritsch) (agate mortar and balls) for 4 hours	[104]
Chloramphenicol palmitate	Milling in a Pulverisette 0 grinder (Fritsch) (agate mortar and balls) for 85 hours	[105]
Aspirin	Grinding with adsorbents under reduced pressure	[106]
	Grinding with β-cyclodextrin	[107]
Ibuprofen	Roll mixing with β-cyclodextrin	[108]
Hydrocortisone acetate	Grinding with crystalline cellulose	[109]

Table 5 Continued

Compound	Method used	Reference
Digoxin	Milling in a Glen Creston Model M270 ball mill for 8 hours	[110]
	Comminution of 1 g at 196°C for 15 minutes in a freezer mill	[111]
Amobarbital	Ball-milling with methylcellulose, microcrystalline cellulose, or dextran 2000	[112,113]
Acetaminophen	Ball milling for 24 hours with α- and β-cyclodextrin	[114]
6-Methyleneandrosta-1, 4-diene-3,17-dione	Co-grinding with β-cyclodextrin for 2 hours	[115]

C. Spray-Drying

In the pharmaceutical industry, spray-drying is used to dry heat-sensitive pharmaceuticals, to change the physical form of materials for use in tablet and capsule manufacture, and to encapsulate solid and liquid particles. This methodology is also used extensively in the processing of foods [116]. In the spray-drying process, a liquid feed stream is first atomized for maximal air spray contact. The particles are then dried in the airstream in seconds owing to the high surface area in contact with the drying gas. Spray-drying can produce spherical particles that have good flow properties, and the process can be optimized to produce particles of a range of sizes required by the particular application. The process can be run using either aqueous or nonaqueous solutions. Examples of pharmaceuticals obtained in the form of amorphous powders by spray-drying are found in Table 6.

D. Lyophilization

Lyophilization (also known as freeze-drying) is a technique that is widely employed for the preparation of dry powders to be reconstituted at the time of administration. It is a particularly useful technique in the

Table 6   Amorphous Pharmaceuticals Obtained by Spray-Drying

Compound	Method used	Reference
YM022	Spray-drying a methanol solution	[117]
α-Lactose monohydrate	Spray-drying in a Buchi 190	[118]
	Spray-drying a solution or suspension	[119]
4″-O-(4-methoxy-phenyl) acetyltylosin	Spray drying a dichloromethane solution	[120]
Salbutamol sulfate	Spray-drying of an aqueous solution in Buchi 90 spray dryer	[121]
Lactose	Spray-drying an aqueous solution	[118,122]
Furosemide	Spray-drying from a 4:1 chloroform: methanol solution at 50 and 150°C inlet temperature	[123,124]
Digoxin	Spray-drying an aqueous solution containing hydroxypropyl methylcellulose	[125]
Cefazolin sodium	Spray-drying from a 25% aqueous solution with an inlet temperature of 150°C and an outlet temperature of 100°C	[126]
9,3″-Diacetyl-midecamycin	Spray-drying of aqueous solution in the presence and absence of ethylcellulose	[127]

case of compounds that are susceptible to decomposition in the presence of moisture but that are more stable as dry solids. The physical form, chemical stability, and dissolution characteristics of lyophilized products can be influenced by the conditions of the freeze-drying cycle. In most pharmaceutical applications, lyophilization is performed on aqueous solutions containing bulking agents, and these often are chosen so as to form a coherent cake after completion of the freeze-drying process. However, lyophilization also can be employed to convert crystalline materials into their amorphous counterparts. The lyophilization process usually consists of the three stages of freezing, primary drying, and secondary drying. For the preparation of amorphous materials, rapid freezing is employed so as to avoid the crystallization process. Both aqueous solutions and solutions containing organic solvents have been lyophilized. The primary drying phase involves sublimation of frozen water or vaporization of another solvent. This step is carried out by reducing the pressure in the chamber and supplying heat to the product. The secondary drying phase consists of the desorption of moisture (or residual solvent) from the solid.

Recently, excipients of various types have been employed in frozen solutions so as to inhibit crystallization. Cyclodextrins appear to be particularly useful for this purpose, although it is generally necessary to employ rapid freezing to liquid nitrogen temperatures to ensure that the freeze-dried product is noncrystalline. When α-cyclodextrin, which has a larger cavity than does β-cyclodextrin, is frozen at a relatively slow rate, it will cocrystallize with compounds such as benzoic acid, salicylic acid, *m*-hydroxybenzoic acid, *p*-hydroxybenzoic acid, and methyl *p*-hydroxybenzoate [128]. However, rapid freezing of a methyl *p*-hydroxybenzoate solution containing α-cyclodextrin at a benzoate/ cyclodextrin ratio of 0.33 yields an amorphous solid after freeze-drying [29].

β-Cyclodextrin and its derivatives have been shown to form amorphous lyophilized products with a number of compounds, principally nonsteroidal antiinflammatory agents. Examples from the literature of excipients and pharmaceuticals prepared as amorphous materials by lyophilization are given in Table 7.

E. Removal of Solvent from a Solvate or Hydrate

Solids can sometimes be rendered amorphous by the simple expedient of allowing solvent molecules of crystallization to evaporate at modest temperatures. If the solvent merely occupies channels in the crystal structure, the structure often remains intact, but when the solvent is strongly bonded to molecules of the host, the structure frequently will collapse when the solvent is removed and one obtains an amorphous powder. A few examples of amorphous solids obtained in this manner are found in Table 8.

Table 7 Amorphous Pharmaceuticals Obtained by Lyophilization

Compound	Method used	Reference
Lactose	Lyophilization of a 5% Aqueous Solution	[130]
MK-0591	Lyophilization	[131]
Raffinose	Lyophilization of a 10% aqueous solution frozen at −45°C	[132]
Sucrose	Lyophilization of 10% aqueous solutions	[133]
Dirithromycin	Freeze-drying from methylene chloride solution	[134]
Cefalexin	Aqueous solution frozen at −196°C, then freeze-dried	[135]
	Lyophilization of a saturated aqueous solution	[136]
Calcium gluceptate	Freeze-drying from 2% aqueous solution	[137]
Griseofulvin	Freeze-drying of solutions of griseofulvin or of solutions of mixtures of griseofulvin and mannitol in dioxane or 1:1 dioxane-water with fast freezing in liquid nitrogen	[138]
Tolobuterol hydrochloride	Freeze-drying of aqueous solution	[139]
E1040	Freeze-drying of aqueous solution	[140]
Glutathione	Freeze-drying of a 5% aqueous solution	[141]
Aspirin	Freeze drying of an aqueous solution in the presence of 1.0% hydroxypropyl-β-cyclodextrin	[142]
Ketoprofen	Freeze-drying in the presence of heptakis-(2,6-O-dimethyl)-β-cyclodextrin	[143]
	Freeze-drying with β-cyclodextrin (rapid freezing with liquid nitrogen)	[144]
Glibenclamide	Freezing at liquid nitrogen temperature, freeze-drying over 24 hours	[145]

Table 7 Continued

Compound	Method used	Reference
Naproxen	Colyophilization (223K and 0.013 torr) of naproxen and hydroxyethyl-β-cyclodextrin, or hydroxypropyl-β-cyclodextrin	[146]
Sodium ethacrynate	Rapid freezing of an aqueous solution to −50°C, followed by freeze-drying	[147]
p-Aminosalicylic acid	Colyophilization of p-aminosalicylic acid in aqueous solution with pullulan	[148]
Ceftazidime	Freeze-drying a nearly saturated aqueous solution of the free acid	[149]
Cefaclor	Freeze-drying from a nearly saturated aqueous solution	[149]
Cephalothin sodium	Freeze-drying from a 25% aqueous solution	[149]
Cefamandol sodium	Freeze-drying from a 25% aqueous solution	[149]
Cefazolin sodium	Freeze-drying an aqueous solution at low temperature	[149]
Nicotinic acid	Freeze-drying in the presence of β-cyclodextrin (fast-freezing); and heptakis (2,6-O-dimethyl)-β-cyclodextrin	[150]

F. Precipitation of Acids or Bases by Change in pH

If the level of supersaturation is carefully controlled, it is often possible to avoid crystallization when a water-soluble salt of a weak acid is precipitated with a base, or when a water-soluble salt of a weak base is precipitated with an acid. When crystalline iopanoic acid is dissolved in 0.1 N NaOH, and 0.1 N HCl is added, an amorphous powder is precipitated [43]. A similar phenomenon is observed in the case of the precipitation of piretanide [155]. Another example in this genre is the

Table 8 Amorphous Pharmaceuticals Obtained by Solvent Removal

Compound	Method used	Reference
Tranilast anhydrate	Dehydration of the monohydrate over P_2O_5	[151]
Raffinose	Lyophilization and heat drying of the pentahydrate	[132]
Erythromycin	Heating the dihydrate for 2 hours at 135°C in an oven, and then cooling to room temperature	[152,153]
Calcium DL-pantothenate	Drying the methanol:water 4:1 solvate *in vacuo* at 50–80°C	[154]

precipitation of amorphous calcium carbonate, which occurs when a calcium chloride solution is combined with a sodium carbonate solution at 283K [156].

G. Miscellaneous Methods

Earlier during the discussion on the preparation of polymorphs, the doping of crystals was mentioned as a technique for encouraging the formation of one type of polymorph over another. Similarly, if a dopant is employed at levels that will disrupt the crystal lattice, the substance can be made to solidify as an amorphous material. Duddu and Grant [157] observed changes in the enthalpy of fusion of (−)-ephedrinium 2-naphthatenesulfonate when the opposite enantiomer, (+)-ephedrinium 2-naphthalenesulfonate, was added as a dopant.

When *m*-cresol was added to a suspension of insulinotropin crystals grown from a normal saline solution, the crystals were immediately rendered amorphous. It was postulated [158] that the *m*-cresol molecules diffused into the crystals through solvent channels and disturbed the lattice interactions that ordinarily maintained the integrity of the crystal. When zinc acetate or zinc chloride was added to the suspension, the zinc ion stabilized the crystal lattice so that the subsequent addition of *m*-cresol did not alter the integrity of the crystals.

Sometimes solvents exert a similar effect. When a small amount of ethyl acetate is added to a calcium chloride solution prior to addition of sodium fenoprofen, the calcium fenoprofen that precipitates has a low degree of crystallinity [159]. Similarly, when calcium DL-pantothenate is precipitated from methanol or ethanol solution by the addition of acetone, ether, ethyl acetate, or other solvents, the precipitate obtained is found to be amorphous [154].

V. SUMMARY

The pharmaceutical development scientist who is assigned the task of demonstrating that a substance exhibits only one crystalline form, or that of discovering whether additional forms exist, can utilize the techniques outlined in this chapter as a starting point. Upon completion of this program, one can certainly conclude that due diligence has been employed to isolate and characterize the various solid-state forms of any new chemical entity. One should always be aware that nuclei capable of initiating the crystallization of previously undiscovered forms might be lurking around the laboratory, ready to confound the investigator should their effects become known. In addition, the phenomenon of ''disappearing polymorphs'' can come into play, and techniques that formerly yielded the same crystals every time may subsequently yield crystals of another, more stable form. In the future, the use of computer simulations of alternative crystallographic structures will suggest how much laboratory work might be required to isolate the polymorphs or solvates of a given compound. Until then, the empirical approach remains superior.

REFERENCES

1. S. R. Byrn, personal communication, October 2, 1996.
2. W. C. McCrone, Jr., *Fusion Methods in Chemical Microscopy*, Interscience, New York, 1957.
3. H. R. Karfunkel, F. J. J. Leusen, and R. J. Gdanitz, *J. Comp.-Aided Mater. Des., 1,* 177 (1993).
4. H. R. Karfunkel, Z. J. Wu, A. Burkhard, G. Rihs, D. Sinnreich, H. M. Buerger, and J. Stanek, *Acta Cryst., B52,* 555 (1996).
5. J. D. Dunitz, and J. Bernstein, *Acc. Chem. Res. 28,* 193 (1995).

6. U. J. Grießer, and A. Burger, *Sci. Pharm., 61*, 133 (1993).
7. M. Kuhnert-Brandstätter, A. Burger, and R. Völlenklee, *Sci. Pharm., 62*, 307 (1994).
8. J. Bernstein, *J. Phys. D: Appl. Phys., 26*, B66 (1993).
9. M. Kuhnert-Brandstätter, *Thermomicroscopy in the Analysis of Pharmaceuticals*, Pergamon Press, Oxford, 1971.
10. S.-Y. Tsai, S.-C. Kuo, and S.-Y. Lin, *J. Pharm. Sci., 82*, 1250 (1993).
11. J. G. Fokkens, J. G. M. van Amelsfoort, C. J. de Blaey, C. G. de Kruif, and J. Wilting, *Int. J. Pharm., 14*, 79 (1983).
12. M. Sakiyama, and A. Imamura, *Thermochim. Acta, 142*, 365 (1989).
13. Y. Suzuki, *Bull. Chem. Soc. Japan, 47*, 2551 (1974).
14. R. J. Behme, T. T. Kensler, D. G. Mikolasek, and G. Douglas, U. S. Patent 4,810,789 (to Bristol-Myers Co.), Mar. 7, 1989.
15. J. W. Mullin, *Crystallization*, 3d ed., Butterworth-Heinemann, Oxford, 1993.
16. R. J. Behme, and D. Brooke, *J. Pharm. Sci., 80*, 986 (1991).
17. W. Ostwald, *Z. Phys. Chem., 22*, 289 (1897).
18. K. K. Nass, "Process Implications of Polymorphism in Organic Compounds," in *Particle Design via Crystallization*, R. Ramanarayanan, W. Kem, M. Larson, and S. Sikdar, eds., AIChE Symposium Series, *284*, 72 (1991).
19. A. J. Kim, and S. L. Nail, Abstract PDD7443, AAPS Annual Meeting: Invited and Contributed Paper Abstracts, Seattle, WA, 1996.
20. F. Kaneko, H. Sakashita, M. Kobayashi, and M. Suzuki, *J. Phys. Chem., 98*, 3801 (1994).
21. N. Garti, E. Wellner, and S. Sarig, *Kristall. Tech., 15*, 1303 (1980).
22. S. R. Byrn, R. R. Pfeiffer, G. Stephenson, D. J. W. Grant, and W. B. Gleason, *Chem. Materials, 6*, 1148 (1994).
23. M. C. Etter, D. A. Jahn, B. S. Donahue, R. B. Johnson, and C. Ojala, *J. Cryst. Growth, 76*, 645 (1986).
24. M. C. Martínez-Ohárriz, M. C. Martin, M. M. Goni, C. Rodriguez-Espinosa, M. C. Tros de Ilarduya-Apaolaza, and M. Sanchez, *J. Pharm. Sci., 83*, 174 (1994).
25. L.-S. Wu, G. Torosian, K. Sigvardson, C. Gerard, and M. A. Hussain, *J. Pharm. Sci., 83*, 1404 (1994).
26. Y. Chikaraishi, A. Sano, T. Tsujiyama, M. Otsuka, and Y. Matsuda, *Chem. Pharm. Bull., 42*, 1123 (1994).
27. W. C. McCrone, "Polymorphism," Chapter 8 in *Physics and Chemistry of the Organic Solid State*, Vol. 11 (D. Fox, M. M. Labes, and A. Weissberger, eds.), Interscience, New York, 1965.

28. S. Khoshkhoo, and J. Anwar, *J. Phys. D: Appl. Phys., 26*, B90 (1993).
29. H. G. Brittain, K. R. Morris, D. E. Bugay, A. B. Thakur, and A. T. M. Serajuddin, *J. Pharm. Biomed. Anal., 11*, 1063 (1993).
30. L.-C. Chang, M. R. Caira, and J. K. Guillory, *J. Pharm. Sci., 84*, 1169 (1995).
31. M. Otsuka, M. Onoe, and Y. Matsuda, *Drug Dev. Ind. Pharm., 20*, 1453 (1994).
32. M. Kitamura, H. Furukawa, and M. Asaeda, *J. Cryst. Growth, 141*, 193 (1994).
33. N. Kaneniwa, M. Otsuka, and T. Hayashi, *Chem. Pharm. Bull., 33*, 3447 (1985).
34. A. Burger, and A. W. Ratz, *Pharm. Ind. 50*, 1186 (1988).
35. R. F. Shanker, "Micro-Techniques for Physicochemical Measurements," presented at the AAPS Symposium on Pharmaceutical Development Contributions During the Drug Discovery Process, Miami Beach, FL, Nov. 9, 1995.
36. M. Otsuka, and Y. Matsuda, *Drug Dev. Ind. Pharm., 19*, 2241 (1993).
37. T. Uchida, E. Yonemochi, T. Oguchi, K. Terada, K. Yamamoto, and Y. Nakai, *Chem. Pharm. Bull., 41*, 1632 (1993).
38. J. Pirttimäki, E. Laine, J. Ketolainen, and P. Paronen, *Int. J. Pharm., 95*, 93 (1993).
39. B. Peffenot, and G. Widmann, *Thermochim. Acta, 234*, 31 (1994).
40. J. Rambaud, A. Bouassab, B. Pauvert, P. Chevallet, J.-P. Declerq, and A. Terol, *J. Pharm. Sci., 82*, 1262 (1993).
41. M. Yoshioka, B. C. Hancock, and G. Zografi, *J. Pharm. Sci., 83*, 1700 (1995).
42. T. P. Shakhtshneider, and V. V. Boldyrev, *Drug Dev. Ind. Pharm., 19*, 2055 (1993).
43. W. C. Stagner, and J. K. Guillory, *J. Pharm. Sci., 68*, 1005 (1979).
44. B. H. Kim, and J. K. Kim, *Arch. Pharm. Res., 7*, 47 (1984).
45. Y. Chikaraishi, M. Otsuka, and Y. Matsuda, *Chem. Pharm. Bull., 44*, 1614 (1996).
46. Stephen R. Byrn, *Solid State Chemistry of Drugs*, Academic Press, New York, 1982.
47. W. L. Rocco, C. Morphet, and S. M. Laughlin, *Int. J. Pharm., 122*, 17 (1995).
48. W. L. Rocco, *Drug Dev. Ind. Pharm., 20*, 1831 (1994).
49. A.-K. Bjerga Bjaen, K. Nord, S. Furuseth, T. Agren, H. Tønnesen, and J. Karlsen, *Int. J. Pharm., 92*, 183 (1993).
50. B. R. Jasti, J. Du, and R. C. Vasavada, *Int. J. Pharm., 118*, 161 (1995).

51. H. Kitaoka, C. Wada, R. Moroi, and H. Hakusui, *Chem. Pharm. Bull.*, *43*, 649 (1995).
52. L. Addadi, Z. Berkovitch-Yellin, I. Weissbuch, J. van Mil, L. J. W. Shimon, M. Lahav, and L. Leiserowitz, *Angew. Chem., Int. Engl. Ed.*, *24*, 466 (1985).
53. I. Weissbuch, L. Addadi, M. Lahav, and L. Leiserowitz, *Science, 253*, 637 (1991).
54. R. J. Davey, S. J. Maginn, S. J. Andrews, S. N. Black, A. M. Buckley, D. Cottler, P. Dempsey, R. Plowman, J. E. Rout, D. R. Stanley, and A. Taylor, *J. Chem. Soc. Far. Trans. II, 40*, 1003 (1994).
55. K. Ikeda, I. Saitoh, T. Oguma, and Y. Takagishi, *Chem. Pharm. Bull.*, *42*, 2320 (1994).
56. I. Weissbuch, L. Leisorowitz, and M. Lahav, *Adv. Materials, 6*, 952 (1994).
57. M. Otsuka, K. Otsuka, and N. Kaneniwa, *Drug Dev. Ind. Pharm.*, *20*, 1649 (1994).
58. R. K. Khankari, and D. J. W. Grant, *Thermochim. Acta, 248*, 61 (1995).
59. J. Pirttimäki, and E. Laine, *Eur. J. Pharm. Sci., 1*, 203 (1994).
60. M. R. Caira, E. W. Pienaar and A. P. Lötter, *Mol. Cryst. Liquid Cryst.*, *279*, 241 (1996).
61. M. J. Jozwiakowski, S. O. Williams, and R. D. Hathaway, *Int. J. Pharm., 91*, 195 (1993).
62. A. Michel, M. Drouin, and R. Glaser, *J. Pharm. Sci., 83*, 508 (1994).
63. D. Giron, *Thermochim. Acta, 248*, 1 (1995).
64. K. Sasaki, H. Suzuki, and H. Nakagawa, *Chem. Pharm. Bull., 41*, 325 (1993).
65. M. R. Caira, L. R. Nassimbeni, and B. van Oudtshoorn, *J. Pharm. Sci., 82*, 1006 (1993).
66. H. K. Cammenga, and I. D. Steppuhn, *Thermochim. Acta, 229*, 253 (1993).
67. W. W. L. Young, and R. Suryanarayanan, *J. Pharm. Sci., 80*, 496 (1991).
68. A. Burger, and A. Lettenbichler, *Pharmazie, 48*, 262 (1993).
69. R. Rajala, E. Laine, and G. Örn, *Eur. J. Pharm. Sci., 1*, 219 (1994).
70. A. Burger, and A. Lettenbichler, *Eur. J. Pharm. Biopharm., 39*, 64 (1993).
71. V. J. van Geerestein, J. A. Kanters, P. van der Stuis, and J. Kroon, *Acta Cryst., C42*, 1521 (1986).
72. S. R. Byrn, P. A. Sutton, B. Tobias, J. Frye, and P. Main, *J. Am. Chem. Soc., 110*, 1609 (1988).
73. P. van der Sluis, and J. Kroon, *J. Crystal Growth, 97*, 645 (1989).

74. M. Shiro, H. Nakai, K. Nagashima, and N. Tsuiji, *J. Chem. Soc. Chem. Comm.*, 682 (1978).
75. O. Dahl, K. H. Ziedrich, G. J. Marek, and H. H. Paradies, *J. Pharm. Sci.*, *78*, 598 (1989).
76. G. A. Stephenson, J. G. Stowell, P. H. Toma, D. E. Dorinan, G. R. Green, and S. R. Byrn, *J. Am. Chem. Soc.*, *116*, 5766 (1994).
77. Y. Fukumori, T. Fukuda, Y. Yamamoto, Y. Shigitani, Y. Hanyu, T. Takeuchi, and N. Sato, *Chem. Pharm. Bull.*, *31*, 4029 (1983).
78. M. R. Caira, and R. Mohamed, *Supramol. Chem.*, *2*, 201 (1993).
79. H. D. Beckstead, G. A. Neville, and H. F. Shurvell, *Fresenius J. Anal. Chem.*, *345*, 727 (1993).
80. G. A. Neville, H. D. Beckstead, and J. D. Cooney, *Fresenius J. Anal. Chem.*, *349*, 746 (1994).
81. W. L. Rocco, *Drug Dev. Ind. Pharm.*, *20*, 1831 (1994).
82. S. Kitamura, L.-C. Chang, and J. K. Guillory, *Int. J. Pharm.*, *101*, 127 (1994).
83. F. Franks, R. H. M. Hatley, and S. F. Mathias, *Biopharm.*, *4*, 38, 40–42, 55, (1991).
84. E. Fukuoka, M. Makita, and S. Yamamura, *Chem. Pharm. Bull.*, *37*, 1047 (1989).
85. B. Perrenot, and G. Widmann, *Thermochim. Acta*, *243*, 31 (1994).
86. M. Yoshioka, B. C. Hancock, and G. Zografi, *J. Pharm. Sci.*, *83*, 1700 (1994).
87. E. Fukuoka, M. Makita and S. Yamamura, *Chem. Pharm. Bull.*, *34*, 4314 (1986).
88. S. Srčič, J. Kerč, U. Urleb, I. Zupančič, G. Lahajnar, B. Kofler, and J. ŠmidKorbar, *Int. J. Pharm.*, *87*, 1 (1992).
89. J. Kerc, M. Mohar, and J. Smid-Korbar, *Int. J. Pharm.*, *68*, 25 (1991).
90. Y. Aso, S. Yoshioka, T. Otsuka, and S. Kojima, *Chem. Pharm. Bull.*, *43*, 300 (1995).
91. A. E. H. Gassim, P. Girgis Takla, and K. C. James, *Int. J. Pharm.*, *34*, 23 (1986).
92. E. Nümberg, and A. Hopp, *Pharm. Ind.*, *44*, 1081 (1982); *45*, 85 (1983).
93. M. W. Gouda, A. R. Ebian, M. A. Moustafa, and S. A. Khalil, *Drug Dev. Ind. Pharm.*, *3*, 273 (1977).
94. J. P. Elder, *Thermochim. Acta*, *166*, 199 (1990).
95. K. Dialer, and K. Kuessner, *Kolloid-S. S. Polymer*, *251*, 710 (1973).
96. A. Bauer-Brandl, *Int. J. Pharm.*, *140*, 195 (1996).
97. A. Miyamae, H. Kema, T. Kawabata, T. Yasuda, M. Otsuka, and Y. Matsuda, *Drug Dev. Ind. Pharm.*, *20*, 2881 (1994).

98. N. Kaneniwa, K. Imagawa, and M. Otsuka, *Chem. Pharm. Bull., 33,* 802 (1985).

99. M. Otsuka, T. Matsumoto, and N. Kaneniwa, *Chem. Pharm. Bull., 34,* 1784 (1986).

100. A. Miyamae, S. Kitamura, T. Tada, S. Koda, and T. Yasuda, *J. Pharm. Sci., 80,* 995 (1991).

101. T. Sato, M. Ishiwata, S. Nemoto, H. Yamaguchi, T. Kobayashi, K. Sekiguchi, and Y. Tsuda, *Yakuzaigaku, 49,* 70 (1989).

102. F. Forni, G. Coppi, V. Iannuccelli, M. A. Vandelli, and R. Cameroni, *Acta Pharm. Suec., 25,* 173 (1988).

103. F. Fomi, G. Coppi, V. Iannucelli, M. A. Vandelli, and M. T. Bemabei, *Drug Dev. Ind. Pharm., 14,* 633 (1988).

104. R. Suryanarayanan, and A. G. Mitchell, *Int. J. Pharm., 24,* 1 (1985).

105. F. Fomi, V. Iannuccelli, and R. Cameroni, *J. Pharm. Pharmacol., 39,* 1041 (1987).

106. T. Konno, K. Kinuno, and K. Kataoka, *Chem. Pharm. Bull., 34,* 301 (1986).

107. Y. Nakai, *Yakugaku Zasshi, 105,* 801 (1985).

108. Y. Nozawa, K. Suzuki, Y. Sadzuka, A. Miyagishima, and S. Hirota, *Pharm. Acta Helv., 69,* 135 (1994).

109. M. Morita, S. Hirota, K. Kinuno, and K. Kataoka, *Chem. Pharm. Bull., 33,* 795 (1985).

110. A. T. Florence, and E. G. Salole, *J. Pharm. Pharmacol., 28,* 637 (1976).

111. D. B. Black, and E. G. Lovering, *J. Pharm. Pharmacol., 29,* 684 (1977).

112. A. Ikekawa, and S. Hayakawa, *Bull. Chem. Soc. Japan, 55,* 1261 (1982).

113. A. Ikekawa, and S. Hayakawa, *Bull. Chem. Soc. Japan, 55,* 3123 (1982).

114. S.-Y. Lin, and C.-S. Lee, *J. Incl. Phen. Mol. Recogn. Chem., 7,* 477 (1989).

115. C. Torricelli, A. Martini, L. Muggetti, and R. De Ponti, *Int. J. Pharm., 71,* 19 (1991).

116. A. S. Rankell, H. A. Liebermann, and R. F. Schiffman, "Drying," in *The Theory and Practice of Industrial Pharmacy,* 3d ed. (L. Lachman, A. A. Lieberman, and J. L. Kanig, eds.), Lea and Febiger, Philadelphia, 1986, p. 47.

117. K. Yano, N. Takamatsu, S. Yamazaki, K. Sako, S. Nagura, S. Tomizawa, J. Shimaya, and K. Yamamoto, *Yakugaku Zasshi, 116,* 639 (1996).

118. L.-E. Briggner, G. Buckton, K. Bystrom, and P. Darcy, *Int. J. Pharm., 105,* 125 (1994).

119. T. Sebhatu, M. Angberg, and C. Ahlneck, *Int. J. Pharm., 104*, 135 (1994).

120. T. Yamaguchi, M. Nishimura, R. Okamoto, T. Takeuchi, and K. Yamamoto, *Int. J. Pharm., 85*, 87 (1992).

121. G. Buckton, P. Darcy, D. Greenleaf, and P. Holbrook, *Int. J. Pharm., 116*, 113 (1995).

122. H. Vromans, G. K. Bolhuis, C. F. Lerk, and K. D. Kussendrager, *Int. J. Pharm., 39*, 201 (1987).

123. Y. Matsuda, M. Otsuka, M. Onoe, and E. Tatsumi, *J. Pharm. Pharmacol., 44*, 627 (1992).

124. Y. Matsuda, M. Otsuka, M. Onoe, and E. Tatsumi, *J. Pharm. Pharmacol., 44*, 627 (1992).

125. E. Nürnberg, B. Dölle, and J. M. Bafort, *Pharm. Ind., 44*, 630 (1982).

126. M. J. Pikal, A. L. Lukes, J. E. Lang, and K. Gaines, *J. Pharm. Sci., 67*, 767 (1978).

127. T. Saito, M. Ishiwata, A. Okada, T. Kobayashi, K. Sekiguchi, and Y. Tsuda, *Yakuzaigaku, 49*, 93 (1989).

128. T. Oguchi, M. Okada, E. Yonemochi, K. Yamamoto, and Y. Nakai, *Int. J. Pharm., 61*, 27 (1990).

129. T. Oguchi, K. Terada, K. Yamamoto, and Y. Nakai, *Chem. Pharm. Bull., 37*, 1881 (1989).

130. L. Figura, *Thermochim. Acta, 222*, 187 (1993).

131. S.-D. Clas, R. Faizer, R. E. O'Connor, and E. B. Vadas, *Int. J. Pharm., 121*, 73 (1995).

132. A. Saleki-Gerhardt, J. G. Stowell, and G. Zografi, *J. Pharm. Sci., 84*, 318 (1995).

133. A. Saleki-Gerhardt, and G. Zografi, *Pharm. Res., 11*, 1166 (1994).

134. G. A. Stephenson, J. G. Stowell, P. H. Toma, D. E. Dorman, J. R. Greene, and S. R. Byrn, *J. Am. Chem. Soc., 116*, 5766 (1994).

135. H. Egawa, S. Maeda, E. Yonemochi, T. Oguchi, K. Yamamoto, and Y. Nakai, *Chem. Pharm. Bull., 40*, 819 (1992).

136. M. Otsuka, and N. Kaneniwa, *Chem. Pharm. Bull., 31*, 4489 (1983).

137. R. Suryanarayanan, and A. G. Mitchell, *Int. J. Pharm., 32*, 213 (1986).

138. K.-H. Frömming, U. Grote, A. Lange, and R. Hosemann, *Pharm. Ind., 48*, 283 (1986).

139. M. Saito, H. Yabu, M. Yamazaki, K. Matsumura, and H. Kato, *Chem. Pharm. Bull., 30*, 652 (1982).

140. K. Ashizawa, K. Uchikawa, T. Hattori, Y. Ishibashi, T. Sato, and Y. Miyake, *J. Pharm. Sci., 78*, 893 (1989).

141. M. Morita, and S. Hirota, *Chem. Pharm. Bull., 30*, 3288 (1982).

142. S. Duddu, and K. Weller, *J. Pharm. Sci., 85*, 345 (1996).
143. O. Funk, L. Schwabe, and K.-H. Frömming, *Drug Dev. Ind. Pharm., 20*, 1957 (1994).
144. O. Funk, L. Schwabe, and K.-H. Frömming, *Pharmazie, 48*, 745 (1993).
145. M. T. Esclusa-Diaz, J. J. Torres-Labandeira, M. Kata, and J. L. Vila-Jato, *Eur. J. Pharm. Sci., 1*, 291 (1994).
146. G. Bettinetti, A. Gazzaniga, P. Mura, F. Giordano, and M. Setti, *Drug Dev. Ind. Pharm., 18*, 39 (1992).
147. R. J. Yarwood, A. J. Phillips, and J. H. Collett, *Drug Dev. Ind. Pharm., 12*, 2157 (1986).
148. T. Oguchi, E. Yonemochi, K. Yamamoto, and Y. Nakai, *Chem. Pharm. Bull., 37*, 3088 (1989).
149. M. J. Pikal, and K. M. Dellerman, *Int. J. Pharm., 50*, 233 (1989).
150. O. Funk, L. Schwabe, and K.-H. Frömming, *J. Incl. Phen. Mol. Recogn. Chem., 16*, 299 (1993).
151. Y. Kawashima, T. Niwa, H. Takeuchi, T. Hino, Y. Itoh, and Y. Furuyama, *J. Pharm. Sci., 80*, 472 (1991).
152. Y. Fukumori, T. Fukuda, Y. Yamamoto, Y. Shigitani, Y. Hanyu, Y. Takeuchi, and N. Sato, *Chem. Pharm. Bull., 31*, 4029 (1983).
153. E. Laine, P. Kahela, R. Rajala, T. Heiklilä, K. Saamivaara, and I. Piippo, *Int. J. Pharm., 38*, 33 (1987).
154. M. Inagaki, *Chem. Pharm. Bull., 25*, 1001 (1977).
155. Y. Chikaraishi, M. Otsuka, and Y. Matsuda, *Chem. Pharm. Bull., 44*, 1614 (1996).
156. F. A. Andersen, and L. Brecevic, *Acta Chem. Scand., 45*, 1018 (1991).
157. S. P. Duddu, and D. J. W. Grant, *Thermochim. Acta, 248*, 131 (1995).
158. Y. Kim, and A. M. Haren, *Pharm. Res., 12*, 1664 (1995).
159. B. A. Hendriksen, *Int. J. Pharm., 60*, 243 (1990).

6

Methods for the Characterization of Polymorphs and Solvates

Harry G. Brittain

Discovery Laboratories, Inc.
Milford, New Jersey

I. INTRODUCTION

Certainly the most important aspect relating to an understanding of polymorphic solid and solvate species is the range of analytical methodology used to perform the characterization studies [1–3]. The importance of this area has been recognized from both scientific and regulatory concerns, so the physical methods have begun to come under the same degree of scrutiny as have the traditional chemical methods of analysis. Byrn et al. have provided a series of useful definitions that concisely give the characteristics of the various solid forms that can be found for a given drug substance [4] and that will be used throughout this chapter. Compounds may be *polymorphs* (forms having the same chemical composition but different crystal structures), *solvates* (forms containing solvent molecules within the crystal structure), *desolvated solvates* (forms when the solvent is removed from a specific solvate while still retaining the original crystal structure), or *amorphous* (solid forms that have no long-range molecular order).

Of all the methods available for the physical characterization of solid materials, it is generally agreed that crystallography, microscopy, thermal analysis, solubility studies, vibrational spectroscopy, and nuclear magnetic resonance are the most useful for characterization of polymorphs and solvates. However, it cannot be overemphasized that the defining criterion for the existence of polymorphic types must always be a nonequivalence of crystal structures. For compounds of pharmaceutical interest, this ordinarily implies that a nonequivalent x-ray powder diffraction pattern is observed for each suspected polymorphic variation. All other methodologies must be considered as sources of supporting and ancillary information; they cannot be taken as definitive proof for the existence of polymorphism by themselves alone.

In the present work, the practice of the most commonly encountered techniques performed for the solid-state characterization of polymorphic or solvate properties will be reviewed. No attempt will be made to summarize every recorded use of these methodologies for such work, but selected examples will be used to illustrate the scope of information that can be extracted from the implementation of each technique.

II. CRYSTALLOGRAPHY: X-RAY DIFFRACTION

The x-ray crystallography technique, whether performed using single crystals or powdered solids, is concerned mainly with structural analysis and is therefore eminently suited for the characterization of polymorphs and solvates. An external examination of crystals reveals that they often contain facets, and that well-formed crystals are completely bounded by flat surfaces. Planarity of this type is not commonly encountered in nature, and it was quickly deduced that the morphological characteristics of a crystal are inherent in its interior structure. In fact, the microscopic form of a crystal depends critically on structural arrangements at the atomic or molecular level; the underlying factor controlling crystal formation is the way in which atoms and molecules can pack together.

A. Single Crystal X-Ray Diffraction

Every crystal consists of exceedingly small fundamental structural units that are repeated indefinitely in all directions. In 1830, Hessel conducted a purely mathematical investigation of the possible types of symmetry for a solid figure bounded by planar faces and deduced that only 32 symmetry groups were possible for such objects. The same conclusion was reached by Bravais in 1949 and Gadolin in 1867. These 32 crystallographic point groups are grouped into six crystal systems, denoted triclinic, monoclinic, orthorhombic, tetragonal, trigonal, hexagonal, and cubic. Each crystal system is characterized by unique relationships existing among the crystal axes and the angles between these, and this information is summarized in Table 1.

One of the characteristics of many crystals is their ability to be split along certain directions, yielding fragments containing smooth faces along the direction of the break. The angular relations between these cleavage planes were found to be the same in every fragment. Ultimately, it was learned that crystal cleavage planes corresponded to planes of atoms or molecules in the crystal, which in turn resulted from the repetition of unit cells. This three-dimensional pattern of atoms in a crystalline solid was shown to be capable of acting as a diffraction grating to light having wavelengths of the same order of magnitude as the translational repeat period of the molecular pattern. This period is

Table 1 Characteristics of the Six Crystal Systems

System	Description
Cubic	Three axes of identical length (identified as a_1, a_2, and a_3) intersect at right angles.
Hexagonal	Four axes (three of which are identical in length, denoted a_1, a_2, and a_3) lie in a horizontal plane and are inclined to one another at 120°. The fourth axis, c, is different in length from the others and is perpendicular to the plane formed by the other three.
Tetragonal	Three axes (two of which are denoted a_1 and a_2 and are identical in length) intersect at right angles. The third axis, c, is different in length with respect to a_1 and a_2.
Orthorhombic	Three axes of different lengths (denoted a, b, and c) intersect at right angles. The choice of the vertical c axis is arbitrary.
Monoclinic	Three axes (denoted a, b, and c) of unequal length intersect so that a and c lie at an oblique angle and the b axis is perpendicular to the plane formed by the other two.
Triclinic	Three axes (denoted a, b, and c) of unequal length intersect at three oblique angles.

of the order of 10^{-10} meters (i.e., angstrom units), and light having wavelengths of this magnitude is called x-ray radiation. The discovery that x-rays could be diffracted by crystalline solids was made by von Laue and his collaborators, and the method was quickly improved by Bragg and subsequently developed by countless others. It must be emphasized, however, that it is the electron density about an atom that is responsible for the scattering of x-rays by matter.

All x-ray diffraction techniques are ultimately based on Bragg's law, which describes the diffraction of a monochromatic x-ray beam impinging on a plane of atoms [5]. Parallel incident rays strike the crystal planes at an angle θ and are then diffracted at the same angle. The observation of reinforcement requires that the path difference of the impinging beam (i.e., the distance between molecular planes) be equal to a whole number of wavelengths. The scattering angles are therefore correlatable to the spacings between planes of molecules in the lattice by means of Bragg's law:

$$n\lambda = 2\,d\,\sin\,\theta \tag{1}$$

where

n = order of the diffraction pattern

λ = wavelength of the incident beam

d = distance between the planes in the crystal

θ = angle of beam diffraction

It should be noted that the Bragg equation yields only the scattering angles with respect to the incident x-ray beam and has nothing to say about the relative intensities of diffracted radiation. To describe scattered intensities, one uses the concept of the scattering power of a sample. This is equal to the number of free and independent electrons, scattering according to Thompson's law governing the scattering by a free electron, which would be required to replace the object in order to obtain the same scattered intensity.

A determination of the internal structure of a crystal requires the specification of the unit cell dimensions (axis lengths, and angles between these) and measurement of the intensities of the diffraction pattern of the crystal. For a given lattice, regardless of the content of the unit cell, the directions of reflection are the same. The experimental determination of these directions is used to deduce the reciprocal lattice of the crystal, which unambiguously yields the crystal lattice. In addition, the relative intensities diffracted by different planes depend on the contents of the unit cell. Their measurement leads to the determination of the crystal structure factor, and these data permit the determination of the atomic structure of crystals. More detailed expositions of the procedures used to obtain the structures of single crystals are available in the literature [5–9] and are beyond the scope of this article.

Genuine polymorphism ordinarily arises either from differences in the packing of conformationally equivalent molecules or from different modes of assembly of conformationally inequivalent molecules. The former situation is well-known for inorganic and geological crystals [10], and the latter has been discussed in detail [11]. One of the best known instances of packing polymorphism are the allotropes of carbon, graphite and diamond. As shown in Fig. 1, in diamond each carbon atom is tetrahedrally surrounded by four equidistant neighbors, and the tetrahedra are arranged to give a cubic unit cell. Graphite is composed of planar hexagonal nets of carbon atoms, which can be arranged to yield either a hexagonal unit cell (the α-form) or a rhombohedral unit cell (the β-form).

Two anhydrous polymorphs of nitrofurantoin have been reported [12], as well as two polymorphic monohydrate forms [13]. In the triclinic α-anhydrate form, the nitrofurantoin molecules were found to be associated in a head-to-head manner, forming centrosymmetric dimer units through two identical intermolecular hydrogen bonds, with these dimer units being further linked into sheets by a system of weaker hydrogen bonds. The hydrogen bonding existing in the molecular planes of the monoclinic β-anhydrate form was found to differ in symmetry, where the key hydrogen bond was seen to link nitrofurantoin molecules by a twofold screw-axis [12]. In each of the two monohydrate forms, the conformations of the nitrofurantoin molecules are essentially equivalent to each other and not significantly different from the conformation observed for the anhydrate forms. However, the hydrogen bonding pattern induced by the presence of lattice water molecules yields two very different packing modes for the two monohydrate polymorphs. In the monoclinic form, virtually all of the atoms lie in

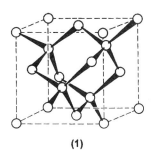

(1)

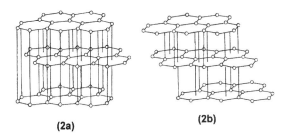

(2a) (2b)

Fig. 1 Crystal structure of (1) diamond, showing the tetrahedral coordination of each carbon atom. Also shown are the crystal structures of the two polymorphs of graphite, specifically (2a) the hexagonal α-form and (2b) the rhombohedral β-form.

a single plane, giving rise to a completely layered structure. In the orthorhombic form, layers of molecules are arranged parallel to different planes, so the overall packing is that of a herringbone arrangement [13].

Probucol is an example of a compound where the polymorphism arises from the packing of different conformers [14]. Although both forms were found to be monoclinic, the unit cells belonged to different space groups and the molecular conformations of the title compound were quite different (Fig. 2). In Form II, the C–S–C–S–C chain is extended, and the molecular symmetry approximates C_{2v}. This symmetry is lost in Form I, where the torsion angles about the two C–S bonds deviate significantly from 180°. The extended conformer was shown to be less stable relative to the bent conformer, as simple grinding was sufficient to convert Form II into Form I.

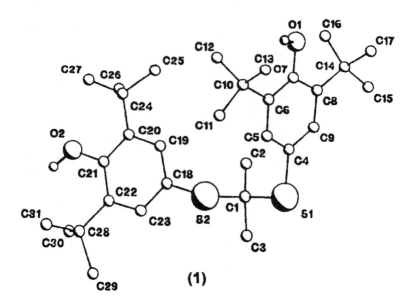

(1)

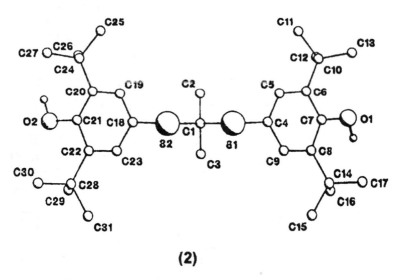

(2)

Fig. 2 Conformation of the probucol molecule existing in (1) Form I and (2) Form II. (The figure was adapted from data contained in Ref. 14).

Not all instances of conformational polymorphism are as dramatic as that just described, and often different conformers of a single side chain are able to pack into different crystalline arrangements. For instance, the two polymorphs of *p*-(1*R*,3*S*)-3-thioanisoyl-1,2,2-trimethyl-cyclopentane carboxylic acid were found to be associated with different conformations of the carboxylate group [15]. Torsion about a single C–N bond was shown to be the origin of the polymorphism detected for lomeridine dihydrochloride [16]. Finally, relatively small differences in molecular conformation were detected for the two polymorphic and four solvated crystalline forms of spironlactone [17].

B. X-Ray Powder Diffraction

Although the solving of a crystal structure provides the greatest understanding of polymorphic solids, the necessity for obtaining suitable single crystals and the degree of complexity associated with the data analysis preclude this technique from being used on a routine basis for batch characterization. In fact, most drug substances are obtained as microcrystalline powders, from which it is often fiendishly difficult to obtain crystallographically adequate crystals. Furthermore, during the most common evaluation of drug substances, it is usually sufficient to establish only the polymorphic identity of the solid and to verify that the isolated compound is indeed of the desired structure. For these reasons, and to its inherent simplicity of performance, the technique of x-ray powder diffraction (XRPD) is the predominant tool for the study of polycrystalline materials [18] and is eminently suited for the routine characterization of polymorphs and solvates.

A correctly prepared sample of a powdered solid will present an entirely random selection of all possible crystal faces at the powder interface, and the diffraction off this surface provides information on all possible atomic spacings in the crystal lattice. To measure a powder pattern, a randomly oriented sample is prepared so as to expose all the planes of a sample and is irradiated with monochromatic x-ray radiation. The scattering angle θ is determined by slowly rotating the sample and using a scintillation counter to measure the angle of diffracted x-rays with respect to the angle of the incident beam. Alternatively, the angle between sample and source can be kept fixed, and the detector moved along a proscribed path to determine the angles of the scattered radiation. Knowing the wavelength of the incident beam, the spacing between the planes (identified as the d-spacings) is calculated using Bragg's law.

The XRPD pattern will therefore consist of a series of peaks detected at characteristic scattering angles. These angles, and their relative intensities, can be correlated with the computed d-spacings to provide a full crystallographic characterization of the powdered sample. After indexing all the scattered lines, it is possible to derive unit cell dimensions from the powder pattern of the substance under analysis [18]. For routine work, however, this latter analysis is not normally performed, and one typically compares the powder pattern of the analyte to that of reference materials to establish the polymorphic identity. Since every compound produces its own characteristic powder pattern owing to the unique crystallography of its structure, powder x-ray diffraction is clearly the most powerful and fundamental tool for a specification of the polymorphic identity of an analyte. The USP general chapter on x-ray diffraction states that identity is established if the scattering angles of the ten strongest reflections obtained for an analyte agree to within ±0.20 degrees with that of the reference material, and if the relative intensities of these reflections do not vary by more than 20 percent [19].

The power of XRPD as a means to establish the polymorphic identity of an analyte can be illustrated by considering the case of the anhydrate and trihydrate phases of ampicillin. The crystal structures of both phases have been obtained, and they differ in the nature of the molecular packing [20]. The amino group in the monoclinic anhydrate is hydrogen bonded to the ionized carboxyl groups of two molecules, while the amino group of the orthorhombic trihydrate is hydrogen bonded to a single carboxylate group and to the waters of hydration that link other molecules in the structure. The powder patterns of these two materials are shown in Fig. 3 and are seen to be readily distinguishable from each other. Amoxycillin trihydrate has been found to crystallize in the same space group as does ampicillin trihydrate, and it exhibits a very similar pattern of hydrogen bonding [21]. However, the dimensions of the two unit cells differ significantly, and this fact is

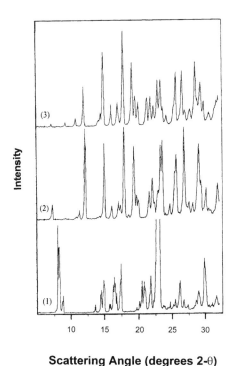

Fig. 3 X-ray powder diffraction patterns of (1) ampicillin anhydrate, (2) ampicillin trihydrate, and (3) amoxicillin trihydrate.

reflected in the differences among the relative intensities of the corresponding peaks contained in Fig. 3. Even though the two structures would be considered as being isostructural, the XRPD patterns of the two trihydrate phases readily permit an unambiguous identification and distinction between these.

X-ray powder diffraction can also be used for the quantitative determination of phase composition, and this approach has been discussed in detail [22]. In one particularly well-developed example, XRPD was used to quantitate the relative amounts of the anhydrate and dihydrate phases existing in carbamazepine samples [23]. The method was based on the observation that the XRPD of each phase featured a scattering peak unique to each form, which was noted at a scattering angle where no scattering was observed for the other phase. Unlike loose powders, compressed samples yielded highly reproducible intensity values, so pelletized materials were used for the data acquisition. Good correlation between sample composition and scattering intensities was obtained in standard materials, permitting the generation of analytical relations suitable for the analysis of analyte samples.

The degree of crystallinity associated with a sample can often be established using powder x-ray diffraction. If the patterns of 100% crystalline and 100% amorphous material can be established, then the integrated peak intensity of the analyte is used to deduce the percent crystallinity. Such methodology has been used to measure the crystallinity of digoxin [24] and calcium gluceptate [25]. The XRPD method is extremely important during the characterization of lyophilized materials, since the stability of a crystalline solid is expected to exceed that of an amorphous or disordered solid. For instance, the technique has been used to study the properties of lyophilized imipenem [26].

III. MORPHOLOGY: MICROSCOPY

An extremely important tool for the characterization of polymorphs and solvates is that of microscopy, since the observable habits of differing crystal structures must necessarily be different and therefore useful for the characterization of such systems [27]. Common sense would dictate that the visual observation of such materials would immediately follow an x-ray crystallographic study, which would in principle make the science of optical crystallography [28–30] an essential aspect of any program of study. A review of crystallography from the pharmaceutical viewpoint is available [31].

As stated in an earlier section, a crystal is a polyhedral solid, bounded by a number of planar faces that are normally identified using the Miller indices. The arrangement of these faces is termed the *habit* of the crystal, and the crystal is built up through the repetition of the unit cell. The three-dimensional basic pattern of molecules in a solid form the space lattice, and the application of simple geometry has shown that only 14 different types of simple space lattices are possible. By taking combinations of the various lattices possible for each crystallographic system, it has also been determined that all solids must belong to one of 230 space groups [28–30].

Both optical and electron microscopies have found widespread use for the characterization of polymorphs and solvates. Although optical microscopy is more limited in the range of magnification suitable for routine work (working beyond 600× being difficult when observing microcrystalline materials), the use of polarizing optics introduces enormous power into the technique not available with other methods. Electron microscopy work can be performed at extraordinarily high magnification levels (up to 90,000× on most units), and the images that can be obtained contain a considerable degree of three-dimensional information. The two methods are complementary in that each can provide information not obtainable by the other. With judicious use of these techniques, one can obtain substantial characterization of a polymorphic system. These data are extremely useful during the early stages of drug development, since often only a limited amount of the drug candidate is available at that time.

The literature pertaining to microscopy and its applications is quite large but fortunately is updated periodically [32]. A number of texts and review articles have been written that cover light or electron microscopy or some combination of the two [33,34]. It is beyond the scope of this chapter to provide a mechanical description of the various instruments; readers are referred to representative references for light microscopy [35–40] and electron microscopy [41–44].

McCrone has provided an excellent discussion of the synergistic aspects of optical and electron microscopies [45]. He concludes that electron microscopy yields excellent topographic and shape information and is most useful in forensic situations involving trace evidence characterization and identification. When polarizing optics are used during a light microscopy study, the optical properties of the crystals under investigation can also be determined. This latter aspect is extremely useful in the characterization of polymorphs and solvates, and consequently polarizing optical microscopy is an extremely important tool for the study of such systems.

A. Polarizing Optical Microscopy

The polarizing microscope is essentially a light microscope equipped with a linear polarizer located below the condenser and an additional polarizer mounted on top of the eyepiece. Full performance of an optical crystallographic study requires the additional use of a rotating stage. Application of this method can yield parameters such as the sign and magnitude of any observed birefringence, knowledge of the refractive indices associated with each crystal direction, what the axis angles are, and what are the relations among the optical axes. To conduct the analysis, the light from the source is rendered linearly polarized by the initial polarizer. The second polarizer mounted above the sample (the analyzer) is oriented so that its axis of transmission is orthogonal to that of the initial polarizer. In this condition of "crossed polarizers," no transmitted light can be perceived by the observer. The passage (or lack thereof) of light though the crystal as a function of the angle between crystal axes and the direction of polarization is of key importance to the method.

The refractive index of light passing through an isotropic crystal must be identical along each of the crystal axes, and such crystals therefore possess *single refraction*. By virtue of their symmetry, crystals within the cubic system are isotropic along all three crystal axes. This property is to be contrasted with that associated with anisotropic substances, which exhibit different refractive indices for light polarized with respect to the crystal axes. This latter property results in the property of *double refraction*, or birefringence. Crystals within the hexagonal and tetragonal systems possess one isotropic direction and are termed *uniaxial*. Anisotropic crystals possessing two isotropic axes are termed *biaxial* and include all crystals belonging to the orthorhombic, monoclinic, or triclinic systems. Biaxial crystals will exhibit unequal indices of refraction along each of the crystal axes.

Isotropic samples will have no effect on the polarized light no matter how the crystal is oriented, since all crystal axes are completely equivalent. This effect is known as complete or *isotropic extinction*. Noncrystalline, amorphous samples yield the same behavior.

When the sample is capable of exhibiting double refraction, the specimen will then appear bright against a dark background. For example, when a uniaxial crystal is placed with the unique c axis horizontal on the stage, it will appear to be alternately dark and bright as the stage is rotated. Furthermore, the crystal will be completely dark when the c axis is parallel to the transmission plane of the polarizer or analyzer. If the crystal has edges or faces parallel to the c axis, then it will be extinguished when such an edge or face is parallel to one of the polarizer directions. This condition is called *parallel extinction*. At intermediate positions, the crystal will appear light and will exhibit colors depending on the thickness of the crystal. A rhombohedral or pyramidal crystal will be extinguished when the bisector of a silhouette angle is parallel to a polarization direction, and this type of extinction is termed *symmetrical extinction*. However, when a uniaxial crystal is mounted with the c axis vertical on the stage, it is found that the crystal remains uniformly dark as the stage is rotated (another case of isotropic extinction).

For biaxial crystals, similar results are obtained to those with uniaxial crystals. The exception to this rule is that in monoclinic and triclinic systems, the polarization directions need not be parallel to faces or to the bisectors of face angles. If the prominent faces or edges of an extinguished crystal are not parallel to the axes of the initial polarizer, the extinction is said to be *oblique*.

A summary of the extinction characters covering biaxial and uniaxial crystals is provided in Table 2. Knowledge of the type of extinction permits a determination of the system to which a given crystal belongs, and such information can be extremely important during the initial evaluation of crystal polymorphs or solvates. For example, identification of the polymorphic or solvated forms of thiamine hydrochloride, bromvalerylurea, ampicillin [46], and the polymorphs of sulfamethoxydiazine [47] is easily done using optical microscopic methods. Application of this procedure requires that one determine the refractive indices along each of the crystal axes, which is done using immersion oils of known refractive indices [48].

Optical microscopy can be a powerful tool in the study of processes associated with phase conversion of one crystal form that is metastable with respect to another form in the presence of an appropriate solvent. For instance, Form II of sulfathiazole can be converted to Form I when suspended in glycerin at 90°C, while at 100°C the

Table 2 Differentiation of Crystal Systems by the Character of Their Extinction

System	Character of observed extinction
Cubic	Isotropic or complete extinction.
Hexagonal	Parallel or symmetrical extinction. Can be isotropic if viewed down the c axis, but then a six-sided silhouette should be observed.
Tetragonal	Parallel or symmetrical extinction. Can be isotropic, but then a four-sided silhouette should be observed.
Orthorhombic	All crystals will show parallel or symmetrical extinction.
Monoclinic	Some crystals will show parallel or symmetrical extinction, and others oblique extinction.
Triclinic	All crystals will show oblique extinction.

reverse process has been found to take place [49]. During the development of a method to obtain solubility data for rapidly reverting solid states, optical microscopy proved to be a useful tool for study of the phase conversion.

The anhydrate/monohydrate equilibrium of theophylline has received a great deal of attention, especially since the monohydrate phase is considerably less soluble than the anhydrate phase [50,51]. This situation has implications for the production and storage of theophylline solid dose forms, since phase interconversion may take place either during processing [52–54] or as a result of environmental exposure [55–56]. Optical microscopy has been used to study the anhydrate-to-monohydrate phase conversion that takes place when the drug substance is exposed to bulk water, and scanning electron microscopy was used to study the surfaces of compressed tablets when these were exposed to high degrees of relative humidity [57].

It has been found that (S)-4-[[[1-(4-fluorophenyl)-3-(1-methylethyl)-1H-indol-2-yl]-ethynyl]-hydroxyphosphinyl]-3]-hydroxybutanoic acid, disodium salt, is capable of existing in three crystalline hydrate and one liquid crystalline phase depending on the relative humidity to which the compound is exposed [58]. Among other things, these have been found to exhibit varying fluorescence properties in their respective solid states [59]. As shown in Fig. 4, the particle morphology of these

(1)   (2)

(3)   (4)

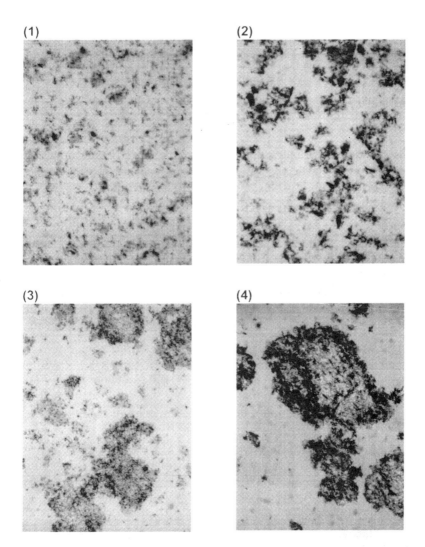

Fig. 4 Particle morphology of (S)-4-[[[1-(4-fluorophenyl)-3-(1-methyl-ethyl)-1H-indol-2-yl]-ethynyl]-hydroxyphosphinyl]-3]-hydroxybutanoic acid, disodium salt, after equilibration at relative humidity values of (1) 11%, (2) 60%, (3) 75%, and (4) 84%. All photomicrographs were taken at a magnification of 100×.

materials is profoundly affected by the hydration state of the drug substance. The birefringence associated with the liquid crystalline state was particularly striking and useful for qualitative identification purposes.

B. Thermal Microscopy

Often referred to as fusion microscopy or hot-stage microscopy, thermal microscopy can be an extremely valuable tool for the characterization of polymorphic or solvate systems. The technique requires that one make observations during the heating and cooling of a few milligrams of substance on a microscope slide, as well as observations made on the crystallized material [60,61]. It is therefore possible to conduct a very rapid analysis using only small quantities of material, and the entire phase diagram of a drug material can be deduced upon the conduct of suitably designed experiments. The most widely used device in the conduct of thermal microscopic studies is the hot stage of Kofler and Kofler [62], which has facilitated an extraordinary number of studies [63,64].

According to the phase rule, a single component system of two phases in equilibrium can only have one degree of freedom and is said to be univariant. If the experiments are conducted at constant and fixed pressure (as would exist on the surface of an open microscope slide), then the temperature of the system would be fixed at the transition point. Nevertheless, the transition point must be influenced by pressure, which may either raise or lower the transition point. The direction of the effect can be predicted on the basis of Le Chatelier's principle, if the change of volume accompanying the passage of one form into the other is known. The magnitude of the pressure effect can be calculated by the Clapeyron equation, and one finds that since the change in volume accompanying a phase transition is small, pressure can only exert a minor effect on the transition point.

The transition point at which a phase transition will take place is defined as the intersection of the pressure–temperature curves associated with each phase. Clearly, the transition point (at atmospheric pressure) can exist at a temperature that either exceeds or is less than the melting point. The former situation is termed *monotropy*, while the latter is termed *enantiotropy*. An enantiotropic substance is distinguished by its ability to change reversibly into one phase or another, while the transformation of a monotropic material is irreversible. The thermal microscope can be used to determine melting points with great accuracy, and through studies where the heating and cooling processes are cycled one can usually deduce the nature of a polymorphic transition. In addition, the loss of solvent molecules from a solvate species can easily be detected by the processes that take place when the sample is immersed in a suitable fluid.

The conduct of a thermal microscopic study may be illustrated through the work conducted by Kuhnert-Brandstätter on the polymorphs and solvates of a series of steroid hormones [65]. For example, if Form II of corticosterone-21-acetate is preheated to 140°C on a hot stage, the solid is observed to melt at 145–148°C. Further heating of the system results in a solidification of the melt, to yield Form I, which then melts at 153–155°C. Similarly, Form II of prednisone acetate is isolated from most solvents and melts over the range of 225–228°C. With continued heating, the melt solidifies, and one observes the melting point of Form I at 232–241°C.

IV. PHASE TRANSITIONS: THERMAL METHODS OF ANALYSIS

Thermal analysis methods are defined as those techniques in which a property of the analyte is determined as a function of an externally applied temperature. Regardless of the observable parameter measured, the usual practice requires that the physical property and the sample temperature be recorded continually and automatically, and that the sample temperature be altered at a predetermined rate. Measurements of thermal analysis are conducted for the purpose of evaluating the physical and chemical changes that may take place in a heated sample, requiring that the operator interpret the events noted in a thermogram in terms of plausible reaction processes. Thermal reactions can be endothermic (melting, boiling, sublimation, vaporization, desolvation, solid–solid phase transitions, chemical degradation, etc.) or exothermic (crystallization, oxidative decomposition, etc.) in nature.

Such methodology has found widespread use in the pharmaceutical industry for the characterization of compound purity, polymorphism, solvation, degradation, and excipient compatibility [66,67]. However, given the utility that thermal microscopy has shown for the characterization of polymorphic systems, it is not surprising that the quantitative applications of thermal analysis have proven to be even more useful. Although a large number of techniques have been developed, the most commonly applied are those of thermogravimetry (TG), differential thermal analysis (DTA), and differential scanning calorimetry (DSC).

A. Thermogravimetry

Thermogravimetry (TG) is a measure of the thermally induced weight loss of a material as a function of the applied temperature [68]. TG analysis is restricted to transitions that involve either a gain or a loss of mass, and it is most commonly used to study desolvation processes and compound decomposition. TG analysis is a useful method for the quantitative determination of the total volatile content of a solid and can be used as an adjunct to Karl Fischer titrations for the determination of moisture. As such it readily permits the distinction between solvates and the anhydrous forms of a given compound.

TG analysis also represents a powerful adjunct to the other methods of thermal analysis, since a combination of either a DTA or a DSC study with a TG determination can be used in the assignment of observed thermal events. Desolvation processes or decomposition reactions must be accompanied by weight changes, and they can be thus identified by a TG weight loss over the same temperature range. On the other hand, solid–liquid and solid–solid phase transformations are not accompanied by any loss of sample mass and would not register in a TG thermogram.

When a solid is capable of decomposing by means of several discrete, sequential reactions, the magnitude of each step can be separately evaluated. The TG analysis of compound decomposition can also be used to compare the stability of similar compounds. In general, the higher the decomposition temperature of a given compound, the greater would be its stability.

The measurement of thermogravimetry is simple in principle; it consists of the continual recording of the mass of the sample as it is heated in a furnace. The weighing device most used is a microbalance, which permits the characterization of milligram quantities of sample. The balance chamber itself is constructed so that the atmosphere may be controlled, which is normally accomplished by means of a flowing gas stream. The furnace must be capable of being totally programmable in a reproducible fashion, and its inside surfaces must be resistant to the gases evolved during the TG study. It is essential in TG design that the temperature readout be that of the sample and not that of the furnace. Thus the thermocouple or resistance thermometer must be mounted as close to the sample pan as possible.

Although the TG methodology is conceptually simple, the accuracy and precision associated with the results are dependent on both instrumental and sample factors [69]. The furnace heating rate used for the determination will greatly affect the transition temperatures, while the atmosphere within the furnace can influence the nature of the thermal reactions. The sample itself can play a role in governing the quality of the data obtained; factors such as the sample size, the nature of the evolved gases, the particle size, the heats of reaction, the sample packing, and the thermal conductivity all influence the observed thermogram.

Other than its ability to demonstrate the anhydrous nature of genuine polymorphic materials, the entire utility of TG analysis is in the differentiation and characterization of solvate species. The multitude of solvate species formed by the disodium salt of (S)-4-[[[1-(4-fluorophenyl)-3-(1-methylethyl)-1H-indol-2-yl]-ethynyl]-hydroxyphosphinyl]-3]-hydroxybutanoic acid was most effectively characterized by TG analysis of materials exposed to various relative humidity values [58]. Representative thermograms of the different hydrates are shown in Fig. 5, where it is evident that the well-defined plateaus observed upon completion of the individual dehydration steps were ideally suited for the evaluation of the hydration state of an isolated material.

Thermogravimetric analysis played an important role during the characterization of the insoluble adducts formed by 5-nitrobarbituric acid (dilituric acid) with alkali metal (Group IA) and alkaline earth (Group IIA) cations [70]. This study was performed to understand the

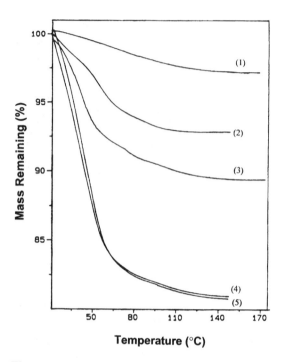

Fig. 5 Thermogravimetric analysis of (S)-4-[[[1-(4-fluorophenyl)-3-(1-methylethyl)-1H-indol-2-yl]-ethynyl]-hydroxyphosphinyl]-3]-hydroxybuta-noic acid, disodium salt. Shown are the thermograms for (1) the initial material containing approximately 3% water, and materials obtained after exposure to relative humidities of (2) 33%, (3) 52%, (4) 60%, and (5) 75%.

scientific foundations that permitted this particular compound to function as a useful analytical reagent for Group IA and IIA cations, which were identified on the basis of the characteristic crystal morphologies associated with the adduct compounds. As is evident in the data collected in Table 3, the origin of the different crystal morphologies associated with each adduct species began with the formation of numerous, crystallographically inequivalent, hydrate species.

B. Differential Thermal Analysis

In the differential thermal analysis (DTA) technique, one monitors the difference in temperature existing between a sample and a reference

Table 3 Thermogravimetric Analysis of the Adducts of Dilituric Acid with Group IA and IIA Cations [70]

Cation	Volatile content, measured at the 150°C plateau (%)	Theoretical volatile content for the n-hydrate species	Deduced hydration state (n)
Free acid	23.8	23.9	3
Lithium	12.3	12.2	2
Sodium	0	0	0
Potassium	0	0	0
Rubidium	0	0	0
Cesium	0	0	0
Beryllium	22.9	23.0	3
Magnesium	12.1	12.1	1.5
Calcium	3.9	4.1	0.5
Strontium	0	0	0
Barium	2.5	2.8	0.50

Source: Ref. 70.

as a function of temperature. Nonequivalences in temperature between the sample and the reference are observed when a process takes place that requires a finite heat of reaction. Typical solid-state changes of this type would be phase transformations, structural conversions, decomposition reactions, or desolvation processes. These processes may require either the input or the release of energy in the form of heat, which in turn translates into events that affect the temperature of the sample relative to a nonreactive reference. The primary applicability of DTA analysis to the study of polymorphs and solvates has been with respect to its ability to yield information about the phase transformations that take place as a function of temperature.

Although it is possible to use DTA as a quantitative tool, such applications require extensive calibration and care in data interpretation. For this reason, DTA has historically been mostly used in a quantitative sense as a means to determine the temperatures at which thermal events take place. Owing to the fortuitous combination of experimental conditions normally used for its measurement, the technique can be used for the characterization of materials that evolve corrosive gases during the heating process. DTA analysis is highly useful as a means for the determination of compound melting points, although in systems capable of undergoing phase changes the analyst must always be aware of such concerns.

Methodology appropriate for the measuring of DTA profiles has been extensively reviewed [71,72] and need only be outlined here. Both the sample and the reference materials are contained within the same furnace, whose temperature program is externally controlled. The outputs of the sensing thermocouples are amplified, electronically subtracted, and finally shown on a suitable display device. If the observed change in enthalpy ΔH is positive as in the case of endothermic reactions, the temperature of the sample will lag behind that of the reference. If the ΔH is negative (exothermic reaction), the temperature of the sample will exceed that of the reference.

Wendlandt has provided an extensive compilation of conditions and requirements that influence the shape of DTA thermograms [73]. These can be divided into instrumental factors (furnace atmosphere, furnace geometry, sample holder material and geometry, thermocouple details, heating rate, and thermocouple location in the sample) and sample characteristics (particle size, thermal conductivity, heat capacity, packing density, swelling or shrinkage of sample, mass of sample taken, degree of crystallinity). A sufficient number of these factors are under the control of the operator, thus permitting selectivity in the methods of data collection. The ability to correlate an experimental DTA thermogram with a theoretical interpretation is profoundly affected by the details of heat transfer between the sample and the calorimeter [74].

The calibration of DTA systems is dependent on the use of appropriate reference materials rather than on the application of electrical heating methods. The temperature calibration is normally accomplished with the thermogram being obtained at the heating rate normally used for analysis [75], and the temperatures known for the thermal events are used to set temperatures for the empirically observed features. Recommended reference materials that span melting ranges of pharmaceutical interest include benzoic acid (melting point 122.4°C), indium (156.4°C), and tin (231.9°C).

The simplest and most straightforward application of DTA analysis is concerned with studies of the relative stability of polymorphic forms. For example, DTA thermograms enabled the deduction that one commercially available form of chloroquine diphosphate was phase pure while another consisted of a mixture of two polymorphs [76]. DTA analysis was used to demonstrate that even though different crystal habits of sulfamethazine could be obtained, these in fact consisted of the same anhydrous polymorph [77]. In a study aimed at profiling the dissolution behavior of the three polymorphs and five solvates of spironlactone, DTA analysis was used in conjunction with powder x-ray diffraction to establish the character of the various materials [78].

In one study, it was found that three different crystalline forms of phenylbutazone could be obtained by the spray-drying of methylene chloride solutions at different temperatures, and that neither of these corresponded to the stable form that was obtained at the highest drying temperatures [79]. In fact, one of these could not be obtained by any conventional crystallization process but was only detected under conditions involving a slower solvent evaporation rate. As shown in Fig. 6, the DTA thermograms of each form are distinct and permit a facile distinction among the isolated forms.

Prior to the widespread use of differential scanning calorimetry, DTA analysis was the only method whereby one could obtain heats of transition or fusion. For example, sulfathiazone was found to undergo a transition from Form I to Form II at 161°C, for which the heat of transition was determined to be 1420 cal/mol [80]. The heat of fusion directly obtained for Form II was found to be 5970 cal/mol, which compared favorably with the heat of fusion determined for the material resulting from the conversion of Form I (5960 cal/mol). In another study, heats of fusion were determined for sixteen sulfonamides, some of which exhibited polymorphism and some of which did not [81]. In this work, a fuller understanding of the thermodynamics was provided, where the entropies as well as the enthalpies of the various processes were deduced.

C. Differential Scanning Calorimetry

In many respects, differential scanning calorimetry (DSC) is similar to the DTA method, and analogous information about the same type of thermal events can be obtained. However, DSC is far easier to use

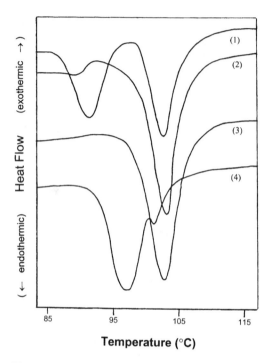

Fig. 6 Differential scanning calorimetry study of various spray-dried forms of phenylbutazone. Shown are the thermograms of (1) Form α, (2) Form β, (3) Form δ, and (4) Form ε. (The figure was adapted from data contained in Ref. 79.)

routinely on a quantitative basis, and for this reason it has become the most widely accepted method of thermal analysis for the pharmaceutical industry. The relevance of the DSC technique as a tool for pharmaceutical scientists has been amply discussed in numerous reviews [66,83–86], and a general chapter on DSC is documented in the *United States Pharmacopoeia* [87].

In the DSC method, the sample and reference materials are maintained at the same temperature, and the heat flow required to keep the equality in temperature is measured. DSC plots are therefore obtained as the differential rate of heating (in units of W/s, cal/s, or J/s) against temperature [88]. The area under a DSC peak is directly proportional to the heat absorbed or evolved by the thermal event, and integration of these peak areas yields the heat of reaction (in units of cal/s · g or J/s · g).

Two types of DSC measurements are possible, which are usually identified as power-compensation DSC and heat-flux DSC, and the details of each have been fully described [88,89]. In power-compensated DSC, the sample and reference materials are kept at the same temperature by the use of individualized heating elements, and the observable parameter recorded is the difference in power inputs to the two heaters. In heat-flux DSC, one simply monitors the heat differential between the sample and the reference materials, with the methodology not being much different from that used for DTA.

In the DTA measurement, an exothermic reaction is plotted as a positive thermal event, while an endothermic reaction is usually displayed as a negative event. Unfortunately, the use of power-compensation DSC results in endothermic reactions being displayed as positive events, a situation counter to the latest IUPAC recommendations [90]. When the heat-flux method is used to detect the thermal phenomena, the signs of the DSC events concur with those obtained using DTA and also agree with the IUPAC recommendations.

The calibration of DSC instruments is normally accomplished through the use of compounds having accurately known transition temperatures and heats of fusion. A list of the standards currently supplied by the National Technical Information Service (NTIS) [91] is provided in Table 2. Once the DSC system is properly calibrated, it is trivial to obtain the melting point and enthalpy of fusion data for any compound upon integration of its empirically determined endotherm and application of the calibration parameters. The current state of methodology is such, however, that unless a determination is repeated a large number of times, the deduced enthalpies must be regarded as being accurate only to within approximately 5%.

As has been noted for DTA analysis, differential scanning calorimetry can also be used to establish the melting points of polymorphic species. For example, gepirone hydrochloride has been obtained in three polymorphic forms, which were found to melt at 180°C, 200°C, and 212°C [92]. In this work, it was shown that Forms I and II, and Forms I and III, were enantiotropic pair systems, but that Forms II and III were monotropic with respect to each other. The two polymorphs of phenylephrine oxazolidine each exhibited well-defined melting points and could easily be distinguished on the basis of their thermograms [93].

Some of the most interesting DSC studies have been conducted on metastable phases that undergo a phase transformation to a more stable phase during the lifetime of the thermal analysis experiment. It was found that Form I of iopanoic acid yielded a single melting endotherm at 154°C, but that the thermogram obtained on Form II was much more complicated [94]. As shown in Fig. 7, Form II exhibits one endotherm at 133°C (the melting transition of Form I), an exotherm at 141°C (crystallization to Form II), and another endotherm at 153°C (melting of the recrystallized Form II). The literature abounds with similar examples where metastable polymorphs have been observed to convert

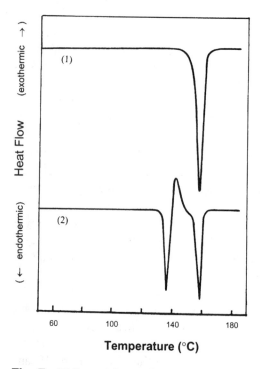

Fig. 7 Differential scanning calorimetry thermograms obtained for the (1) Form I and (2) Form II polymorphs of iopanoic acid. (The figure was adapted from data contained in Ref. 94.)

into more stable forms, with carbamazepine [95], piroxicam [96], and pireanide [97] being quoted as recent examples of work in this genre.

When studying the stability relationships between different polymorphic forms, the use of temperature cycling experiments has often proven to be very useful. Samples are heated to a preset temperature (but not high enough to induce a thermal decomposition event), cooled back to a lower temperature, and then reheated. Alterations in the recorded thermograms resulting from the cycling process can provide information about the ease of phase conversion. For example, such work enabled deductions to be made regarding the relative stabilities of the polymorphs of 1,2-dihydro-6-neopentyl-2-oxonicotinic acid [98]. The polymorphism associated with glyceryl monostearate was found to exert a profound effect upon the stability of formulations containing raw materials obtained from different sources, but the system could be understood in terms of phase transformations brought about by repeated melting and congealing cycles [99]. In fact, the use of cycling experiments may be essential to eliminate artifacts from entering into the construction of phase diagrams [100].

DSC techniques may be used to determine the kinetics of solid-state transformations as well. The kinetics associated with the transformation of disopyramide Form I to Form II were followed by changes in the DSC thermograms, and an activation energy of 144 kJ/mol was calculated for the system [101]. According to this work, the model proposed by Prout and Tompkins [102] for decomposition in solid phases (with no prior melting or liquefaction) appears to be appropriate for solid-state polymorphic transitions. As anticipated for any solid-state reaction, the phase change would initiate at defect sites, producing energetically favorable configurations that would serve as templates for continued phase transformation. In a similar work, a combination of DSC and XRPD studies was used to evaluate the kinetics associated with the various polymorphic transitions of phenylbutazone [103].

DSC analysis is often used in conjunction with structural techniques during the characterization of hydrate and solvate systems, with the thermal method being used to pinpoint the transition temperature range over which the bound water or solvent can be liberated. For instance, although a number of solvate forms could be crystallized for ethynylestradiol, the different solvent molecules were found incapable of exerting any effect on the conformation of the drug [104]. DSC studies permitted the deduction of the relative stabilities of the three solvates of alprazolam [105] and contributed to the characterization of the solvate systems identified for norfloxacin [106] and dehydroepiandrosterone [107].

The character of the anhydrous phase resulting after the desolvation process has taken place can be effectively understood using DSC analysis. In many cases, loss of solvation or hydration molecules leads to the formation of an amorphous material, but this is not always the case. It was established that the acetone, isopropanol, and ethyl acetate solvates of cyclopenthiazide all reverted to Form I (a known anhydrous phase) above the endotherm corresponding to loss of solvent [108]. In other systems, loss of solvation results in the formation of a metastable phase, which undergoes a phase conversion to the stable anhydrate phase before overall melting takes place. Such behavior is illustrated in Fig. 8, where it was reported that the *tert*-butanol and dioxane solvates of piretanide exhibited a recrystallization after desolvation, but that the propylene glycol and dimethylformamide solvates did not [109].

V. MOLECULAR MOTION: VIBRATIONAL SPECTROSCOPY

The energies associated with the vibrational modes of a chemical compound lie within the range of 400–4000 cm^{-1}. These modes can be observed directly through their absorbance in the infrared region of the spectrum, or through the observation of the low-energy scattered bands that accompany the passage of an intense beam of light through the sample (the Raman effect). In either case, the use of Fourier-transform methodology has vastly improved the quality of data that can be obtained [110]. Most workers are familiar with the use of mid-infrared spectra for identity purposes, where the pattern of absorption bands is taken to be diagnostic for a given compound. However, it has come to be recognized that the vibrational spectra of solid materials will reflect details of the crystal structure, and hence these methods can be used in the spectroscopic investigation of polymorphs and solvates [111,112].

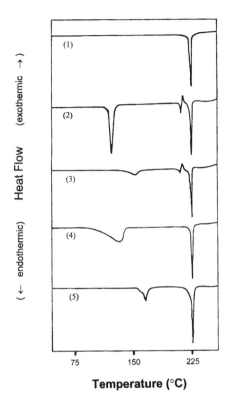

Fig. 8 Differential scanning calorimetry study of various solvates of piretan-ide. Shown are the thermograms of (1) the anhydrate phase, (2) the *tert*-butanol solvate, (3) the dioxane solvate, (4) the propylene glycol solvate, and (5) the dimehylformamide solvate. (The figure was adapted from data contained in Ref. 109.)

In the best-designed studies of polymorphic or solvate systems, the purpose of the vibrational spectroscopy investigation should be to gather information from the observed pattern of vibrational frequencies and to use these data to understand the structural aspects that yield crystallographic differences. Once suitable spectral features are identi-fied from this work, they can be used to develop easily performed meth-ods for the quantitative analysis of one polymorph (or solvate) in the presence of the other. Unfortunately, all too many workers are merely satisfied to obtain the spectra of the various polymorphs and/or solvates and simply to display the differences. In doing so, they miss a great opportunity to gain additional understanding about the system they are trying to characterize.

A. Infrared Absorption Spectroscopy

The acquisition of high-quality infrared absorption spectra appropriate for the characterization of polymorphs and solvates is best performed using Fourier transform technology (the FTIR method), since this approach minimizes transmission and beam attenuation problems. Essentially all FTIR spectrometers use a Michelson interferometer. Radiation entering the interferometer is split into two beams by means of a beam splitter. One beam follows a path of fixed distance before being reflected back into the beam splitter, while the other beam travels a variable distance before being recombined with the first beam. The recombination of these two beams yields an interference pattern, and the time-dependent constructive and destructive interferences have the effect of forming a cosine signal.

Each component wavelength of the source will yield a unique cosine wave having a maximum at the zero path length difference (ZPD) that decays with increasing distance from the ZPD. The detector is placed so that radiation in the central image of the interference pattern will be incident upon it, and therefore intensity variations in the recombined beam are manifest as phase differences. The observed signal at the detector is a summation of all the cosine waves, having a maximum at the ZPD that decays rapidly with increasing distance from the ZPD. If the component cosine waves can be resolved, then the contribution from individual wavelengths can be observed. The frequency domain spectrum is obtained from the interferogram by performing the Fourier transformation mathematical operation.

The solid-state FTIR spectra of many polymorphic systems often are found to be only slightly different, indicating that the pattern of molecular vibrations is not grossly affected by differences in crystal structure. Examples of this type of behavior are typified in studies conducted on etoposide [113], tegafur [114], lomeridine dihydrochloride [115], and carbamazepine [116]. In other instances, the FTIR spectra of polymorph systems differ substantially, permitting the ready identification of a given form. For instance, the two forms of ranitidine hydrochloride yielded spectra that differed in the region above 3000 cm^{-1} and in the regions spanning 2300–2700 cm^{-1} and 1570–1620 cm^{-1} [117].

Mexiletine hydrochloride has been found to crystallize in six polymorphic forms, each of which yields a characteristic IR spectrum [118]. As shown in the data collected in Table 4, the very intense bands due to the C–N(H) stretching mode (1020–1060 cm^{-1}) and the aromatic C–H out-of-plane deformation mode (760–780 cm^{-1}) were found to be particularly sensitive to the structural differences. In fact, the splitting of the latter mode observed in four of the six polymorphs suggests the existence of different types of molecules within the unit cell. The data in Table 4 also illustrate the typical magnitudes in band energy differences encountered within the vibrational spectra of polymorphic systems.

When solvent molecules are incorporated in a crystal lattice, the new structure is often sufficiently different from that of the anhydrous phase so that many of the molecular vibrational modes are altered [119]. Not surprisingly, the most pronounced effects are associated with modes expressing the motion of atoms involved in hydrogen bonding with the solvent. For this reason, studies carried out in the high-

Table 4 Effect of Crystal Polymorphism on Selected Infrared Bands of Mexiletine Hydrochloride

Polymorph	Energy of band(s) associated with the C–N(H) stretching mode (cm^{-1})	Energy of band(s) associated with the aromatic C–H out-of-plane deformation mode (cm^{-1})
I	1037	772
II	1033	770, 765
III	1039	769, 760
IV	1027	775, 760
V	1047	769, 760
VI	1052, 1034	766

Source: Ref. 118.

energy region of the infrared spectrum (2000–4000 cm $^{-1}$) often yield the most striking spectral differences between solvate phases and the corresponding anhydrous phase.

When water acts as the solvation agent, observations within the - OH stretching region (3100–3600 cm^{-1}) are most fruitful for identification purposes. For example, the anhydrate and hydrate phases of oxyphenbutazone [120], digoxin [121], trazodone hydrochloride [122], and ampicillin [123] are readily distinguished on the basis of their IR absorption spectra. Organic solvents are equally amenable to study within the high-energy spectral region. As illustrated in Fig. 9, the - OH stretch of bound methanol in the solvate of 9,10-anthraquinone is observed at 3440 cm^{-1} [124], which represents a shift from the frequency of 3400 cm^{-1} noted for the same mode in liquid methanol [125].

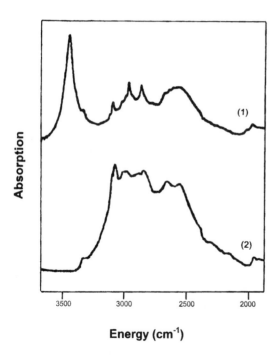

Fig. 9 Infrared absorption spectra obtained within the high-energy vibrational region for the (1) methanol solvate and (2) anhydrous Form I polymorphs of 9,10-anthraquinone. (The figure was adapted from data contained in Ref. 124.)

When systems can form multiple anhydrate, hydrate, and solvate phases, the use of infrared spectroscopy can be extremely valuable. For example, depending on the recrystallization solvent, delavirdine mesylate has been found to form two anhydrous phases, two hydrates, an ethanol solvate, and an acetonitrile solvate, as well as six other phases resulting as the products of solid-state transformations of the hydrated and solvated phases [126]. In this work, FTIR spectroscopy played an important role in working out the characterization of this system.

However, it is clear that vibrational spectroscopy has considerable use beyond the identification of polymorphs and solvates. The infrared spectra obtained on the polymorphs of acetohexamide and selected derivatives has been used to study the tautomerism of the drug compound [127]. It was deduced in this work that Form A existed in the enol form, stabilized by the intramolecular bonding between the O—H and S=O groups that produces a six-membered ring. Form B was characterized by the existence of the keto form, with the urea carbonyl group being intermolecular bonding to a sulfonamide N—H functionality. This behavior can be contrasted with that noted for spironolactone, where no evidence was found for the existence of enolic tautomers in any of the four polymorphs [128].

Variable-temperature vibrational spectroscopy can be a powerful tool for the study of phase transitions and/or desolvation processes. The technique has been combined with factor analysis to deduce the three phase transitions and four conformational changes associated with pentaerythritol tetrastearate [129]. In this particular work, each thermal event was substantiated by analogous DSC studies. The dihydrates prepared from the two polymorphs of carbamazepine were also studied using variable-temperature techniques [130].

When a nonequivalence in absorption bands is identified for a polymorphic or solvate system, the band intensities can be used for a quantitative analysis of mixtures containing the various phases. Such work requires the nontrivial preparation of homogeneous calibration samples containing known amounts of the phases in question, but most workers underestimate the difficulty associated with mixing powders of potentially different morphologies. For example, the monohydrate phase of cefepime dihydrochloride consists of rods, while the dihydrate phase consists of needles, and the mixing of these could only be properly effected using a solvent slurrying technique [131]. But with this superior method of mixing the two phases, workers were able to obtain superior analytical performance parameters for their validated assay method. Quantitative analytical methods have also been reported for sulfamethoxazole [132], chlorpropamide [133], and 7-[3-(4-acetyl-3-methoxy-2-propylphenoxy)-propoxy-3,4-dihydro-8-propyl-2H-1-benzo-pyran-2-carboxylic acid [134], illustrating the generality of the approach.

B. Raman Spectroscopy

The vibrational modes of a compound may also be studied using Raman spectroscopy, where one measures the inelastic scattering of radiation by a nonabsorbing medium [135]. When a beam of light is passed through a material, approximately one in every million incident photons is scattered with a loss or gain of energy. The inelastically scattered radiation can occur at lower (Stokes lines) and higher (anti-Stokes lines) frequencies relative to that of the incident (or elastically scattered) light, and the energy displacements relative to the energy of the incident beam correspond to the vibrational transition frequencies of both mediums. The actual intensities of the Stokes and anti-Stokes lines are determined by the Boltzmann factor characterizing the vibrational population. For high-frequency vibrations, the Stokes lines are intense relative to the anti-Stokes lines, so conventional Raman spectroscopy makes exclusive use of the Stokes component.

The Raman effect originates from the interaction of the oscillating induced polarization or dipole moment of the medium with the electric field vector of the incident radiation. Raman spectra are measured by passing a laser beam through the sample and observing the scattered light either perpendicular to the incident beam or through backscatter detection. The scattered light is analyzed at high resolution by a monochromator and ultimately detected by a suitable device. One way of obtaining good spectra is to use a notch filter that will eliminate the exciting line, since this is required to obtain acceptable signal-to-noise ratios.

Although both infrared absorption and Raman scattering yield information on the energies of the same vibrational bands, the different selection rules governing the band intensities for each type of spectros-copy can yield useful information. For the low-symmetry situations presented by the structures of molecules of pharmaceutical interest, every vibrational band will be active to some degree in both infrared absorption and Raman scattering spectroscopies. The relative intensi-ties of analogous bands will differ, however, when observed by infrared absorption or Raman spectroscopy. In general, symmetric vibrations and nonpolar groups yield the most intense Raman scattering bands, while antisymmetric vibrations and polar groups yield the most intense infrared absorption bands. Both types of vibrational spectroscopy were used to investigate the polymorphism of nimodipine, and the data were contrasted in a particularly illustrative series of figures [136]. It was evident from the intensity relations that although each technique yielded a summary of the vibrational transitions, substantial differences in band intensity were readily discernible.

Raman scattering bands are often quite sharp, and consequently Raman spectra often contain significantly less spectral overlap than in-frared absorption spectra. As shown in Fig. 10, the high-energy spectral region obtained for the two polymorphs of fluconazole consists of a series of well-resolved spectral features, which permits a more facile characterization of the structural differences between the two systems [137]. Raman spectra are generally not complicated by contributions from adventitious moisture, owing to the weak scattering nature of the water molecule.

Even when Raman spectroscopy is used primarily as a means for the differentiation of polymorphs or solvates, the degree of spectral simplification associated with Raman data permits a more facile gener-ation of band assignments. This feature has been successfully exploited in the cases of spironolactone [128], losartan [138], 3-(p-thioanisoyl)-1,2,2-trimethyl-cyclopentanecarboxylic acid [139], and *trans*-3,4-dichloro-*N*-methyl-*N*-[1,2,3,4-tetrahydro-5-methoxy-2-(pyrrolidin-1-yl)]-naphth-1-ylbenzeneacetamide [140].

Owing to its ability to provide data on very low frequency vibra-tional modes, Raman spectroscopy can yield information on the lattice vibrations of a crystal. Such work has been performed on the various crystal forms of ampicillin and griseofulvin [141,142], and useful infor-mation has been obtained on the nature of the solvate species formed.

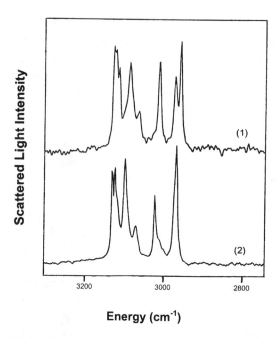

Fig. 10 Raman spectra obtained within the high-energy vibrational region for the (1) Form I and (2) Form II polymorphs of fluconazole. (The figure was adapted from data contained in Ref. 137.)

Most interesting was the use of variable-temperature Raman studies, where the characteristics of the crystal lattice were followed during the desolvation process [142]. Variable-temperature Raman spectroscopy was also used to study the phase transformations of 2-(*S*)-hydroxy-3-(*R*)-(2-carboxyethyl)thiol-3-[2-(phenyloctyl)-phenyl]-propionic acid [143].

VI. CHEMICAL ENVIRONMENT: NUCLEAR MAGNETIC RESONANCE SPECTROMETRY

One technique that is becoming increasingly important for the characterization of materials is that of solid-state nuclear magnetic resonance (NMR) spectroscopy [144]; the application of this methodology to topics of pharmaceutical interest has been amply demonstrated [112,145–146]. Although any nucleus that can be studied in the solution phase can also be studied in the solid state, most of the work has focused on ^{13}C studies. As mentioned in the case of vibrational spectroscopy, the ability of solid-state NMR to differentiate between a system of polymorphs or solvates requires that individual nuclei exist in nonequivalent magnetic environments within the two crystal structures. If the structural variations do not lead to a magnetic nonequivalence for a given nucleus, then the resonances obtained for the nucleus will not differ. Powerful as the technique has proven to be, one must remember that the ultimate arbiter of polymorphism is crystallography and not spectroscopy.

^{1}H-NMR remains an extremely difficult measurement in the solid state, and the data obtained from such work can only be obtained at medium resolution. The field acting at the nucleus is affected by the magnetic dipoles of neighboring nuclei, and the local fields thus generated are sensitive to both the internuclear distances and their orientation relative to the external field. Since protons are abundantly present in organic compounds, the removal of proton–proton dipolar interactions is necessary to obtain high-resolution ^{1}H spectra in solids. Although this is possible, the resulting ^{1}H-NMR spectra are still inferior to those obtained in the solution phase. The primary reason for this is that ^{1}H-NMR has one of the smallest isotropic chemical shift ranges (12 ppm), but with peak broadening effects that can span several ppm in magnitude. Other nuclei yield far better data, with ^{13}C and ^{31}P solid-state NMR studies being very useful to the physical characterization of all pharmaceutical solids.

The local magnetic field B_{loc} at a ^{13}C nucleus in an organic solid is given by

$$B_{loc} = \pm \left\{ \frac{h\gamma_H}{4\pi} \right\} \left\{ \frac{3\cos^2 \theta - 1}{r^3} \right\} \tag{2}$$

where γ_H is the magnetogyric ratio of the proton, r is the internuclear C–H distance to the bonded proton, and θ is the angle between the C–H bond and the external applied field (B_0). The $\pm$ sign indicates that the local field may add to or subtract from the applied field depending on whether the neighboring proton dipole is aligned with or against the direction of B_0. In a microcrystalline organic solid, there is a summation over many values of θ and r, resulting in a proton dipolar broadening of many kilohertz. A rapid reorientation of the C–H internuclear vectors (such as those associated with the random molecular motions that take place in the liquid phase) would result in reduction of the dipolar broadening. In solids, such rapid isotropic tumbling is not possible, but since $3 \cos^2 \theta - 1$ equals zero if θ equals $\cos^{-1} 3^{-1/2}$ (approximately 54°44′), spinning the sample at the so-called magic angle of 54°44′ with respect to direction of the applied magnetic field results in an averaging of the chemical shift anisotropy. In a solid sample, the anisotropy reflects the chemical shift dependence of chemically identical nuclei on their spatial arrangement with respect to the applied field. Since it is this anisotropy that is primarily responsible for the spectral broadening associated with ^{13}C samples, spinning at the magic angle makes it possible to obtain high-resolution ^{13}C-NMR spectra of solid materials.

An additional method for the removal of $^{13}C-^{1}H$ dipolar broadening is to use a high-power proton decoupling field. This is often referred to as dipolar decoupling. One irradiates the sample using high power at an appropriate frequency, which results in the complete collapse of all $^{13}C-^{1}H$ couplings. With proton dipolar coupling alone, the resonances in a typical solid-state ^{13}C spectrum will remain very broad (on the order of 10–200 ppm). This broadening arises because the chemical shift of a particular carbon is directional, depending on the orientation of the molecule with respect to the magnetic field.

Even though high-resolution spectra can be obtained on solids using the magic angle spinning (MAS) technique, the data acquisition time is lengthy due to the low sensitivity of the nuclei and the long relaxation times exhibited by the nuclei. This problem is circumvented through the use of cross polarization (CP), where spin polarization is transferred from the high-abundance, high-frequency nucleus (^{1}H) to the rare, low-frequency nucleus (^{13}C). This process results in up to a fourfold enhancement of the normal ^{13}C magnetization and permits a shortening of the waiting periods between pulses. The CP experiment also allows the measurement of several relaxation parameters that can be used to study the dynamic properties of the solid under investigation.

It is often observed that the NMR spectra of compound polymorphs or solvates contain nonequivalent resonance peaks for analogous nuclei. This effect arises because the intimate details of the molecular environments associated with differing crystal structures can yield a nonequivalent relationship with respect to the applied magnetic field of the NMR experiment, which in turn causes the analogous nuclei to resonate at different energies. As has been already noted for the infrared spectra of polymorphs or solvates, it is not uncommon for certain resonance peaks to be observed at identical chemical shifts, while other resonances are significantly shifted [145]. Since it is not difficult to assign organic functional groups to observed resonances, solid-state NMR spectra can be used to deduce the nature of polymorphic variations. This technique is especially valuable when the crystal polymorphism is conformational in origin. Such information is extremely valuable at the early stages in drug development when solved single crystal structures for each polymorph or solvate may not be available.

In their simplest application, solid-state NMR spectra can be used to differentiate qualitatively between polymorphs or solvates, much in the manner described for vibrational spectroscopy. Such data have been reported for the polymorphs of sulfathiazole [147], cyclopentthiazide [148], and indomethacin [149]. The technique can also be profitably used to differentiate between anhydrate and solvate phases, as has been reported for ampicillin [123], androstanolone [150], and dirithromycin [151].

When more detailed interpretation of the results is required, solution-phase NMR spectra can often be useful in the assignment of resonances observed in the solid-state NMR spectrum of the same compound. At the same time, the effects of magnetic nonequivalence associated with details of the crystallography can also be evaluated. This situation has been illustrated in Fig. 11, where the solid-state spectrum reported for benoxaprofen Form I is found to exhibit both similarities and differences relative to the solution-phase spectrum [152]. Contrasts between solution-phase and solid-state NMR spectra have also been drawn for mofebutazone, phenylbutazone, and oxyphenbutazone [153].

When detailed assignments of solid-state spectra have been made, the technique can be used to deduce differences in molecular conformation that cause crystallographic variations to exist. During the development of fosinopril sodium, a crystal structure was solved for the most stable phase, but no such structure could be obtained for its metastable phase [154]. The compound contains three carbonyl groups, but the

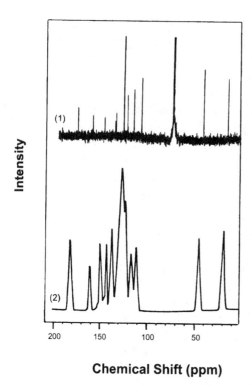

Fig. 11 (1) Completely decoupled solution-phase and (2) solid-state ^{13}C nuclear magnetic resonance spectra obtained for benoxaprofen. The solution-phase spectrum is compared with the solid-state spectrum of Form I. (The figure was adapted from data contained in Ref. 152.)

solid-state ^{13}C-NMR spectra of two of these were essentially equivalent. The third carbonyl, located on the acetal side chain, was found to resonate at different chemical shifts in the two structures. When combined with the observations obtained using vibrational spectroscopy, these results permitted the deduction that the solid-state polymorphism was associated with different conformations of this side chain. The NMR data also suggested that additional conformational differences between the two polymorphs were associated with cis–trans isomerization along the peptide bond, which in turn results in the presence of nonequivalent molecules existing in the unit cell. In the absence of solved crystal structures for the two polymorphs, this information would not have been otherwise obtainable.

The acquisition of solid-state ^{13}C-NMR spectra at various temperatures can be a powerful approach to the study of molecular motion in solids and for the study of phase conversion. Such methodology was used to study the two polymorphs of 2,2-bis(p-acetoxyphenyl)propane, where markedly different molecular mobilities were found to exist in the two crystal structures [155]. Below the glass transition temperature, the important active mode correlates with the flipping of the phenylene rings, and this effect was evident in the NMR spectra obtained at different temperatures. Some of the resonance bands associated with Form II were found to coalesce upon sample heating from $-10°C$ to ambient probe temperature.

The solid-state NMR technique can be used to deduce quantitative measurements of phase composition, as has been reported for the anhydrate and dihydrate phases of carbamazepine [156]. In the solid-state NMR spectra of delavirdine mesylate, Form VIII shows a unique resonance at 17.3 ppm and Form XI a unique resonance at 20.2 ppm, while a resonance at 23.9 ppm is shared by both forms [157]. These spectral characteristics have been exploited for the development of a method to determine the Form VIII content in bulk Form XI. As evident from the spectra shown in Fig. 12, the empirical limit of detection for the determination of Form VIII was approximately 2%.

The applications of solid-state ^{13}C-NMR spectra for the study of polymorphs and solvates can go beyond evaluations of resonance band positions and make use of additional spectral characteristics. For instance, studies of $T_{1\rho}$ relaxation times of furosemide polymorphs were used to show the presence of more molecular mobility and disorder in Form II, while the structure of Form I was judged to be more rigid and uniformly ordered [158]. The analysis of the solid-state ^{13}C-NMR spectra of (1R,3S)-3-p-thioanisoyl-1,2,2-trimethylcyclopentanecarboxylic acid was facilitated by the J-modulated spin-echo technique, which was used to deduce the number of protons bound to each carbon atom [159]. Differences in the dipolar dephasing behavior between the two polymorphs of ($\pm$)-$trans$-3,4-dichloro-N-methyl-N-[1,2,3,4-tetrahydro-5-methoxy-2-(pyrrolidin-1-yl)]naphth-1-yl-benzeneacetamide were noted and ascribed to motional modulation of the carbon–hydro-

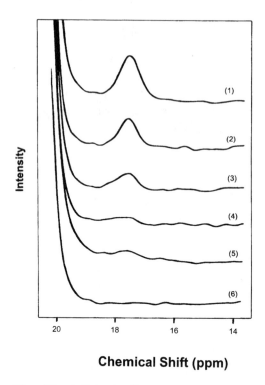

Chemical Shift (ppm)

Fig. 12 Solid-state ^{13}C nuclear magnetic resonance spectra obtained for Form XI of delavirdine mesylate, spiked with various levels of Form VIII. Spectra are shown for spiking levels of (1) 1%, (2) 2%, (3) 3%, (4) 5%, (5) 10%, and (6) 15%. (The figure was adapted from data contained in Ref. 157.)

gen dipolar interaction [160]. This added degree of molecular motion was used to deduce a loosely packed crystal structure for Form II of this compound.

VII. SUMMARY

The study of polymorphs and solvates begins with the methods used to elucidate the nature of the system in question. Beginning with the basic arsenal of crystallography, microscopy, thermal analysis, solubility studies, vibrational spectroscopy, and nuclear magnetic resonance, one can do significant work on the characterization of polymorphs and solvates. In spite of the power associated with the thermal or analytical techniques, one must always remember that the defining criterion for the existence of different crystal forms is a nonequivalence of crystal structures. All other methodologies must be considered as sources of supporting and ancillary information; they cannot be taken as definitive proof for the existence of polymorphism by themselves.

It is hoped that the range of studies discussed in the present work sheds sufficient light onto the practice of the most commonly encountered techniques for the solid-state characterization of polymorphic or solvate properties. Every system presents its own range of challenges, requiring an entire program of study to comprehend the system. Through suitably designed multidisciplinary work, any clever investigator will be able to obtain information at whatever level of complexity might be required. It is also anticipated that continued developments in methodology will enable the design of even better studies and ultimately result in the generation of even more useful data.

REFERENCES

1. H. G. Brittain, S. J. Bogdanowich, D. E. Bugay, J. DeVincentis, G. Lewen, and A. W. Newman, *Pharm. Res., 8*, 963 (1991).
2. H. G. Brittain, *Physical Characterization of Pharmaceutical Solids*, Marcel Dekker, New York, 1995.
3. T. L. Threlfall, *Analyst, 120*, 2435 (1995).
4. S. R. Byrn, R. R. Pfeiffer, M. Ganey, C. Hoiberg, and G. Poochikian, *Pharm. Res., 12*, 945 (1995).
5. R. W. James, *The Optical Principles of the Diffraction of X-Rays*, G. Bell, London, 1954.
6. M. J. Buerger, *X-Ray Crystallography*, John Wiley, New York, 1942.
7. G. H. Stout, and L. H. Jensen, *X-Ray Structure Determination*, Macmillan, New York, 1968.
8. M. M. Woolfson, *An Introduction to X-Ray Crystallography*, Cambridge Univ. Press, Cambridge, 1970.
9. J. P. Glusker, and K. N. Trueblood, *Crystal Structure Analysis*, Oxford Univ. Press, New York, 1972.

10. A. F. Wells, *Structural Inorganic Chemistry*, 5th ed., Clarendon Press, Oxford, 1984.
11. J. Bernstein,"Conformational Polymorphism," Chapter 13 in *Organic Solid State Chemistry*, Elsevier, Amsterdam, 1987, pp. 471–518.
12. E. W. Pienaar, M. R. Caira, and A. P. Lötter, *J. Cryst. Spect. Res., 23*, 785 (1993).
13. E. W. Pienaar, M. R. Caira, and A. P. Lötter, *J. Cryst. Spect. Res., 23*, 739 (1993).
14. J. J. Gerber, M. R. Caira, and A. P. Lötter, *J. Cryst. Spect. Res., 23*, 863 (1993).
15. J. Rambaud, A. Bouassab, B. Pauvert, P. Chevallet, J.-P. Declercq, and A. Terol, *J. Pharm. Sci., 82*, 1262 (1993).
16. Y. Hiramatsu, H. Suzuki, A. Kuchiki, H. Nakagawa, and S. Fuji, *J. Pharm. Sci., 85*, 761 (1996).
17. V. Agafonov, B. Legendre, N. Rodier, D. Wouessidjewe, and J.-M. Cense, *J. Pharm. Sci., 80*, 181 (1991).
18. H. P. Klug, and L. E. Alexander, *X-Ray Diffraction Procedures for Polycrystalline and Amorphous Materials*, 2d ed., John Wiley, New York, 1974.
19. "X-Ray Diffraction," General Test <941>, *The United States Pharmacopeia*, 23d ed., United States Pharmacopeial Convention, Rockville, MD, 1995, pp. 1843–1844.
20. M. O. Boles, and R. J. Girven, *Acta Cryst., B32*, 2279 (1976).
21. M. O. Boles, R. J. Girven, and P.A.C. Gane, *Acta Cryst., B34*, 461 (1978).
22. R. Suryanarayanan,"X-Ray Powder Diffractometry," Chapter 7 in *Physical Characterization of Pharmaceutical Solids* (H. G. Brittain, ed), Marcel Dekker, New York, 1995, pp. 199–216.
23. R. Suryanarayanan, *Pharm. Res., 6*, 1017 (1989).
24. D. B. Black, and E. G. Lovering, *J. Pharm. Pharmacol., 29*, 684 (1977).
25. R. Suryanarayanan, and A. G. Mitchell, *Int. J. Pharm., 24*, 1 (1985).
26. L. S. Crocker, and J. A. McCauley, *J. Pharm. Sci., 84*, 226 (1995).
27. J. K. Haleblian, *J. Pharm. Sci., 64*, 1269 (1975).
28. A. N. Winchell, *The Optical Properties of Organic Compounds*, 2d ed., Academic Press, New York, 1954.
29. N. H. Hartshorne, and A. Stuart, *Practical Optical Crystallography*, Elsevier, New York, 1964.
30. E. E. Wahlstrom, *Optical Crystallography*, 4th ed., John Wiley, New York, 1969.
31. J. A. Biles, *J. Pharm. Sci., 51*, 499–509; 601 (1962).

32. P. M. Cooke, *Anal. Chem., 68*, 333R (1996).
33. T. G. Rochow, and E. G. Rochow, *An Introduction to Microscopy by Means of Light, Electrons, X-Rays, or Ultrasound*, Plenum Press, New York, 1978.
34. A. W. Newman, and H. G. Brittain, "Particle Morphology: Optical and Electron Microscopies," Chapter 5 in *Physical Characterization of Pharmaceutical Solids* (H. G. Brittain, ed), Marcel Dekker, New York, 1995, pp. 127–156.
35. J. Belling, *The Use of The Microscope*, McGraw-Hill, New York, 1930.
36. G. Needham, *The Practical Use of The Microscope*, Charles C. Thomas, Springfield, IL, 1958.
37. E. M. Chamot, and C. W. Mason, *Handbook of Chemical Microscopy*, volume 1, 3d ed., John Wiley, New York, 1958.
38. A. F. Hallimond, *The Polarizing Microscope*, 3d ed., Vickers, New York, 1970.
39. N. H. Hartshorne, and A. Stuart, *Crystals and the Polarizing Microscope*, 4th ed., Edward Arnold, London, 1970.
40. W. C. McCrone, L. B. McCrone, and J. G. Delly, *Polarized Light Microscopy*, Ann Arbor Science, Ann Arbor, Michigan, 1978.
41. A. W. Agar, R. H. Alderson, and D. Chescoe, *Principles and Practice of Electron Microscopy*, North Holland, Amsterdam, 1974.
42. O. C. Wells, *Scanning Electron Microscopy*, McGraw-Hill, New York, 1974.
43. J. I. Goldstein, D. E. Newbury, P. Echlin, D. C. Joy, C. Fiori, and E. Ligshin, *Scanning Electron Microscopy and X-Ray Microanalysis*, Plenum Press, New York, 1981.
44. E. E. Hunter, *Practical Electron Microscopy*, Praeger Scientific, New York, 1984.
45. W. C. McCrone, *Scanning Microscopy, 7*, 1 (1993).
46. A. Watanabe, Y. Tanaka, and Y. Tanaka, *Chem. Pharm. Bull., 25*, 2239 (1977).
47. T. Yokoyama, T. Umeda, K. Kuroda, and A. Watanabe, *Chem. Pharm. Bull., 26*, 1044 (1978).
48. A. Watanabe, Y. Yamaoka, and K. Kuroda, *Chem. Pharm. Bull., 28*, 372 (1980).
49. G. Misosovich, *J. Pharm. Sci., 53*, 484 (1964).
50. E. Shefter, and T. Higuchi, *J. Pharm. Sci., 52*, 781 (1963).
51. J. B. Bogardus, *J. Pharm. Sci., 72*, 837 (1983).
52. J. Herman, J. P. Remon, N. Visavarungroj, J. B. Schwartz, and G. H. Klinger, *Int. J. Pharm., 42*, 15 (1988).

53. J. Herman, N. Visavarungroj, and J. P. Remon, *Int. J. Pharm., 55*, 143 (1989).

54. M. Otsuka, N. Kaneniwa, K. Otsuka, K. Kawakami, and O. Umezawa, *Drug Dev. Indust. Pharm., 19*, 541 (1993).

55. M. Otsuka, N. Kaneniwa, K. Kawakami, and O. Umezawa, *J. Pharm. Pharmacol., 43*, 226 (1991).

56. M. Otsuka, N. Kaneniwa, K. Otsuka, K. Kawakami, O. Umezawa, and Y. Matsuda, *J. Pharm. Sci., 81*, 1189 (1992).

57. H. Ando, M. Ishii, M. Kayano, and H. Ozawa, *Drug Dev. Indust. Pharm., 18*, 433 (1992).

58. K. R. Morris, D. E. Bugay, A. W. Newman, S. A. Ranadive, A. K. Singh, M. Szyper, S. A. Varia, H. G. Brittain, and A. T. M. Serajuddin, *Int. J. Pharm., 108*, 195 (1994).

59. H. G. Brittain, S. A. Ranadive, and A. T. M. Serajuddin, *Pharm. Res., 12*, 556 (1995).

60. W. C. McCrone, *Anal. Chem. 21*, 436 (1949).

61. W. C. McCrone, *Fusion Methods in Chemical Microscopy*, Interscience, New York, 1957.

62. L. Kofler, and A. Kofler, *Thermomikromethoden*, Wagner, Innsbruck, 1954.

63. M. Kuhnert-Brandstätter, *Thermomicroscopy in the Analysis of Pharmaceuticals*, Pergamon Press, Oxford, 1971.

64. M. Kuhnert-Brandstätter, "Thermomicroscopy of Organic Compounds," Chapter 2 in *Comprehensive Analytical Chemistry*, Volume XVI (G. Svehla, ed.), Elsevier, Amsterdam, 1982.

65. M. Kuhnert-Brandstätter, and P. Gasser, *Microchem. J., 16*, 419, 577, 590 (1971).

66. D. Giron, *J. Pharm. Biomed. Anal., 4*, 755 (1986).

67. J. A. McCauley, and H. G. Brittain, "Thermal Methods of Analysis," Chapter 8 in *Physical Characterization of Pharmaceutical Solids* (H. G. Brittain, ed.), Marcel Dekker, New York, 1995, pp. 223–251.

68. C. J. Keattch, and D. Dollimore, *Introduction of Thermogravimetry, 2d* ed., Heyden, London, 1975.

69. C. Duval, *Inorganic Thermogravimetric Analysis*, 2d ed., Elsevier, Amsterdam, 1963.

70. H. G. Brittain, *J. Pharm. Biomed. Anal., 15*, 1143 (1997).

71. W. Smykatz-Kloss, *Differential Thermal Analysis*, Springer-Verlag, Berlin, 1974.

72. M. I. Pope, and M. D. Judd, *Differential Thermal Analysis*, Heyden, London, 1977.

73. W. W. Wendlandt, *Thermal Analysis*, 3d ed., Wiley-Interscience, New York, 1986.
74. G. M. Lukaszewski, *Lab. Pract., 15*, 664 (1966).
75. M. J. Richardson, and P. Burrington, *J. Therm. Anal., 6*, 345 (1974).
76. Ph. Van Aerde, J. P. Remon, D. De Rudder, R. Van Severen, and P. Braeckman, *J. Pharm. Pharmacol., 36*, 190 (1984).
77. L. Maury, J. Rambaud, B. Pauvert, Y. Lasserre, G. Bergé, and M. Audran, *J. Pharm. Sci., 74*, 422 (1985).
78. E. G. Salole, and H. Al-Sarraj, *Drug Dev. Indust. Pharm., 11*, 855 (1985).
79. H. Matsuda, S. Kawaguchi, H. Kobayashi, and J. Nishijo, *J. Pharm. Sci., 73*, 173 (1984).
80. J. K. Guillory, *J. Pharm. Sci., 56*, 72 (1967).
81. S. S. Yang, and J. K. Guillory, *J. Pharm. Sci., 61*, 26 (1972).
82. D. Ghiron-Forest, C. Goldbronn, and P. Piechon, *J. Pharm. Biomed. Anal., 7*, 1421 (1989).
83. I. Townsend, *J. Therm. Anal., 37*, 2031 (1991).
84. A. F. Barnes, M. J. Hardy, and T. J. Lever, *J. Therm. Anal., 40*, 499 (1993).
85. J. L. Ford, and P. Timmins, *Pharmaceutical Thermal Analysis*, Ellis Horwood, Chichester, 1989.
86. D. Giron, *Acta Pharm. Jugosl., 40*, 95 (1990).
87. *United States Pharmacopeia* 23, general test <891>, 1995, pp. 1837–1838.
88. W. W. Wendlandt, *Thermal Analysis*, 3d ed., Wiley-Interscience, New York, 1986.
89. D. Dollimore, ''Thermoanalytical Instrumentation,'' Chapter 25 in *Analytical Instrumentation Handbook* (G. W. Ewing, ed.), Marcel Dekker, New York, 1990, pp. 905–960.
90. R. C. Mackenzie, *Pure Appl. Chem., 57*, 1737 (1985).
91. E. L. Charsley, J. A. Rumsey, and S. B. Warrington, *Anal. Proc.*, 5 (1984).
92. R. J. Behme, D. Brooke, R. F. Farney, and T. T. Kensler, *J. Pharm. Sci., 74*, 1041 (1985).
93. Y. Qui, R. D. Schownwald, and J. K. Guillory, *Pharm. Res., 10*, 1507 (1993).
94. W. C. Stagner, and J. K. Guillory, *J. Pharm. Sci., 68*, 1005 (1979).
95. R. J. Behme, and D. Brooke, *J. Pharm. Sci., 80*, 986 (1991).
96. F. Vrecer, S. Srcic, and J. Smid-Korbar, *Int. J. Pharm., 68*, 35 (1991).
97. Y. Chikaraishi, M. Otsuka, and Y. Matsuda, *Chem. Pharm. Bull., 44*, 1614 (1996).

98. A. C. Shad, and N. J. Britten, *J. Pharm. Pharmacol., 39*, 736 (1987).
99. R. O'Laughlin, C. Sachs, H. G. Brittain, E. Cohen, P. Timmins, and S. Varia, *J. Soc. Cosmet. Chem., 40*, 215 (1989).
100. A. Gonthier-Vassal, H. Szwarc, and F. Romain, *Thermochim. Acta, 202*, 87 (1992).
101. S. R. Gunning, M. Freeman, and J. A. Stead, *J. Pharm. Pharmacol., 28*, 758 (1976).
102. E. G. Prout, and F. C. Tompkins, *Trans. Far. Soc., 40*, 488 (1944).
103. Y. Matsuda, E. Tatsumi, E. Chiba, and Y. Miwa, *J. Pharm. Sci., 73*, 1453 (1984).
104. T. Ishida, M. Doi, N. Shimamoto, N. Minamino, K. Nonaka, and M. Inoue, *J. Pharm. Sci., 78*, 274 (1989).
105. M. R. Caira, B. Easter, S. Honiball, A. Horne, and L. R. Nassimbeni, *J. Pharm. Sci., 84*, 1379 (1995).
106. A. V. Katdare, J. A. Ryan, J. F. Bavitz, D. M. Erb, and J. K. Guillory, *Mikrochim. Acta*(Wien), *3*, 1 (1987).
107. L.-C. Chang, M. R. Caira, and J. K. Guillory, *J. Pharm. Sci., 84*, 1169 (1995).
108. J. J. Gerber, J. G. van der Watt, and A. P. Lötter, *Int. J. Pharm., 69*, 265 (1991).
109. Y. Chikaraishi, A. Sano, T. Tsujiyama, M. Otsuka, and Y. Matsuda, *Chem. Pharm. Bull., 42*, 1123 (1994).
110. R. J. Markovich, and C. Pidgeon, *Pharm. Res., 8*, 663 (1991).
111. D. E. Bugay, and A. C. Williams, "Vibrational Spectroscopy," Chapter 3 in *Physical Characterization of Pharmaceutical Solids* (H. G. Brittain, ed.) Marcel Dekker, New York, 1995, pp. 59–91.
112. H. G. Brittain, *J. Pharm. Sci., 86*, 405 (1997).
113. B. R. Jasti, J. Du, and R. C. Vasavada, *Int. J. Pharm., 118*, 161 (1995).
114. T. Uchida, E. Yonemochi, T. Oguchi, K. Terada, K. Yamamoto, and Y. Nakai, *Chem. Pharm. Bull., 41*, 1632 (1993).
115. Y. Hiramatsu, H. Suzuki, A. Kuchiki, H. Nakagawa, and S. Fujii, *J. Pharm. Sci., 85*, 761 (1996).
116. M. M. J. Lowes, M. R. Caira, A. P. Lötter, and J. G. van der Watt, *J. Pharm. Sci., 76*, 744 (1987).
117. T. J. Cholerton, J. H. Hunt, G. Klinkert, and M. Martin-Smith, *J. Chem. Soc. Perkin Trans. 2*, 1761 (1984).
118. A. Kiss, and J. Repasi, *Analyst, 118*, 661 (1993).
119. R. K. Khankari, and D. J. W. Grant, *Thermochim. Acta, 248*, 61 (1995).
120. M. Stoltz, M. R. Caira, A. P. Lötter, and J. G. van der Watt, *J. Pharm. Sci., 78*, 758 (1989).

121. S. A. Botha, and D. R. Flanagan, *Int. J. Pharm., 82*, 195 (1992).
122. K. Sasaki, H. Suzuki, and H. Nakagawa, *Chem. Pharm. Bull., 41*, 325 (1993).
123. H. G. Brittain, D. E. Bugay, S. J. Bogdanowich, and J. DeVincentis, *Drug Dev. Indust. Pharm., 14*, 2029 (1988).
124. S.-Y. Tsai, S.-C. Kuo, and S.-Y. Lin, *J. Pharm. Sci., 82*, 1250 (1993).
125. G. Herzberg, *Infrared and Raman Spectra of Polyatomic Molecules*, Van Nostrand Reinhold, New York, 1945, pp. 334–335.
126. M. S. Bergren, R. S. Chao, P. A. Meulman, R. W. Sarver, M. A. Lyster, J. L. Havens, and M. Hawley, *J. Pharm. Sci., 85*, 834 (1986).
127. P. G. Takla, and C. J. Dakas, *J. Pharm. Pharmacol., 41*, 227 (1989).
128. G. A. Neville, H. D. Beckstead, and H. F. Shurvell, *J. Pharm. Sci., 81*, 1141 (1992).
129. W. Gu, *Anal. Chem, 65*, 827 (1993).
130. L. E. McMahon, P. Timmins, A. C. Williams, and P. York, *J. Pharm. Sci., 85*, 1064 (1996).
131. D. E. Bugay, A. W. Newman, and W. P. Findlay, *J. Pharm. Biomed. Anal., 15*, 49 (1996).
132. K. J. Hartauer, E. S. Miller, and J. K. Guillory, *Int. J. Pharm., 85*, 163 (1992).
133. A. M. Tudor, S. J. Church, P. J. Jendra, M. C. Davies, and C. D. Melia, *Pharm. Res., 10*, 1772 (1993).
134. D. A. Roston, M. C. Walters, R. R. Rhinebarger, and L. J. Ferro, *J. Pharm. Biomed. Anal., 11*, 293 (1993).
135. J. G. Grasselli, M. K. Snavely, and B. J. Bulkin, *Chemical Applications of Raman Spectroscopy*, John Wiley, New York, 1981.
136. A. Grunenberg, B. Keil, and J.-O. Henck, *Int. J. Pharm., 118*, 11 (1995).
137. X. J. Gu, and W. Jiang, *J. Pharm. Sci., 84*, 1438 (1995).
138. K. Raghavan, A. Dwivedi, G. C. Campbell, E. Johnston, D. Levorse, J. McCauley, and M. Hussain, *Pharm. Res., 10*, 900 (1993).
139. A. Terol, G. Cassanas, J. Nurit, B. Pauvert, A. Bouassab, J. Rambaud, and P. Chevallet, *J. Pharm. Sci., 83*, 1437 (1994).
140. K. Raghavan, A. Dwivedi, G. C. Campbell, G. Nemeth, and M. Hussain, *J. Pharm. Biomed. Anal., 12*, 777 (1994).
141. J. C. Bellows, F. P. Chem, and P. N. Prasad, *Drug Dev. Indust. Pharm., 3*, 451 (1977).
142. B. A. Bolton, and P. N. Prasad, *J. Pharm. Sci., 70*, 789 (1981).
143. C. S. Randall, B. K. DiNenno, R. K. Schultz, L. Dayter M. Konieczny, and S. L. Wunder, *Int. J. Pharm., 120*, 235 (1995).

144. C. A. Fyfe, *Solid State NMR for Chemists*, C.F.C. Press, Guelph, 1983.
145. D. E. Bugay, *Pharm. Res., 10*, 317 (1993).
146. D. E. Bugay, "Magnetic Resonance Spectrometry," Chapter 4 in *Physical Characterization of Pharmaceutical Solids* (H. G. Brittain, ed), Marcel Dekker, New York, 1995, pp. 93–125.
147. J. Anwar, S. E. Tarling, and P. Barnes, *J. Pharm. Sci., 78*, 337 (1989).
148. J. J. Gerber, J. G. van der Watt, and A. P. Lötter, *Int. J. Pharm., 73*, 137 (1991).
149. S.-Y. Lin, *J. Pharm. Sci., 81*, 572 (1992).
150. R. K. Harris, B. J. Say, R. Y. Yeung, R. A. Fletton, and R. W. Lancaster, *Spectrochim. Acta, 45A*, 465 (1989).
151. G. A. Stephenson, J. G. Stowell, P. H. Toma, D. E. Dorman, J. R. Greene, and S. R. Byrn, *J. Am. Chem. Soc, 116*, 5766 (1994).
152. S. R. Byrn, G. Gray, R. R. Pfeiffer, and J. Frye, *J. Pharm. Sci., 74*, 565 (1985).
153. M. Stoltz, D. W. Oliver, P. L. Wessels, and A. A. Chalmers, *J. Pharm. Sci., 80*, 357 (1991).
154. H. G. Brittain, K. R. Morris, D. E. Bugay, A. B. Thakur, and A. T. M. Serajuddin, *J. Pharm. Biomed. Anal., 11*, 1063 (1993).
155. D. Casarini, R. K. Harris, and A. M. Kenwright, *Mag. Reson. Chem., 31*, 540 (1993).
156. R. Suryanarayanan, and T. S. Wiedmann, *Pharm. Res., 7*, 184 (1990).
157. P. Gao, *Pharm. Res., 13*, 1095 (1996).
158. C. Doherty, and P. York, *Int. J. Pharm., 47*, 141 (1988).
159. A. Terol, G. Cassanas, J. Nurit, B. Pauvert, A. Bouassab, J. Rambaud, and P. Chevallet, *J. Pharm. Sci., 83*, 1437 (1994).
160. K. Raghavan, A. Dwivedi, G. C. Campbell, G. Nemeth, and M. Hussain, *J. Pharm. Biomed. Anal., 12*, 777 (1994).

7

Effects of Polymorphism and Solid-State Solvation on Solubility and Dissolution Rate

Harry G. Brittain

Discovery Laboratories, Inc.
Milford, New Jersey

David J. W. Grant

College of Pharmacy
University of Minnesota
Minneapolis, Minnesota

I. INTRODUCTION

Earlier chapters have amply demonstrated that the different lattice energies (and entropies) associated with different polymorphs or solvates give rise to measurable differences in the physical properties (density, color, hardness, refractive index, conductivity, melting point, enthalpy of fusion, vapor pressure, etc.—see Chapter 1, Table 3). Even the explosive power of cyclotetramethylene-tetranitramine depends on which of its four polymorphs is being used [1]. We have seen in Chapter 1 that the different lattice energies of polymorphs or solvates give rise to different solubilities and dissolution rates. If the solubilities of the various solid forms are sufficiently different, they can be very important during the processing of drug substances into drug products [2] and may have implications for the adsorption of the active drug from its dosage form [3]. These concerns have led to an increased regulatory interest in the solid-state properties and behavior of drug substances and in their characterization [4].

That the crystal structure can have a direct effect on the solubility of a solid can be understood using a simple model. For a solid to dissolve, the forces of attraction between solute and solvent molecules must overcome the attractive forces holding together the solid and the liquid solvent. In other words, for the process to proceed spontaneously, the solvation free energy released upon dissolution must exceed the sum of the lattice free energy of the solid plus the free energy of cavity formation in the solvent. The balance of the attractive and disruptive forces will determine the equilibrium solubility of the solid in question (which is an exponential function of the free energy change of the system—see Chapter 1, Equation 15). The enthalpy change and the increase in disorder of the system (i.e., the entropy change) determine the Gibbs free energy change. Since different lattice energies (and enthalpies) characterize different crystal structures (as discussed in Chapter 1), the solubility of different crystal polymorphs (or solvate species) must differ as well. Finally, the act of dissolution may be endothermic or exothermic in nature, so that measurements of solution calorimetry can be used to provide important information about the substance under study. The most common solvent media used in the characterization of polymorphs or solvates are liquids or liquid mixtures that give rise to liquid solutions of the solute [5], and which constitute the focus of this chapter.

As explained in Chapter 1, the solubility differences between polymorphs or solvates enable a less stable form to convert to the most stable form. When such a conversion can take place, the measured solubility of each form will approach a common value, namely that of the most stable form at the temperature of measurement.

The effect of polymorphism on solubility becomes especially critical because the rate of compound dissolution must also be dictated by the balance of attractive and disruptive forces existing at the crystal–solvent interface. A solid having a higher lattice free energy (i.e., a less stable polymorph) will tend to dissolve faster, because the release of a higher amount of stored lattice free energy will increase the solubility and hence the driving force for dissolution. At the same time, each species would liberate (or consume) the same amount of solvation energy, because all dissolved species (of the same chemical identity) must be thermodynamically equivalent. The varying dissolution rates possible for different structures of the same drug entity can in turn lead to varying degrees of bioavailability for different polymorphs or solvates.

To achieve bioequivalence for a given drug compound usually requires equivalent crystal structures in the drug substance, although exceptions are known to exist.

The dissolution rate and solubility in a solvent medium are two of the most important characteristics of a drug substance, because these quantities determine the bioavailability of the drug for its intended therapeutic use. Solubility is defined as the equilibrium concentration of the dissolved solid (the solute) in the solvent medium and is ordinarily a function of temperature and pressure.

II. EQUILIBRIUM SOLUBILITY

The capacity of any system to form solutions has limits imposed by the phase rule of Gibbs:

$$F + P = C + 2 \tag{1}$$

where F is the number of degrees of freedom in a system consisting of C components with P phases. For a system of two components and two phases (e.g., solid and liquid) under the pressure of their own vapor and at constant temperature, F equals zero. If one of the phases consists solely of one component (a pure substance), the equilibrium solubility at constant temperature and pressure is a fixed quantity that is given as the amount of solute contained in the saturated solution in a unit amount of the solvent or solution.

For any case in which F is zero, a definite reproducible solubility equilibrium can be reached. Complete representation of the solubility relations is accomplished in the phase diagram, which gives the number, composition, and relative amounts of each phase present at any temperature in a sample containing the components in any specified proportion. Solubilities may therefore be expressed in any appropriate units of concentration, such as the quality of the solute dissolved (defined mass, number of moles) divided by the quantity either of the solvent (defined mass, volume, or number of moles) or of the solution (defined mass, volume, or number of moles). Jacques et al. have provided a compilation of the expressions for concentration and solubility [6].

A. Determination of Equilibrium Solubility

Methods for the determination of solubility have been thoroughly reviewed [5,7,8], especially with respect to the characterization of pharmaceutical solids [9]. Solubility is normally highly dependent on temperature, so the temperature must be recorded for each solubility measurement in addition to the precise nature of the solvent and the solid phase at equilibrium. Plots of solubility against temperature are commonly used for characterizing pharmaceutical solids and have been extensively discussed [5,10]. Frequently (especially over a relatively narrow temperature range), a linear relationship can be given either by a van't Hoff plot:

$$\ln X_2^{sat} = \frac{-a}{RT} + c' \tag{2}$$

or by a Hildebrand plot:

$$\ln X_2^{sat} = \frac{b}{R} \ln T + c'' \tag{3}$$

In Eqs. (2) and (3), X_2^{sat} is the mole fraction solubility of the solid solute at an absolute temperature T, a is the apparent molar enthalpy of solution, b is the apparent molar entropy of solution, and c' and c'' are constants. The combined equation, attributed to Valentiner, has been used by Grant et al. [10] in the form

$$\ln X_2^{sat} = \frac{-a}{RT} + \frac{b}{R} \ln T + c''' \tag{4}$$

This three-parameter equation enables solubility to be simulated and correlated quite accurately over a wide temperature range (e.g., 60 K).

As implied in the previous paragraph, the validity of the aforementioned equations requires that each crystal phase be stable, as indicated by the absence of any phase conversion during the determination of equilibrium solubility. Systems characterized by the presence of metastable phases constitute special cases that have been discussed in Chapters 1 and 2.

Two general methods, the analytical method and the synthetic method [9], are available for determining solubility. In the analytical method, the temperature of equilibration is fixed, while the concentration of the solute in a saturated solution is determined at equilibrium by a suitable analytical procedure. The analytical method can be either the traditional, common batch agitation method, or the more recent flow column method. In the synthetic method, the composition of the solute–solvent system is fixed by appropriate addition and mixing of the solute and solvent, and then the temperature at which the solid solute just dissolves or just crystallizes is carefully bracketed.

It is usually not difficult to determine the solubility of solids that are moderately soluble (greater than 1 mg/mL), but the direct determination of solubilities much less than 1 mg/mL is not straightforward. Problems such as slow equilibrium resulting from a low rate of dissolution, the influence of impurities, and the apparent heterogeneity in the energy content of the crystalline solid [11] can lead to large discrepancies in reported values. For example, reported values of the aqueous solubility of cholesterol range from 0.066 to 3000 mg/L at 30°C [12].

B. Studies of the Equilibrium Solubility of Polymorphic and Solvate Systems

The thermodynamic relationships examined in Chapter 1 involving polymorphism and solubility have been applied to the methylprednisolone system [13]. The solubilities of the two polymorphs of this steroid were determined at various temperatures in water, decyl alcohol, and dodecyl alcohol. Because the chemical potential and thermodynamic activity of the drug in the solid state and in each saturated solution are constant, the solubility ratios for the two forms (which can be found in Table 1) were found to be independent of the solvent. The enthalpy, entropy, and temperature of transition calculated from the data were 1600 cal/mol, 4.1 cal/K · mol, and 118°C, respectively.

Solubility determinations were used to characterize the polymorphism of 3-(((3-(2-(7-chloro-2-quinolinyl)-(E)-ethenyl)phenyl)-((3-dimethylamino-3-oxopropyl)thio)methyl)thio)propanoic acid [14]. The solubility of Form II was found to be higher than that of Form I in both isopropyl alcohol (IPA, solubility ratio approximately 1.7 from 5 to 55°C) and in methyl ethyl ketone (MEK, solubility ratio approximately 1.9 from 5 to 55°C), indicating that Form I is the thermodynamically stable form in the range of 5 to 55°C. An analysis of the entropy contributions to the free energy of solution from the solubility results

Table 1 Equilibrium Solubility and Solubility Ratios of
the Polymorphs of Methylprednisolone in Different
Solvent Systems

Temperature (°C)	Solubility in water (mg/mL)		
	Form I	Form II	Solubility ratio, II/I
30	0.09	0.15	1.67
39	0.12	0.20	1.67
49	0.16	0.26	1.63
60	0.21	0.33	1.57
72	0.30	0.43	1.43
84	0.43	0.55	1.28
Temperature (°C)	Solubility in decyl alcohol (mg/mL)		
	Form I	Form II	Solubility ratio, II/I
30	2.9	4.8	1.66
39	3.5	5.7	1.63
49	4.3	6.9	1.60
60	5.5	8.6	1.56
72	8.3	11.9	1.43

Source: Ref. 13.

implied that the saturated IPA solutions were more disordered than
were the corresponding MEK solutions, in turn indicating the existence
of stronger solute–solvent interactions in the MEK solution. This finding corroborated results determined for the enthalpy with respect to the
deviations of the saturated solutions from ideality.

Phenylbutazone has been found capable of existing in five different polymorphic structures, characterized by different X-ray powder
diffraction patterns and melting points [15]. The equilibrium solubilities of all five polymorphs in three different solvent systems are summarized in Table 2. Form A exhibits the highest melting point (suggesting
the least energetic structure at the elevated temperature), while its solubility is the lowest in each of the three solvent systems studied (actually
demonstrating the lowest free energy). These findings indicate that
Form A is the thermodynamically most stable polymorph both at room

Table 2 Equilibrium Solubility of Phenylbutazone Polymorphs at Ambient Temperature in Different Solvent Systems

	Solubility (mg/mL)				
Solvent system	A	B	D	E	C
pH 7.5 phosphate buffer	4.80	5.10	5.15	5.35	5.9
Above buffer with 0.05% Tween 80	4.50	4.85	4.95	5.10	5.52
First buffer with 2.25% PEG 300	3.52	5.77	5.85	6.15	6.72

Note: The polymorphs are listed in order of increasing free energy at ambient temperature.
Source: Ref. 15.

temperature and at the melting point (105°C). However, identification of the sequence of stability for the other forms at any particular temperature was not straightforward. Following one common convention, the polymorphs were numbered in the order of decreasing melting points, but the solubility data of Table 7 did not follow this order. This finding implies that the order of stability at room temperature is not the same as that at 100°C and emphasizes that only measurements of solubility can predict the stability order at room temperature. If different polymorphs are not discovered in the same study, they are ordinarily numbered according to the order of discovery to avoid renumbering those discovered earlier. Ostwald's rule of stages, discussed in Chapter 1, explains why metastable forms are often discovered first.

Gepirone hydrochloride was found to exist in at least three polymorphic forms, whose melting points were 180°C (Form I), 212°C (Form II), and 200°C (Form III) [16]. Forms I and II, and Forms I and III, were deduced to be enantiotropic pairs in the sense that their G vs. T curves crossed. Form III was found to be monotropic with respect to Form II, since the G vs. T curves did not cross below their melting points, and since there was no temperature at which Form III was the most stable polymorph. The solubility data illustrated in Fig. 1 were used to estimate a transition temperature of 74°C for the enantiotropic Forms I and II, while the reported enthalpy difference was 4.5 kcal/mol at 74°C and 2.54 kcal/mol at 25°C. The most stable polymorph below 74°C was Form I, whereas Form II was the most stable above 74°C.

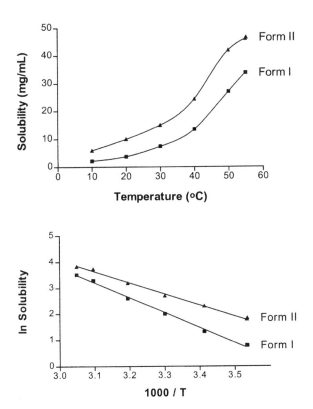

Fig. 1. Temperature dependence of the equilibrium solubilities of two poly-morphic forms of gepirone hydrochloride. (The plots were adapted from data originally presented in Ref. 16.)

The effect of solvent composition on the solubility of polymorphs was investigated with cimetidine [17]. Both forms exhibited almost identical melting points, but Form B was found to be less soluble than Form A, identifying it as the most stable polymorph at room tempera-ture. The two forms were more soluble in mixed water–isopropanol solvents than in either of the pure solvents, reflecting the balance be-tween the solvation of the molecules by water and isopropanol in de-termining the activity coefficient of the solute and hence the solubility. At constant temperature, the difference in the Gibbs free energy and the solubility ratio were constant, independent of the solvent system.

The equilibrium solubilities of two polymorphs of an experimental antiviral compound were used to verify the results of solubility ratio predictions made on the basis of melting point and heat of fusion data [18]. Even though the solubilities of Forms I and III were almost equal in three different solvent systems, the theoretically calculated solubility ratio agreed excellently with the experimentally derived ratios in all of the solvent systems studied. The highest melting form (Form I) was found to be more soluble at room temperature, indicating that an enantiotropic relationship existed between Forms I and III.

It is well established that the temperature range of thermodynamic stability (and certain other quantities) can be determined from measurements of the equilibrium solubilities of the individual polymorphs [19]. In one such study, the two polymorphic forms of 2-[[4-[[2-(1H-tetrazol-5-ylmethyl)phenyl]methoxy]phenoxy]methyl]quinoline were found to exhibit an enantiotropic relationship, because their G vs. T curves intersected with Form I melting at a lower temperature than did Form II [20]. Form I was determined to be the more thermodynamically stable form at room temperature, although the solubility of the two forms was fairly similar. The temperature dependence of the solubility ratio of the two polymorphs afforded the enthalpy of transition (Form II to Form I) as $+0.9$ kcal/mol, while the free energy change of this transition was -0.15 kcal/mol.

When the hydrates or solvates of a given compound are stable with respect to phase conversion in a solvent, the equilibrium solubility of these species can be used to characterize these systems. For instance, the equilibrium solubility of the trihydrate phase of ampicillin at 50°C is approximately 1.3 times that of the more stable anhydrate phase at room temperature [21]. However, below the transition temperature of 42°C, the anhydrate phase is more soluble and is therefore less stable. These relationships are illustrated in Fig. 2.

Amiloride hydrochloride can be obtained in two polymorphic dihydrate forms, A and B [22]. However, each solvate dehydrates around 115–120°C, and the resulting anhydrous solids melt at the same temperature. However, form B was found to be slightly less soluble than form A between 5 and 45°C, indicating that it is the thermodynamically stable form at room temperature. The temperature dependencies of the solubility data were processed by the van't Hoff equation to yield the apparent enthalpies of solution of the two polymorphic dihydrates.

The solubility of polymorphic solids derived from the anhydrate

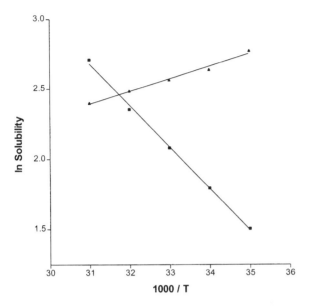

Fig. 2. The van't Hoff plot for the anhydrate (-▲-) and trihydrate (-■-) phases of ampicillin in water. (The relations were adapted from data originally presented in Ref. 21.)

and monohydrate phases of tranilast crystals were evaluated, as were materials processed from them to enhance in vitro availability and micromeritic properties [23]. Agglomerates of monohydrate phases I, II, or III were produced using different crystallization solvents and procedures. Monohydrate Form I transformed directly to the stable α-form upon dehydration, while Forms II and III dehydrated to the amorphous and β phases, respectively. The apparent equilibrium solubilities of monohydrate Form II and the amorphous form were much higher than those of the α and β forms due to their high surface energies. The solubilities of tranilast hydrate phases exceeded those of the anhydrate phases, which runs counter to the commonly observed trend and suggests that the anhydrate/hydrate transition temperatures are below the temperature of measurement. An analogous situation applies to the anhydrate and trihydrate phases of ampicillin [21] discussed above. The trihydrate phase is more soluble than the anhydrate phase at 50°C, because the transition temperature (42°C) is lower.

C. Metastable Solubility

Because only one member of a family of polymorphs or solvates can be the most thermodynamically stable form under a given set of environmental conditions defined by the phase rule, one frequently finds that one form spontaneously converts to another form during the time required to establish an equilibrium solubility. The existence of an unexpected metastable solubility can lead to important (and possible undesirable) consequences. For digoxin, unexpectedly high solubility values and abnormally high dissolution rates have resulted in the overdosing of patients before the phenomenon of its solid-state phase conversion was properly understood and controlled [24]. This phenomenon was caused by higher-energy crystals resulting from a greater density of crystal defects rather than by polymorphism.

Any metastable phase will have a higher free energy than would a thermodynamically more stable phase, and it will undergo a phase transformation to the more stable phase once the activation energy barrier is overcome. Often the barrier to phase transformation is merely the improbability of a suitable nucleation step. Hence, only fortuitously unfavorable kinetics permitted a determination of the equilibrium solubility of the various higher-energy phases discussed in the preceding section.

Conversions of a metastable phase into a more stable phase may include the transformation of one polymorphic phase into another, the solvation of an anhydrous phase, the desolvation of a solvate phase, the transformation of an amorphous phase into a crystalline anhydrate or solvate phase, the degradation of a crystalline anhydrate or solvate phase to an amorphous phase, or in the case of digoxin, the conversion of imperfect (less crystalline, more amorphous) crystals with a high density of defects into more perfect (more crystalline) crystals with a lower density of defects. While it is straightforward to determine the equilibrium solubility of a phase that is stable with respect to conversion, the measurement of solubilities of metastable phases that are susceptible to conversion is not a trivial matter.

Because determinations of the solubility of solid materials are often made by suspending an excess of the compound in question in the chosen solvent or other dissolution medium, the application of this equilibrium method to a metastable phase will result in a determination of the solubility of the stable phase. One of the attempts to measure the solubility of a metastable polymorph was made by Milosovich, who developed a method based on the measurement of the intrinsic dissolution rates (IDR) and used it to deduce the relative solubilities of sulfathiazole Forms I and II [25]. This method assumes that the IDR is proportional to the solubility, the proportionality constant being the transport rate constant, which is constant under constant hydrodynamic conditions in a transport-controlled dissolution process.

Ghosh and Grant have developed an extrapolation technique to determine the solubility of a crystalline solid that undergoes a phase change upon contact with a solvent medium [26]. They proposed a thermodynamic cycle analogous to Hess's law, but based on free energies, and used this cycle to predict the theoretical solubility of solvates in water. In the model systems to which the technique was applied, good agreement was obtained between the solubility values measured by equilibration (and derived from an extrapolation method in a mixed solvent system) and those derived from the extrapolation method and calculated by means of the thermodynamic cycle.

A light scattering method has recently been described for the determination of the solubility of drugs, and its application to solubility evaluations of metastable phases has also been demonstrated [27]. Using this technique to deduce solubility data for theophylline anhydrate (metastable with respect to the monohydrate phase in bulk water at room temperature), agreement with the most reliable literature data was excellent. The light scattering method appears to be most useful in the determination of solubility data for metastable crystal phases in dissolution media in which they spontaneously and rapidly convert into a more stable crystal phase.

D. Studies of the Solubility of Metastable Polymorphic and Solvate Phases

Aqueous suspensions of tolbutamide were reported to thicken to an unpourable state after several weeks of occasional shaking, while samples of the same suspensions that were not shaken showed excellent stability after years of storage at ambient and elevated temperature [28].

Examination by microscopy revealed that the thickening was due to partial conversion of the original platelike tolbutamide crystals to very fine needle-shaped crystals. The new crystals were identified as a different polymorphic form and did not correspond either to a solvate species or to crystals of a different habit. The crystalline conversion was observed to take place in a variety of solvents, the rate of conversion being faster in solvents where the drug exhibited appreciable solubility. Because the conversion rate in 1-octanol was relatively slow, use of this solvent permitted an accurate solubility ratio of 1.22 to be obtained (Form I being more soluble than Form III).

The polymorphism and phase interconversion of sulfamethoxydiazine (sulfameter) have been studied in detail [29]. This compound can be obtained in three distinct crystalline polymorphs, with the metastable Form II being suggested for use in solid dosage forms on the basis of its greater solubility and bioavailability [30]. However, the formulation of Form II in aqueous suspensions was judged inappropriate because of the fairly rapid rate of transformation to Form III. This behavior is illustrated in Fig. 3, which shows that seeding of a Form II suspension with Form III crystals greatly accelerates the phase conversion. It was subsequently learned that phase conversion could be retarded by prior addition of various formulation additives, possibly permitting the development of a suspension containing the metastable Form II [31]. Although there are many examples of the conversion of a metastable polymorph to a stable polymorph during the dissolution process, some of them seminal [32], the use of tailor-made additives to inhibit the crystallization of a more stable polymorph is relatively recent [33].

Carbamazepine is known to exist in both an anhydrate and a dihydrate form, with the anhydrate spontaneously transforming to the dihydrate upon contact with bulk liquid water [34]. The anhydrous phase is reported to be practically insoluble in water, but this observation is difficult to confirm owing to its rapid transition to the dihydrate phase. The rates associated with the phase transformation process have been studied and appear to follow first-order kinetics [35]. Interestingly, the only difference in pharmacokinetics between the two forms was a slightly higher absorption rate for the dihydrate [36]. The slower absorption of anhydrous carbamazepine was attributed to the rapid trans-

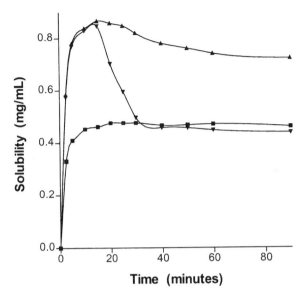

Fig. 3. Effect on the solubility of sulfamethoxydiazine Form II by seeding with crystals of Form III. Shown are the dissolution profiles of Form II (-▲-), Form III (-■-), and Form II seeded with Form III after 20 minutes elapsed time (-▼-). (The plots were adapted from data originally presented in Ref. 31.)

formation to the dihydrate, accompanied by a fast growth in particle size. Comparison of the bioavailabilities of different polymorphs of a given drug suggest that significant differences are found only when the polymorphs differ significantly in Gibbs free energy deduced from the ratio of solubilities or intrinsic dissolution rates, as in the case of chloramphenicol palmitate.

A monohydrate phase of metronidazole benzoate exhibited solubility properties different from those of the commercially available anhydrous form [37]. The monohydrate was found to be the thermodynamically stable form in water below 38°C. The enthalpy and entropy changes of transition for the conversion of the anhydrate to the monohydrate were determined to be −1200 cal/mol and −3.7 cal/K · mol, respectively. This transition was accompanied by a drastic increase in particle size and caused physical instability of oral suspension formulations. These findings were taken to imply that any difference in bio-availability between the two forms could be attributed to changes in particle size distribution and not to an inherent difference in the in vivo activity at body temperature.

Recognizing that the hydration state of a hydrate depends on the water activity in the crystallization medium, Zhu and Grant investigated the influence of solution media on the physical stability of the anhydrate, trihydrate, and amorphous forms of ampicillin [38]. The crystalline anhydrate was found to be kinetically stable in the sense that no change was detected by powder x-ray diffraction for at least 5 days in methanol+water solutions over the whole range of water activity ($a_w = 0$ for pure methanol to $a_w = 1$ for pure liquid water). However, addition of trihydrate seeds to ampicillin anhydrate suspended in methanol+water solutions at $a_w \geq 0.381$ resulted in the conversion of the anhydrate to the thermodynamically stable trihydrate. The trihydrate converted to the amorphous form at $a_w \leq 0.338$ in the absence of anhydrate seeds, but it converted to the anhydrate phase at $a_w \leq 0.338$ when the suspension was seeded with the anhydrate. These trends are illustrated in Fig. 4. The metastable amorphous form took up water progressively with increasing a_w from 0 to 0.338 in the methanol+water mixtures. The most significant finding of this work was that water activity was the major thermodynamic factor determining the nature of the solid phase of ampicillin that crystallized from methanol+water mixtures.

Perhaps the most studied example of phase conversion in the presence of water concerns the anhydrate-to-monohydrate transition of theophylline. It had been noted in a very early work that the anhydrous phase would convert to the monohydrate phase within seconds of exposure of the former to bulk water [39]. The conversion to the monohydrate phase was also demonstrated to take place during wet granulation [40] and could even occur in processed tablets stored under elevated humidity conditions [41]. The difficulty in determining the equilibrium solubility of theophylline anhydrate is evident in the literature, which reports a wide range of values [42–44]. Better success has been obtained in mixed solvent systems, as in the work of Zhu and Grant [45], analogous to the experiments with ampicillin [39]. However, the data obtained in water-rich solutions was distorted by the formation of the monohydrate phase [46,47].

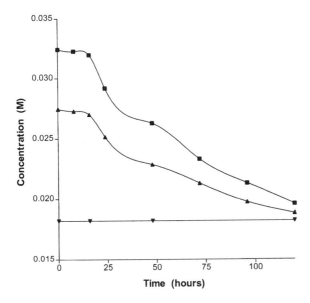

Fig. 4. Conversion of ampicillin anhydrate to the trihydrate phase at various water activities after seeding with the trihydrate. Shown are the concentration-time data at $a_w = 1.0$ (-■-), $a_w = 0.862$ (-▲-), and $a_w = 0.338$ (-▼-). (The curves were adapted from data originally presented in Ref. 38.)

III. SOLUTION CALORIMETRY

The practice of thermochemistry involves measurement of the heat absorbed or evolved during a chemical reaction or physical process. Such a measurement determines the amount of heat q according to the first law of thermodynamics:

$$\Delta E = q + w \tag{5}$$

where w is the work done on the system (negative for work done by the system), and ΔE is the corresponding change in internal energy of the system. Under conditions of constant volume, $\Delta V = 0$, so no mechanical work is done, $w = 0$, and $q = \Delta E$ is the heat of the reaction (or process) at constant volume. Calorimeters, such as bomb calorimeters, that operate under constant volume conditions are not commonly used for studies of polymorphs or solvates. The usual practice for solution calorimetry is to conduct the studies at constant pressure, p, so that the only work done on or by the system is that due to the change in volume, ΔV. The heat of the reaction (or process) at constant pressure is the enthalpy change, ΔH, which is positive for heat absorbed (an endothermic change) and negative for heat evolved (an exothermic change). The following equation shows the relationship between the heat of reaction at constant pressure, ΔH, and the heat of reaction at constant volume, ΔE.

$$\Delta H = \Delta E + p \cdot \Delta V \tag{6}$$

The principal requirement for calorimetry is that the measured heat change must be assignable to a definite process, such as the dissolution of a solute in a solvent medium.

A. Enthalpies of Solution

When a solute is dissolved in a solvent to form a solution, there is almost always absorption or evolution of heat. According to the principle of Le Chatelier, substances that absorb heat as they dissolve must show an increase in solubility with an increase in temperature. Those which evolve heat upon dissolution must become less soluble at higher temperatures.

The heat change per mole of solute dissolved varies with the concentration c of the solution that is formed. It is useful to plot the total enthalpy change ΔH at constant temperature against the final molar concentration. This type of curve increases rapidly at low solute concentrations but levels off at the point when the solution is saturated at the temperature of the experiment. The magnitude of the enthalpy change at a given concentration of solute divided by the corresponding number of moles of that solute dissolved represents the increase in enthalpy per mole of solute when it dissolves to form a solution of a particular concentration. This quantity is called the *molar integral heat of solution* at the given concentration. The integral heat of solution is approximately constant in dilute solution but decreases as the final dissolved solute concentration increases.

For hydrated salts and salts that do not form stable hydrates, the integral heat of solution is ordinarily positive, meaning that heat is absorbed when these substances dissolve. When the anhydrous form of a salt capable of existing in a hydrated form dissolves, there is usually a liberation of heat energy. This difference in behavior between hydrated and anhydrous forms of a given salt is attributed to the usual negative change in enthalpy (evolution of heat) associated with the hydration reaction.

Because the heat of solution of a solute varies with its final concentration, there must be a change of enthalpy when a solution is diluted by the addition of solvent. The *molar integral heat of dilution* is the change in enthalpy resulting when a solution containing one mole of a solute is diluted from one concentration to another. According to Hess's law, this change in enthalpy is equal to the difference between the integral heats of solution at the two concentrations.

The increase of enthalpy that takes place when one mole of solute is dissolved in a sufficiently large volume of solution (which has a particular composition), such that there is no appreciable change in the concentration, is the *molar differential heat of solution*. When stating a value for this quantity, the specified concentration and temperature must also be quoted. Because the differential heat of solution is almost constant in very dilute solutions, the molar differential and integral heats of solution are equal at infinite dilution. At higher concentrations, the differential heat of solution generally decreases as the concentration increases.

The *molar differential heat of dilution* can be defined as the heat change when one mole of solvent is added to a large volume of the solution at the specified concentration. The difference between the integral heats of solution at two different concentrations corresponds to the heat of dilution between these two concentrations. The heat of dilution at a specified concentration is normally obtained by plotting the molar heat of solution at various concentrations against the number of moles of solvent associated with a definite quantity of solute and finding the slope of the curve at the point corresponding to that particular concentration. Because of the approximate constancy of the molar integral heat of solution at small concentrations, such a curve flattens out at high dilutions, and the differential heat of dilution then approaches zero.

The molar differential heats of solution and dilution are examples of partial molar quantities, which are of such importance that they must be used whenever systems of variable composition, such as solutions, are involved.

B. Principles Underlying Partial Molar Quantities

A solution is deduced to be ideal if the chemical potential μ_i of every component is a linear function of the logarithm of its mole fraction X_i according to the relation

$$\mu_i = \mu_i^* + RT \ln X_i \tag{7}$$

where μ_i^* is the (hypothetical or actual) value of μ_i when X_i equals unity and is a function of temperature and pressure. A solution is termed ideal only if Eq. (7) applies to every component in a given range of composition (usually corresponding to dilute solutions), but it is not necessary that the relation apply to the whole range of composition. Any solution that is approximately ideal over the entire composition range is termed a perfect solution, although relatively few such solutions are known. However, because a given solution may approach ideality over a limited composition range, it is worthwhile to develop the equations further.

   When substance i is present both as a pure solid and as a component of an ideal solution, the condition of equilibrium may be stated as

$$\mu_i^s = \mu_i^* + RT \ln X_i \tag{8}$$

where μ_i^s is the chemical potential of the pure solid and X_i is the mole fraction in the solution. Rearranging, one finds

$$\ln X_i = \frac{\mu_i^s}{RT} - \frac{\mu_i^*}{RT} \tag{9}$$

According to the phase rule, this two-component, two-phase system is characterized by two degrees of freedom. One concludes that both the temperature and the pressure of the solution can be varied independently. Since the pressure on the system is normally held fixed as that of the atmosphere during solubility studies, differentiation of Eq. (9) yields

$$\left(\frac{\delta \ln X_i}{\delta T} \right)_p = \frac{H_i - H_i^s}{RT^2} \tag{10}$$

H_i is the partial molar enthalpy of the component in the ideal solution, and H_i^s is its enthalpy per mole as the pure solid. The equation can therefore be rewritten as

$$\left(\frac{\delta \ln X_i}{\delta T} \right)_p = \frac{\Delta H_i}{RT^2} \tag{11}$$

where ΔH_i is the heat absorbed (at constant temperature and pressure) when one mole of the component dissolves in the ideal solution. As stated above, this quantity is the differential heat of solution and is given by

$$\Delta H_i = H_i - H_i^s \tag{12}$$

Provided that the solution remains ideal up to $X_i = 1$, and because H_i is independent of composition in the region of ideality, H_i is the same as the enthalpy per mole of the pure liquid component. ΔH_i is equal to its molar heat of fusion, which was formerly termed the molar latent heat of fusion. It is noted, however, that these quantities refer to the temperature at which the solution having mole fraction X_i is in equilibrium with the pure solid.

If one now assumes that ΔH_i is independent of temperature over a narrow temperature range, then Eq. (11) can be integrated at constant pressure to yield

$$\ln \frac{X_1}{X_2} = \frac{\Delta H_i}{R} \left(\frac{1}{T_2} - \frac{1}{T_1} \right) \tag{13}$$

where X_1 and X_2 refer to the solubilities (expressed as mole fractions) of the solute at temperatures T_1 and T_2, respectively. If Eq. (13) remains approximately valid up to a mole fraction of unity, this situation corresponds to that where pure liquid solute is in equilibrium with its own solid at the melting point. In that case, Eq. (13) yields

$$\ln X = \frac{\Delta H_i}{R} \left(\frac{1}{T_m} - \frac{1}{T} \right)$$ (14)

where X is the solubility at temperature T, and T_m is the melting point of the solute. ΔH_i is the heat of solution but, by the nature of the assumptions that have been made, it is also equal to the latent heat of fusion (ΔH^f) of the pure solute.

Because the number of energy levels available to take up thermal energy is greater in the liquid state than in the solid state, the heat capacity of a liquid frequently exceeds that of the same substance in the solid state. As a result, ΔH^f must be assumed to be a function of temperature. If one assumes the change in heat capacity to be constant over the temperature range of interest, then one can use the relation

$$\Delta H_i = \Delta H_R + \Delta C_p (T_i - T_R)$$ (15)

where ΔH_R is the heat of fusion at some reference temperature T_R and ΔC_p is the difference in heat capacity between the liquid and the solid state. This situation has been treated by Grant and coworkers [10], who have provided the highly useful Eq. (4) for the treatment of solubility data over a wide range of temperature values. Equation (4) may be derived by substituting the expression for ΔH_i of Eq. (15) into the differential form of the van't Hoff Eq. (11) and integrating. In Eq. (4), a is equal to ΔH_R when T_R equals 0 K, while b is equal to ΔC_p.

The determination of solubility data over a defined temperature range can therefore be used to calculate the differential heat of solution of a given material. For instance, the data illustrated in the bottom half of Fig. 1 indicate that Eq. (14) can be used to deduce a value for the molar differential heat of solution of gepirone. In addition, the fact that forms I and II yield lines of different slopes indicates the existence of unique values of the molar differential heat of solution for the two polymorphs. One can subtract the differential heats of solution obtained for the two polymorphs to deduce the heat of transition ΔH_T between the two forms:

$$\Delta H_t = \Delta H_s^B - \Delta H_s^A$$ (16)

where ΔH_s^A and ΔH_s^B denote the differential heats of solution for polymorphs A and B, respectively. For gepirone, A $\equiv$ I, B $\equiv$ II.

The validity of the assumption regarding constancy in the heats of solution for a given substance with respect to temperature can be made by determining the enthalpy of fusion ΔH_f for the two forms and then taking the difference between these:

$$\Delta H_t' = \Delta H_f^B - \Delta H_f^A \qquad (17)$$

where $\Delta H_t'$ represents the heat of transition between forms A and B at the melting point. The extent of agreement between ΔH_t and $\Delta H_t'$ can be used to estimate the validity of the assumptions made.

For example, the heats of fusion and solution have been reported for the polymorphs of auranofin [48], and these are summarized in Table 3. The similarity of the heats of transition deduced in 95% ethanol (2.90 kcal/mol) and dimethylformamide (2.85 kcal/mol) with the heat of transition calculated at the melting point (3.20 kcal/mol) indicates that the difference in heat capacity between the two polymorphs is relatively small.

Because of the temperature dependence of the various phenomena under discussion, and because of the important role played by entropy, discussions based purely on enthalpy changes are necessarily incomplete. One can rearrange Eq. (7) to read

Table 3 Heats of Solution and Fusion for the Polymorphs of Auranofin

	Heat of solution, 95% ethanol (kcal/mol)	Heat of solution, dimethylformamide (kcal/mol)
Form A	12.42	5.57
Form B	9.52	2.72
Difference	2.90	2.85

	Heat of fusion (kcal/mol)
Form A	9.04
Form B	5.85
Difference	3.20

Source: Ref. 48.

$$\mu_i^* - \mu_i^s = -RT \ln X_i \qquad (18)$$

where the left hand side of the equation represents the difference in chemical potential between the chemical potential of i in its pure solid state and the chemical potential of this species in the solution at a defined temperature and pressure. This difference in chemical potential is by definition the molar Gibbs free energy change associated with the dissolution of compound i, so one can write

$$\Delta G_s = -RT \ln X_i \qquad (19)$$

where ΔG_s is the molar Gibbs free energy of solution. By analogy with Eq. (16), the molar Gibbs free energy associated with the transformation of polymorph A to B is given by

$$\Delta G_t = \Delta G_s^B - \Delta G_s^A \qquad (20)$$

$$= RT \ln \frac{X_A}{X_B} \qquad (21)$$

where X_A and X_B are the equilibrium solubilities of polymorphs A and B, respectively, expressed in units of mole fraction.

Finally, the entropy of solution ΔS_s is obtained from the relation

$$\Delta S_s = \frac{\Delta H_s - \Delta G_s}{T} \qquad (22)$$

For basic thermodynamic understanding of the solubility behavior of a given substance, ΔG_s, ΔH_s, and ΔS_s must be determined. Similarly, a basic thermodynamic understanding of a polymorphic transition requires an evaluation of the quantities ΔG_t, ΔH_t, and ΔS_t associated with the phase transition.

To illustrate the importance of free energy changes, consider the solvate system formed by paroxetine hydrochloride, which can exist as a nonhygroscopic hemihydrate or as a hygroscopic anhydrate [49]. The heat of transition between these two forms was evaluated both by differential scanning calorimetry ($\Delta H_t' = 0.0$ kJ/mol) and by solution calorimetry ($\Delta H_t = 0.1$ kJ/mol), which indicates that these two forms are almost isoenthalpic. However, the free energy of transition (-1.25 kJ/mol) favors conversion of the anhydrate to the hemihydrate, and such phase conversion can be initiated by crystal compression or by seeding techniques. Since the two forms are essentially isoenthalpic, the entropy increase that accompanies the phase transformation is responsible for the decrease in free energy and can therefore be viewed as the driving force for the transition.

C. Methodology for Solution Calorimetry

Any calorimeter with a suitable mixing device and designed for use with liquids can be applied to determine heats of solution, dilution, or mixing. To obtain good precision in the determination of heats of solution requires careful attention to detail in the construction of the calorimeter. The dissolution of a solid can sometimes be a relatively slow process and requires efficient and uniform stirring. Substantial experimental precautions are ordinarily made to ensure that heat input from the stirrer mechanism is minimized.

Most solution calorimeters operate in the batch mode, and descriptions of such systems are readily found in the literature [50,51]. The common practice is to use the batch solution calorimetric approach, in which mixing of the solute and the solvent is effected in a single step. Mixing can be accomplished by breaking a bulb containing the pure solute, by allowing the reactants to mix by displacing the seal separating the two reactants in the calorimeter reaction vessel, or by rotating the reaction vessel and allowing the reactants to mix [51]. Although the batch calorimetric approach simplifies the data analysis, there are design problems associated with mixing of the reactants. Guillory and coworkers have described the use of a stainless steel ampoule whose design greatly facilitates batch solution calorimetric analyses [52]. This device was validated by measurement of the enthalpy of solution of potassium chloride in water, and the reproducibility of the method was demonstrated by determination of the enthalpy of solution of the two common polymorphic forms of chloramphenicol palmitate in 95% ethanol.

D. Applications of Solution Calorimetry

Solution calorimetric investigations can be classified into studies that focus entirely on enthalpic processes and studies that seek to understand the contribution of the enthalpy change to the free energy change of the system. Although the former can prove to be quite informative, only the latter permit the deduction of unequivocal thermodynamic conclusions about relative stability.

While heats of solution data are frequently used to establish differences in enthalpy within a polymorphic system, they cannot be used to deduce accurately the relative phase stability. According to Eq. (16), the difference between the differential heats of solution of two polymorphs is a measure of the heat of transition ΔH_t between the two forms. Because enthalpy is a state function (Hess's law), this difference must necessarily be independent of the solvent system used. However, conducting calorimetric measurements of the heats of solution of the polymorphs in more than one solvent provides an empirical verification of the assumptions made. For instance, ΔH_t values of two losartan polymorphs were found to be 1.72 kcal/mol in water and 1.76 kcal/mol in dimethylformamide [53]. In a similar study with moricizine hydrochloride polymorphs, ΔH_t values of 1.0 kcal/mol and 0.9 kcal/mol were obtained from their dissolution in water and dimethylformamide, respectively [54]. These two systems, which show good agreement, can be contrasted with that of enalapril maleate, where ΔH_t was determined to be 0.51 kcal/mol in methanol and 0.69 kcal/mol in acetone [55]. Disagreements of this order (about 30%) suggest that some process, in addition to dissolution, is taking place in one or both solvents.

In systems characterized by the existence of more than one polymorph, the heats of solution have been used to deduce the order of stability. As explained above, the order of stability cannot be deduced from enthalpy changes but only from free energy changes. If the enthalpy change reflects the stability, then the polymorphic change is not driven by an increase in entropy but by a decrease in enthalpy. The heat of solution measured for cyclopenthiazide Form III (3.58 kcal/mol) was significantly greater than the analogous values obtained for Form I (1.41 kcal/mol) or Form II (1.47 kcal/mol); thus Form III is the polymorph with the lower enthalpy but not necessarily the most stable polymorph at ambient temperature [56]. Other examples follow.

In the case of the anhydrate and hydrate phases of norfloxacin [57], the dihydrate phase was found to exhibit a relatively large endothermic heat of solution relative to either the anhydrate or the sesquihydrate. Both urapidil [58] and dehydroepiandrosterone [59] were found to exhibit complex polymorphic/solvate systems, but the relative enthalpy of these could be deduced through the use of solution calorimetry. As an example, the data reported for urapidil [58], which have been collected into Table 4, show that the form with the lowest heat of solution consequently has the highest enthalpy. In this particular case, the rank order of enthalpy changes corresponds to that of the free energy changes.

It is invariably found that the amorphous form of a compound is less stable than its crystalline modification, in the sense that the amorphous form tends to crystallize spontaneously, indicating that the amorphous form has the greater Gibbs free energy. As discussed in Chapter 1, the amorphous form is more disordered and must therefore have a greater entropy than does the crystalline form. Hence the enthalpy of the amorphous form is also greater. The heat of solution of amorphous piretanide in water was found to be 12.7 kJ/mol, while the heat of solution associated with Form C was determined to be 32.8 kJ/mol [60]. The authors calculated the heat of transformation associated with the amorphous-to-crystalline transition to be −20.1 kJ/mol. Any facile transformation of the two phases was obstructed by the significant activation energy (145.5 kJ/mol).

As emphasized above, a basic thermodynamic understanding of

Table 4   Heats of Solution in Water for the Various Polymorphs and Solvates of Urapidil

Crystalline form	Heat of solution (kJ/mol)
Form I	21.96
Form II	24.26
Form III	22.98 (estimated)
Monohydrate	44.28
Trihydrate	53.50
Pentahydrate	69.16
Methanol solvate	48.39

Source: Ref. 58.

a polymorphic system requires a determination of the free energy difference between the various forms. The two polymorphs of 3-amino-1-(m-trifluoromethlyphenyl)-6-methyl-1H-pyridazin-4-one have been characterized by a variety of methods, among which solubility studies were used to evaluate the thermodynamics of the transition from Form I to Form II [61]. At a temperature of 30°C, the enthalpy change for the phase transformation was determined to be -5.64 kJ/mol. From the solubility ration of the two polymorphs, the free energy change was then calculated as -3.67 kJ/mol, which implies that the entropy change accompanying the transformation was -6.48 cal/K·mol. In this system, one encounters a phase change that is favored by the enthalpy term but not favored by the entropy term. However, since the overall free energy change ΔG_T is negative, the process takes place spontaneously, provided that the molecules can overcome the activation energy barrier at a significant rate.

A similar situation has been described for the two polymorphic forms of 2-[[4-[[2-(IH-tetrazol-5-ylmethyl)phenyl]methoxy]phenoxy]methyl]quinoline [20]. The appreciable enthalpic driving force for the transformation of Form II to Form I (-0.91 kcal/mol) was found to be partially offset by the entropy of transformation (-2.6 cal/K · mol), resulting in a modest free energy difference between the two forms (-0.14 kcal/mol).

In other instances, an unfavorable enthalpy term was found to be compensated by a favorable entropy term, thus rendering negative the free energy change associated with a particular phase transformation. Lamivudine can be obtained in two forms, of which one is a 0.2-hydrate obtained from water or from methanol that contains water, and the other nonsolvated and obtained from many nonaqueous solvents [62]. Form II was determined to be thermodynamically favored in the solid state. Solubility studies of both forms as a function of solvent and temperature were used to determine whether entropy or enthalpy was the driving force for solubility. Solution calorimetric data indicated that Form I would be favored in all solvents studied on the basis of enthalpy alone (see Table 5). In higher alcohols and other organic solvents, Form I exhibited a larger entropy of solution than did Form II, compensating for the unfavorable enthalpic factors and yielding an overall negative free energy for the phase change.

Table 5 Thermodynamic Quantities of
Solution for Lamivudine in Various Solvents

	Solvent = water	
	Form I	Form II
ΔG_{Sol}(cal/mol)	2990	2950
ΔH_{Sol}(cal/mol)	5720	5430
ΔS_{Sol}(cal/K · mol)	9.2	8.3
	Solvent = ethanol	
	Form I	Form II
ΔG_{Sol}(cal/mol)	3180	3460
ΔH_{Sol}(cal/mol)	5270	4740
ΔS_{Sol}(cal/K·mol)	7.0	4.3
	Solvent = n-propanol	
	Form I	Form II
ΔG_{Sol}(cal/mol)	3120	3610
ΔH_{Sol}(cal/mol)	5350	5000
ΔS_{Sol}(cal/K·mol)	7.5	4.7

Source: Ref. 62.

Shefter and Higuchi considered the thermodynamics associated with the anhydrate/hydrate equilibrium of theophylline and glutethimide [40]. For both compounds, the free energy change for the transformation from the anhydrate to the hydrate was negative (hence indicating a spontaneous process), the favorable enthalpy changes being mitigated by the unfavorable entropy changes. In this work, the free energy was calculated from the solubilities of the anhydrate and hydrate forms, while the enthalpy of solution was calculated from the temperature dependence of the solubility ratio using the van't Hoff equation. The entropy of solution was evaluated using Eq. (22).

A similar conclusion was reached regarding the relative stability of the monohydrate and anhydrate phases of metronidazole benzoate [37]. The enthalpy term (-1.20 kcal/mol) favored conversion to the monohydrate, but the strong entropy term (-3.7 cal/K · mol) essentially offset this enthalpy change. At 25°C, the overall ΔG_t of the transition was still negative, favoring the monohydrates, but only barely so (-0.049 kcal/mol). This difference was judged to be too small to result in any detectable bioavailability differences.

IV. KINETICS OF SOLUBILITY: DISSOLUTION RATES

Evaluation of the dissolution rates of drug substances from their dosage forms is extremely important in the development, formulation, and quality control of pharmaceutical agents [9,63–65]. Such evaluation is especially important in the characterization of polymorphic systems owing to the possibility of bioavailability differences that may arise from differences in dissolution rate that may themselves arise from differences in solubility [4]. The wide variety of methods for determining the dissolution rates of solids may be categorized either as batch methods or as continuous flow methods, for which detailed experimental protocols have been provided [66].

A. Factors Affecting Dissolution Rates

The dissolution rate of a solid may be defined as dm/dt, where m is the mass of solid dissolved at time t. To obtain dm/dt, the following equation, which defines concentration, must be differentiated:

$$m = Vc_b \tag{23}$$

In a batch dissolution method the analyzed concentration c_b of a well-stirred solution is representative of the entire volume V of the dissolution medium, so that

$$\frac{dm}{dt} = V \frac{dc_b}{dt} \tag{24}$$

In a dissolution study, c_b will increase from its initial zero value until a limiting concentration is attained. Depending on the initial amount of solute presented for dissolution, the limiting concentration will be at the saturation level, or less than this.

Batch dissolution methods are simple to set up and to operate, are widely used, and can be carefully and reproducibly standardized. Nevertheless, they suffer from several disadvantages [9]. The hydrodynamics are usually poorly characterized, a small change in dissolution rate will often create an undetectable and immeasurable perturbation in the dissolution time curve, and the solute concentration may not be uniform throughout the solution volume.

In a continuous flow method, the volume flow rate over the surface of the solid is given by dV/dt, so that differentiation of Eq. (23) leads to

$$\frac{dm}{dt} = c_b \frac{dV}{dt} \tag{25}$$

where c_b is the concentration of drug dissolved in the solvent that has just passed over the surface of the solid drug.

Continuous flow methods have the advantages that sink conditions can be easily achieved, and that a change in dissolution rate is reflected in a change in c_b [9]. At the same time, they require a significant flow rate that may require relatively large volumes of dissolution medium. Should the solid be characterized by a low solubility and a slow dissolution rate, c_b will be small, and a very sensitive analytical method would be required.

The diffusion layer theory is the most useful and best known model for transport-controlled dissolution and satisfactorily accounts for the dissolution rates of most pharmaceutical solids. In this model, the dissolution rate is controlled by the rate of diffusion of solute molecules across a thin diffusion layer. With increasing distance from the surface of the solid, the solute concentration decreases in a nonlinear manner across the diffusion layer. The dissolution process at steady state is described by the Noyes–Whitney equation:

$$\frac{dm}{dt} = k_D A(c_s - c_b) \tag{26}$$

where dm/dt is the dissolution rate, A is the surface area of the dissolving solid, c_s is the saturation solubility of the solid, and c_b is the concentration of solute in the bulk solution. The dissolution rate constant k_D

is given by D/h, where D is the diffusivity. The hydrodynamics of the dissolution process have been fully discussed by Levich [67].

It has been shown [9] that the dissolution rates of solids are determined or influenced by a number of factors, which may be summarized as follows:

1. Solubility of the solid, and the temperature
2. Concentration in the bulk solution, if not under sink conditions
3. Volume of the dissolution medium in a batch-type apparatus, or the volume flow rate in a continuous flow apparatus
4. Wetted surface area, which consequently is normalized in measurements of intrinsic dissolution rate
5. Conditions in the dissolution medium that, together with the nature of the dissolving solid, determine the dissolution mechanism

The conditions in the dissolution medium that may influence the dissolution rate can be summarized as

1. The rate of agitation, stirring, or flow of solvent, if the dissolution is transport-controlled, but not when the dissolution is reaction-controlled.
2. The diffusivity of the dissolved solute, if the dissolution is transport-controlled. The dissolution rate of a reaction-controlled system will be independent of the diffusivity.
3. The viscosity and density influence the dissolution rate if the dissolution is transport-controlled, but not if the dissolution is reaction-controlled.
4. The pH and buffer concentration (if the dissolving solid is acidic or basic), and the pK_a values of the dissolving solid and of the buffer.
5. Complexation between the dissolving solute and an interactive ligand, or solubilization of the dissolving solute by a surface-active agent in solution. Each of these phenomena tends to increase the dissolution rate.

B. Applications of Dissolution Rate Studies to Polymorphs and Hydrates

Historically, batch-type dissolution rate studies of loose powders and compressed discs have played a major role in the characterization of essentially every polymorphic or solid-state solvated system [13,21,39]. Stagner and Guillory used these two methods of dissolution to study the two polymorphs and the amorphous phase of iopanoic acid [68]. As is evident in the loose powder dissolution data illustrated in the upper half of Fig. 5, the two polymorphs were found to be stable with respect to phase conversion, but the amorphous form rapidly converted to Form I under the dissolution conditions. In the powder dissolution studies, the initial solubilities of the different forms followed the same rank order as did their respective intrinsic dissolution rates, but the subsequent phase conversion of the amorphous form to the stable Form I appeared to change the order. The amorphous form demonstrated a 10-fold greater intrinsic dissolution rate relative to Form I, while the intrinsic dissolution rate of Form II was 1.5 times greater than that of Form I.

The nature of the dissolution medium can profoundly affect the shape of a dissolution profile. The relative rates of dissolution and the solubilities of the two polymorphs of 3-(3-hydroxy-3-methylbutylamino)-5-methyl-*as*-triazino[5,6-*b*]indole were determined in USP artificial gastric fluid, water, and 50% ethanol solution [69]. In the artificial gastric fluid, both polymorphic forms exhibited essentially identical dissolution rates. This behavior has been contrasted in Fig. 6 with that observed in 50% aqueous ethanol, in which Form II has a significantly more rapid dissolution rate than Form I. If the dissolution rate of a solid phase is determined by its solubility, as predicted by the Noyes–Whitney equation, the ratio of dissolution rates would equal the ratio of solubilities. Because this type of behavior was not observed for this triazinoindole drug, the different effects of the dissolution medium on the transport rate constant can be suspected.

The solubilities of the two polymorphs of difenoxin hydrochloride have been studied, as well as the solubility of tablets formed from mixtures of these polymorphs [70]. Form I was found to be more soluble than was Form II, and the solubilities of materials containing known

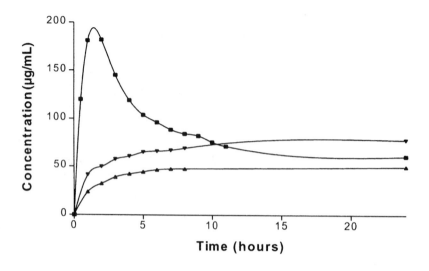

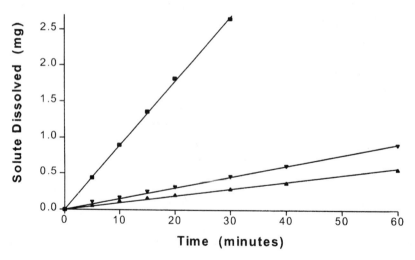

Fig. 5. Loose powder dissolution (upper family of traces) and intrinsic dissolution (lower family of traces) profiles of iopanoic acid. Shown are the profiles of Form I (-▲-), Form II (-▼-), and the amorphous form (-■-). (The plots were adapted from data originally presented in Ref. 68.)

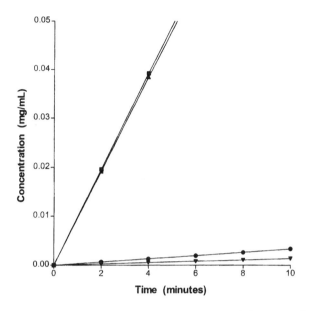

Fig. 6. Initial stages of the dissolution of the two polymorphs of 3-(3-hydroxy-3-methylbutylamino)-5-methyl-*as*-triazino[5,6-*b*]indole in different media. Shown are the profiles obtained for Form I (-■-) and Form II (-▲-) in simulated gastric fluid, as well as the profiles of Form I (-○-) and Form II (-▼-) in 50% aqueous ethanol. (The traces were adapted from data originally presented in Ref. 69.)

proportions of Form I and II reflected the differences in the solubilities of the pure forms. Likewise, the dissolution rate of difenoxin hydrochloride from tablets was determined by the ratio of Form I to Form II. In these studies, no solid-state transformation of the more soluble form to the less soluble form was observed. In addition, micronization proved to be a successful method for improving the dissolution of tablets prepared from the less soluble polymorph.

Stoltz and coworkers have conducted extensive studies on the dissolution properties of the hydrates and solvates of oxyphenbutazone [71,72]. They compared the dissolution properties of the benzene and cyclohexane solvates with those of the monohydrate, hemihydrate, and anhydrate forms, and then compared their findings with results reported in the literature. The powder dissolution rates of the solvates proved to be comparable with those of the hemihydrate and the anhydrate but superior to that of the monohydrate. This trend is illustrated in Fig. 7, which confirms the usual observation that increasing degrees of hydration result in slower dissolution rates. This observation differed from that previously described by Matsuda and Kawaguchi who reported powder dissolution rates in simulated intestinal fluid that were in the sequence: hemihydrate > monohydrate > anhydrate [73]. The reversed order in the dissolution rates of the former work [71] was attributed to the presence of a surfactant in the dissolution medium, which apparently overcame the hydrophobicity of the crystal surfaces of the anhydrate form. In terms of the Noyes–Whitney equation, these results can

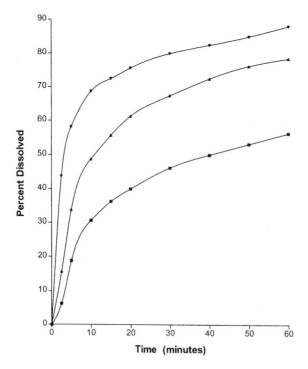

Fig. 7. Powder dissolution profiles obtained for oxyphenbutazone anhydrate (-▼-), hemihydrate (-▲-), and monohydrate (-■-). (The curves were adapted from data originally presented in Ref. 71.)

be explained by the influence of the surface active agent in increasing either the wetted surface area, or the transport rate constant, or both quantities.

It has been noted from the earliest dissolution work [39] that, for many substances, the dissolution rate of an anhydrous phase usually exceeds that of any corresponding hydrate phase. These observations were explained by thermodynamics, where it was reasoned that the drug in the hydrates possessed a lower activity and would be in a more stable state relative to their anhydrous forms [74]. This general rule was found to hold for the previously discussed anhydrate/hydrate phases of theophylline [42,44,46], ampicillin [38], metronidazone benzoate [37], carbamazepine [34,36], glutethimide [75], and oxyphenbutazone [72], as well as for many other systems not mentioned here. In addition, among the hydrates of urapidil, the solubility decreases with increasing crystal hydration [58].

Since the mid-1970s, a number of exceptions to the general rule have been found. For example, Fig. 8 shows that the hydrate phases of erythromycin exhibit a reverse order of solubility where the dihydrate phase exhibits the fastest dissolution rate and the highest equilib-

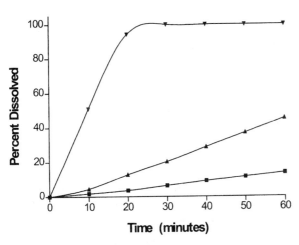

Fig. 8. Dissolution profiles of erythromycin anhydrate (-■-), monohydrate (-▲-), and dihydrate (-▼-). (The plots were adapted from data originally presented in Ref. 76.)

rium solubility [76]. More recent examples include the magnesium, zinc, and calcium salts of nedocromil, for which the intrinsic dissolution rate increases with increasing water stoichiometry of their hydrates [77]. The explanation for this behavior is that the transition temperatures between the hydrates are below the temperature of the dissolution measurements and decrease with increasing water stoichiometry of the hydrates. Consequently, the solubilities, and hence the intrinsic dissolution rates, increase with increasing stoichiometry of water in the hydrates. Acyclovir was recently found to be capable of forming a 3:2 drug/water hydrate phase that exhibited an almost instantaneous dissolution relative to the more slowly dissolving anhydrous form [78]. This latter finding implies a substantial difference in Gibbs free energy between the two forms.

C. Intrinsic Dissolution Rates: Principles and Practice

It should be recognized that the final concentration measured using the loose powder dissolution method is the equilibrium solubility, and that the initial stages of this dissolution are strongly affected by the particle size and surface area of the dissolving solids. For this reaction, many workers have chosen to study the dissolution of compacted materials, by which the particle size and surface area are regulated by the process of forming the compact.

In the disc method for conducting intrinsic dissolution studies, the powder is compressed in a die to produce a compact. One face of the disc is exposed to the dissolution medium and rotated at a constant speed without wobble. The dissolution rate is determined as for a batch method, while the wetted surface area is simply the area of the disc exposed to the dissolution medium.

It is good practice to compare the powder X-ray diffraction patterns of the compacted solid and of the residual solid after the dissolution experiment with that of the original powder sample. In this manner, one can test for possible phase changes during compaction or dissolution.

The dissolution rate of a solid from a rotating disc is governed by the controlled hydrodynamics of the system and has been treated theoretically by Levich [67]. In this system, the intrinsic dissolution rate J can be calculated using either of the following relations:

$$J = 0.620 \, D^{2/3} \nu^{-1/6}(c_s - c_b)\omega^{1/2} \tag{27}$$

or

$$J = 1.555 \, D^{2/3} \nu^{-1/6}(c_s - c_b)W^{1/2} \tag{28}$$

where D is the diffusivity of the dissolved solute, ω is the angular velocity of the disc in radians per second (W revolutions per second, Hz), ν is the kinematic viscosity of the fluid, c_b is the concentration of solute at time t during the dissolution study, and c_s is the equilibrium solubility of the solute. The dependence of J on $\omega^{1/2}$ has been verified experimentally [79].

Equations (27) and (28) enable the diffusivity of a solute to be measured. These relations assume the dissolution of only one diffusing species, but since most small organic molecules exhibit a similar diffusivity (of the order 10^{-5} cm^2/s in water at 25°C), it follows that J depends on the 2/3 power of D. Consequently, the errors arising from several diffusing species only become significant if one or more species exhibit abnormal diffusivities. In fact, diffusivity is only weakly dependent on the molecular weight, so it is useful to estimate the diffusivity of a solute from that of a suitable standard of known diffusivity under the same conditions. In most cases, the diffusivity predictions agree quite well with those obtained experimentally [80].

D. Intrinsic Dissolution Rate Studies of Polymorphic and Hydrate Systems

Under constant hydrodynamic conditions, the intrinsic dissolution rate is usually proportional to the solubility of the dissolving solid. Consequently, in a polymorphic system, the most stable form will ordinarily exhibit the slowest intrinsic dissolution rate. For example, a variety of high-energy modifications of frusemide were produced, but the commercially available form was found to exhibit the longest dissolution times [81]. Similar conclusions were reached for the four polymorphs of tegafur [82] and (R)-N-[3-[5-(4-fluorophenoxy)-2-furanyl]-1-methyl-2-propynyl]-N-hydroxyurea [83]. However, it is possible that one of the less stable polymorphs of a compound can exhibit the slowest dissolution rate, as was noted in the case of diflunisal [84].

Intrinsic dissolution rate studies proved useful during the characterization of the two anhydrous polymorphs and one hydrate modification of alprazolam [85]. The equilibrium solubility of the hydrate phase was invariably less than that of either anhydrate phase, although the actual values obtained were found to be strongly affected by pH. Interestingly, the intrinsic dissolution rate of the hydrate phase was higher than that of either anhydrate phase, with the anhydrous phases exhibiting equivalent dissolution rates. The IDR data of Table 6 reveal an interesting phenomenon, where discrimination between some polymorphs was noted at slower spindle speeds, but not at higher rates. Thus if one is to use IDR values as a means to determine the relative rates of solution of different solids, the effect of stirring speed must be investigated before the conclusions can be judged genuine.

Intrinsic dissolution rate investigations can become complicated when one or more of the studied polymorphs interconverts to another during the time of measurement. Sulfathiazole has been found to crystallize in three distinct polymorphic forms, two of which are unstable in contact with water [86] and convert only slowly to the stable form (i.e., are kinetically stable) in the solid state. As can be seen in Fig. 9, the initial intrinsic dissolution rates of these are all different, but as Forms I and II convert into Form III, the dissolved concentrations converge. Only the dissolution rate of Form III remains constant, which suggests that it is the thermodynamically stable form at room tempera-

Table 6 Intrinsic Dissolution Rates (IDR) for the Various Polymorphs of Aprazolam at Two Different Spindle Speeds

Crystalline form	IDR, 50 RPM (μ/min · cm^2)	IDR, 75 RPM (μ/min · cm^2)
Form I	15.8	21.8
Form II	18.4	21.9
Form V	20.7	27.3

Source: Ref. 85.

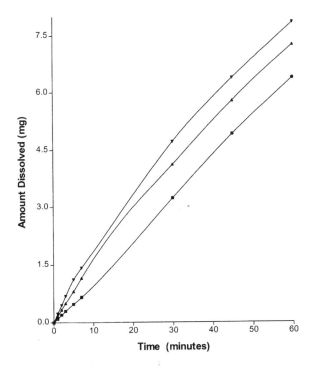

Fig. 9.   Dissolution profiles obtained for sulfathiazole Form I (-▲-), Form II (-▼-), and Form III (-■-) in water at 37°C. (The figure has been adapted from data originally presented in Ref. 86.)

ture. Aqueous suspensions of Forms I or II each converted into Form III over time, supporting the conclusions of the dissolution studies.

Suitable manipulation of the dissolution medium can sometimes inhibit the conversion of one polymorph to another during the dissolution process, thus permitting the measurement of otherwise unobtainable information. In studies on the polymorphs of sulfathiazole and methylprednisolone, Higuchi, who used various alcohols and additives in the dissolution medium to inhibit phase transformations, first employed this approach [87]. Aguiar and Zelmer were able to characterize thermodynamically the polymorphs formed by chloramphenicol palmitate and mefenamic acid by means of dissolution modifiers [88]. Furthermore, the use of an aqueous ethanol medium containing 55.4%

v/v ethanol yielded adequate solubility and integrity of the dissolving disc during studies conducted on digoxin [89].

One area of concern associated with intrinsic dissolution measurements is associated with the preparation of the solid disc by compaction of the drug particles. If a phase transformation is induced by compression, one might unintentionally measure the dissolution rate of a polymorph different from the intended one. This situation was encountered with phenylbutazone, where Form III was transformed to the most stable modification (Form IV) during the initial compression step [90].

One interesting note concerns the aqueous dissolution rates of solvate forms, where the solvent bound in the crystal lattice is not water. As noted earlier, the dissolution rate of an anhydrous phase normally exceeds that of any corresponding hydrate phase, but this relation is not usually applicable to other solvate species. It has been reported that the methanol solvate of urapidil exhibits a heat of solution approximately twice that of any of the anhydrate phases and that it also exhibits the most rapid dissolution rate [91]. Similarly, the pentanol and toluene solvates of glibenclamide (glyburide) exhibit significantly higher aqueous dissolution rates and aqueous equilibrium solubility values when compared to either of the two anhydrous polymorphs [92]. The acetone and chloroform solvates of sulindac yielded intrinsic dissolution rates that were double those of the two anhydrate phases [93]. These trends would imply that a nonaqueous solvate phase could be considered as being a high-energy form of the solid with respect to dissolution in water.

The most usual explanation of these phenomena is that the negative Gibbs free energy of mixing of the organic solvent, released during the dissolution of the solvate, contributes to the Gibbs free energy of solution, increasing the thermodynamic driving force for the dissolution process [39]. This explanation, due to Shefter and Higuchi, was originally derived from observations on the higher dissolution rate of the pentanol solvate of succinylsulfathiazole than of the anhydrate [39]. Prior addition of increasing concentrations of pentanol to the aqueous dissolution medium reduced the initial dissolution rate of the pentanol solvate. This reduction was attributed to a less favorable (less negative) Gibbs free energy of mixing of the released pentanol in the solution that already contained pentanol. In this way, the Gibbs free energy of solution was rendered less favorable (less negative), reducing the thermodynamic driving force for dissolution of the solvate. Thermodynamic characterization of the various steps in the dissolution of solvates and evaluation of their respective Gibbs free energies (and enthalpies) has been carried out by Ghosh and Grant [94].

V. CONSEQUENCES OF POLYMORPHISM AND SOLVATE FORMATION ON THE BIOAVAILABILTY OF DRUG SUBSTANCES

In those specific instances where the absorption rate of the active ingredient in a solid dosage form depends upon the rate of drug dissolution, the use of different polymorphs would be expected to affect the bioavailability. One can imagine the situation in which the use of a metastable polymorph would yield higher levels of a therapeutically active substance after administration owing to its higher solubility. This situation may be either advantageous or disadvantageous depending on whether the higher bioavailability is desirable or not. On the other hand, unrecognized polymorphism may result in unacceptable dose-to-dose variations in drug bioavailability and certainly represents a drug formulation not under control.

The trihydrate/anhydrate system presented by ampicillin has received extensive attention, with conflicting conclusions from several investigations. In one early study, Poole and coworkers reported that the aqueous solubility of the anhydrate phase was 20% higher than that of the trihydrate form at 37°C [95]. They also found that the time for 50% of the drug to dissolve in vitro was 7.5 and 45 minutes for the anhydrate and trihydrate forms, respectively [96]. Using dogs and human subjects, these workers then determined in vivo blood levels, following separate administration of the two forms of the drug in oral suspensions or in capsules. The anhydrous form produced a higher maximum concentration of ampicillin C_{max} and an earlier time to reach maximum concentration T_{max} in the blood serum relative to the trihydrate form. This behavior was more pronounced in the suspension formulations. In addition, the area under the curve (AUC) was found to be greater with the anhydrous form, implying that the anhydrous form was more efficiently absorbed.

Since the early works just discussed, an interesting discussion on the comparative absorption of ampicillin has arisen. Some workers have concluded that suspensions and capsules containing ampicillin anhydrate exhibit superior bioavailabilities to analogous formulations made from the trihydrate [97,98]. For instance, in a particularly well-controlled study, Ali and Farouk [98] obtained the clear-cut distinction between the anhydrate and the trihydrate that is illustrated in Fig. 10. However, others have found that capsules containing either form of ampicillin yielded an essentially identical bioavailability [99–101]. These conflicting observations indicate that the problem is strongly affected by the nature of the formulation used, and that the effects of compounding can overshadow the effects attributed to the crystalline state.

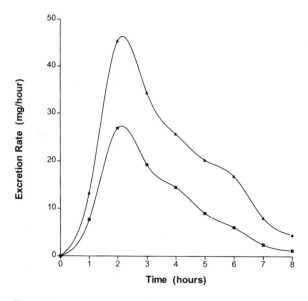

Fig. 10. Ampicillin urinary excretion rates at various times after separate administration of the two forms. Shown are the profiles obtained for the anhydrate (-▲-) and trihydrate phases (-■-). (The figure has been adapted from data provided in Ref. 98.)

Chloramphenicol palmitate has been shown to exist in four crystal modifications, and the effect of two of these on the degree of drug absorption has been compared [102]. After oral ingestion of Forms A and B, the highest mean blood levels were obtained with suspensions containing only Form B. In mixed dosage forms, the blood levels of the drug were found to bear an inverse relationship with the fraction of Form A. This finding explained the previous report, which noted that a particular suspension formulation of chloramphenicol palmitate exhibited an unsatisfactory therapeutic effect [103]. A study of various commercial products indicated that the polymorphic state of the drug in this formulation was uncontrolled, consisting of mixtures of the active polymorph B and the inactive polymorph A.

Sulfamethoxydiazine has been shown to exist in a number of polymorphic forms, which exhibit different equilibrium solubilities and dissolution rates [104]. Form II, the polymorph with the greater thermo-dynamic activity, was found to yield higher blood concentrations than Form III which is stable in water [105]. This relationship is illustrated in Fig. 11. Although the urinary excretion rates during the absorption phase confirmed the different drug absorption of the two forms as previously observed, the extent of absorption (as indicated by 72-hour excretion data) of the two forms was ultimately shown to be equivalent [106].

Fluprednisolone has been shown to exist in seven different solid phases, of which six were crystalline and one was amorphous [107]. Of the crystalline phases, three were anhydrous, two were monohydrates, and one was a *tert*-butylamine solvate. The in vitro dissolution rates of the six crystalline phases of fluprednisolone were determined and compared with in vivo dissolution rates derived from pellet implants in rats [108]. The agreement between the in vitro and in vivo dissolution rates was found to be quite good, but the correlation with animal weight loss and adrenal gland atrophy was only fair. These results can be interpreted to indicate that, for fluprednisolone, differences in dissolution rates of the drug did not lead to measurable biological differences.

Erythromycin base is reported to exist in a number of structural forms, including an anhydrate, a dihydrate, and an amorphous form [109, 110]. The commercially available product appears to be a partially crystalline material, containing a significant amount of amor-

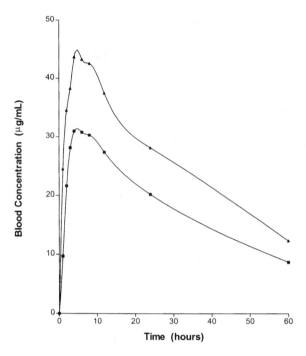

Fig. 11. Mean concentrations of sulfamethoxydiazine in blood as influenced by the polymorphic form of the drug substance. Shown are the profiles of Form II (-▲-) and Form III (-■-). (The figure has been adapted from data provided in Ref. 105.)

phous drug [111]. From studies conducted in healthy volunteers, it was learned that the anhydrate and dihydrate phases were absorbed faster and more completely than was either the amorphous form or the commercially available form [112]. These observations were reflected in the two pharmacokinetic parameters (C_{max} and AUC).

Azlocillin sodium can be obtained either as a crystalline form or as an amorphous form, depending on the solvent and method used for its isolation [113]. The antibacterial activity of this agent was tested against a large number of reference strains, and, in most cases, the crystalline form exhibited less antibacterial activity than did the amorphous form. Interestingly, several of the tested microorganisms also proved to be resistant to the crystalline form.

Whether the different polymorphs or solvates of a given drug substance will lead to the existence of observable differences in the adsorption, metabolism, distribution, or elimination of the compound clearly cannot be predicted a priori at the present time. It is certainly likely that different crystal forms of highly soluble substances ought to be roughly bioequivalent, owing to the similarity of their dissolution rates. An effect associated with polymorphism that leads to a difference in bioavailability would be anticipated only for those drug substances whose absorption is determined by the dissolution rate. However, the literature indicates that, even in such cases, the situation is not completely clear. Consequently, when the existence of two or more polymorphs or solvates is demonstrated during the drug development process, wise investigators will determine those effects that could be associated with the drug crystal form and will modify their formulations accordingly.

REFERENCES

1. W. C. McCrone, *Fusion Methods in Chemical Microscopy*, Interscience, New York, 1957, p. 133.
2. J. K. Haleblian and W. C. McCrone, *J. Pharm. Sci., 58,* 911 (1969).
3. W. I. Higuchi, P. K. Lau, T. Higuchi, and J. W. Shell, *J. Pharm. Sci., 52,* 150 (1963).
4. S. Byrn, R. Pfeiffer, M. Ganey, C. Hoiberg, and G. Poochikian, *Pharm. Res., 12,* 945 (1995).
5. D. J. W. Grant and T. Higuchi, *Solubility Behavior of Organic Compounds*, Vol. 21, Techniques of Chemistry (W. H. Saunders, Jr., series ed.), John Wiley, New York, 1990.
6. J. Jacques, A. Collet, and S. H. Wilen, *Enantiomers, Racemates and Resolutions*, John Wiley, New York, 1981, p. 168.
7. W. J. Mader, R. D. Vold, and M. J. Vold, in *Physical Methods of Organic Chemistry*, 3d ed., Vol. 1, Part I (A. Weissberger, ed.), Interscience, New York, 1959, pp. 655–688.
8. S. H. Yalkowsky and S. Banerjee, *Aqueous Solubility Methods of Estimation for Organic Compounds*, Marcel Dekker, New York, 1992, pp. 149–154.
9. D. J. W. Grant and H. G. Brittain, "Solubility of Pharmaceutical Sol-

ids'', Chapter 11 in *Physical Characterization of Pharmaceutical Solids* (H. G. Brittain, ed.) Marcel Dekker, New York, 1995, pp. 321–386.

10. D. J. W. Grant, M. Mehdizadeh, A. H.-L. Chow, and J. E. Fairbrother, *Int. J. Pharm., 18*, 25 (1984).

11. T. Higuchi, F. -M. L. Shih, T. Kimura, and J. H. Rytting, *J. Pharm. Sci., 68*, 1267 (1979).

12. S. H. Yalkowsky, S. C. Valvani, W.-Y. Kuu, and R.-M. Dannenfelser, *Arizona Database of Aqueous Solubility*, 2nd Edn., Samuel Yalkowsky, 1987.

13. W. I. Higuchi, P. K. Lau, T. Higuchi, and J. W. Shell, *J. Pharm. Sci., 52*, 150 (1963).

14. S. Ghodbane and J. A. McCauley, *Int. J. Pharm., 59*, 281 (1990).

15. M. D. Tuladhar, J. E. Carless, and M. P. Summers, *J. Pharm. Pharmacol., 35*, 208 (1983).

16. R. J. Behme, D. Brooke, R. F. Farney, and T. T. Kensler, *J. Pharm. Sci., 74*, 1041 (1985).

17. S. Sudo, K. Sato, and Y. Harano, *J. Chem. Eng. Japan., 24*, 237 (1991).

18. W. L. Rocco and J. R. Swanson, *Int. J. Pharm., 117*, 231 (1995).

19. A. J. Aguiar and J. E. Zelmer, *J. Pharm. Sci., 58*, 983 (1969).

20. R. A. Carlton, T. J. Difeo, T. H. Powner, I. Santos, and M. D. Thompson, *J. Pharm. Sci., 85*, 461 (1996).

21. J. W. Poole and C. K. Bahal, *J. Pharm. Sci., 57*, 1945 (1968).

22. M. J. Jozwiakowski, S. O. Williams, and R. D. Hathaway, *Int. J. Pharm., 91*, 195 (1993).

23. Y. Kawashima, T. Niwa, H. Takeuchi, T. Hino, Y. Itoh, and S. Furuyama, *J. Pharm. Sci., 80*, 472 (1991).

24. W. L. Chiou and L. E. Kyle, *J. Pharm. Sci., 68*, 1224 (1979).

25. G. Milosovich, *J. Pharm. Sci., 53*, 781 (1964).

26. S. Ghosh and D. J. W. Grant, *Int. J. Pharm., 114*, 185 (1995).

27. H. G. Brittain, *Langmuir, 12*, 601 (1996).

28. E. L. Chiou and B. D. Anderson, *J. Pharm. Sci., 73*, 1673 (1984).

29. M. A. Moustafa, A. R. Ebian, S. A. Khalil, and M. M. Motawi, *J. Pharm. Pharmacol., 23*, 868 (1971).

30. M. A. Moustafa, S. A. Khalil, A. R. Ebian, and M. M. Motawi, *J. Pharm. Pharmacol., 24*, 921 (1972).

31. A. R. Ebian, M. A. Moustafa, S. A. Khalil, and M. M. Motawi, *J. Pharm. Pharmacol., 25*, 13 (1973).

32. G. Levy and J. A. Procknal, *J. Pharm. Sci., 53*, 656 (1964); D. E. Wurster and P. W. Taylor, *J. Pharm. Sci., 54*, 670 (1965); W. I. Higuchi, P. D. Bernardo, and S. C. Mehta, *J. Pharm. Sci., 56*, 200 (1967).

33. L. Addadi, Z. Berkovitch-Yellin, I. Weissbuch, J. van Mil, L. J. W. Shimon, M. Lahav, and L. Leiserowitz, *Angew. Chem. Int. Ed., 24*, 466 (1985); J. Hulliger, *Angew. Chem. Int. Ed., 33*, 143 (1994).

34. E. Laine, V. Tuominen, P. Ilvessalo, and P. Kahela, *Int. J. Pharm., 20*, 307 (1984).

35. W. W. L. Young and R. Suryanarayanan, *J. Pharm. Sci., 80*, 496 (1991).

36. P. Kahela, R. Aaltonen, E. Lewing, M. Anttila, and E. Kristoffersson, *Int. J. Pharm., 14*, 103 (1983).

37. Annie Hoelgaard and Niels Moller, *Int. J. Pharm., 15*, 213 (1983).

38. H. Zhu and D. J. W. Grant, *Int. J. Pharm., 139*, 33 (1996).

39. E. Shefter and T. Higuchi, *J. Pharm. Sci., 52*, 781 (1963).

40. J. Herman, J. P. Remon, N. Visavarungroj, J. B. Schwartz, and G. H. Klinger, *Int. J. Pharm., 42*, 15 (1988).

41. H. Ando, M. Ishii, M. Kayano, and H. Ozawa, *Drug Dev. Indust. Pharm., 18*, 453 (1992).

42. J. G. Fokkens, J. G. M. van Amelsfoort, C. J. de Blaey, C. C. de Kruif, and J. Wilting, *Int. J. Pharm., 14*, 79 (1983).

43. J. B. Bogardus, *J. Pharm. Sci., 72*, 837 (1983).

44. J. H. de Smidt, J. G. Fokkens, H. Grijseels, and D. J. A. Crommelin, *J. Pharm. Sci., 75*, 497 (1986).

45. P. L. Gould, J. R. Howard, and G. A. Oldershaw, *Int. J. Pharm., 51*, 195 (1989).

46. H. Zhu, C. Yuen, and D. J. W. Grant, *Int. J. Pharm., 135*, 151 (1996).

47. N. Otsuka, N. Kaneniwa, K. Otsuka, K. Kawakami, O. Omezawa, and Y. Matsuda, *J. Pharm. Sci., 81*, 1189 (1992).

48. S. Lindenbaum, E. S. Rattie, G. E. Zuber, M. E. Miller, and L. J. Ravin, *Int. J. Pharm., 26*, 123 (1985).

49. P. C. Buxton, I. R. Lynch, and J. A. Roe, *Int. J. Pharm., 42*, 135 (1988).

50. J. M. Sturtevant, "Calorimetry", Chapter 14 in *Physical Methods of Organic Chemistry*, 2d ed. (A. Weisssberger, ed), Interscience, New York, 1949, pp. 817–827.

51. L. D. Hansen, E. A. Lewis, and D. J. Eatough, in *Analytical Solution Calorimetry*, (J. K. Grime, ed) John Wiley, New York, 1985.

52. R. A. Winnike, D. E. Wurster, and J. K. Guillory, *Thermochim. Acta, 124*, 99 (1988).

53. L.-S. Wu, C. Gerard, and M. A. Hussain, *Pharm. Res., 10*, 1793 (1993).

54. L.-S. Wu, G. Torosian, K. Sigvardson, C. Gerard, and M. A. Hussain, *J. Pharm. Sci., 83*, 1404 (1994).

55. D. P. Ip, G. S. Brenner, J. M. Stevenson, S. Lindenbaum, A. W. Douglas, S. D. Klein, and J. A. McCauley, *Int. J. Pharm., 28*, 183 (1986).

56. J. J. Gerber, J. G. vanderWatt, and A. P. Lötter, *Int. J. Pharm.*, *73*, 137 (1991).

57. A. V. Katdare, J. A. Ryan, J. F. Bavitz, D. M. Erb, and J. K. Guillory, *Mikrochim. Acta*, *3*, 1 (1986).

58. S. A. Botha, M. R. Caira, J. K. Guillory, and A. P. Lötter, *J. Pharm. Sci.*, *77*, 444 (1988).

59. L.-C. Chang, M. R. Caira, and J. K. Guillory, *J. Pharm. Sci.*, *84*, 1169 (1995).

60. Y. Chikaraishi, M. Otsuka, and Y. Matsuda, *Chem. Pharm. Bull.*, *44*, 2111 (1996).

61. A. Chauvet, J. Masse, J.-P. Ribet, D. Bigg, J.-M. Autin, J.-M. Maurel, J.-F. Patoiseau, and J. Jaud, *J. Pharm. Sci.*, *81*, 836 (1992).

62. M. J. Jozwiakowski, N.-A. Nguyen, J. M. Sisco, and C. W. Spancake, *J. Pharm. Sci.*, *85*, 193 (1996).

63. L. J. Leeson and J. T. Carstensen, *Dissolution Technology*, American Pharmaceutical Association, Washington, DC, 1974.

64. W. A. Hanson, *Handbook of Dissolution Testing*, Pharmaceutical Technology Publication, Springfield, OR, 1982.

65. H. M. Abdou, *Dissolution, Bioavailability and Bioequivalence*, Mack, Easton, PA, 1989.

66. *United States Pharmacopoeia 23*, United States Pharmacopoeial Convention, Inc., Rockville, MD, 1994, pp. 1791–1793.

67. V. G. Levich, *Physicochemical Hydrodynamics*, Prentice-Hall, Englewood Cliffs, NJ, 1962.

68. W. C. Stagner and J. K. Guillory, *J. Pharm. Sci.*, *68*, 1005 (1979).

69. L. J. Ravin, E. G. Shami, and E. Rattie, *J. Pharm. Sci.*, *59*, 1290 (1970).

70. W. D. Walkling, H. Almond, V. Paragamian, N. H. Batuyios, J. A. Meschino, and J. B. Arpino, *Int. J. Pharm.*, *4*, 39 (1979).

71. M. Stoltz, A. P. Lötter, and J. G. van der Watt, *J. Pharm. Sci.*, *77*, 1047 (1988).

72. M. Stoltz, M. R. Caira, A. P. Lötter, and J. G. van der Watt, *J. Pharm. Sci.*, *78*, 758 (1989).

73. Y. Matsuda and S. Kawaguchi, *Chem. Pharm. Bull.*, *34*, 1289 (1986).

74. S. H. Yalkowsky, *Techniques of Solubilization of Drugs*, Marcel Dekker, New York, 1981, pp. 160–180.

75. R. K. Khankari and D. J. W. Grant, *Thermochim. Acta*, *248*, 61 (1995).

76. P. V. Allen, P. D. Rahn, A. C. Sarapu, and A. J. Vanderwielen, *J. Pharm. Sci.*, *67*, 1087 (1978).

77. H. Zhu, R. K. Khankari, B. E. Padden, E. J. Munson, W. B. Gleason, and D. J. W. Grant, *J. Pharm. Sci.*, *85*, 1026 (1996); H. Zhu, B. E.

Padden, E. J. Munson, and D. J. W. Grant, *J. Pharm. Sci., 86*, 418 (1997); H. Zhu, J. A. Halfen, V. G. Young, B. E. Padden, E. J. Munson, V. Menon, and D. J. W. Grant, *J. Pharm. Sci., 86*, 1439 (1997).

78. A. Kristl, S. Srcic, F. Vrecer, B. Sustar, and D. Vojinovic, *Int. J. Pharm., 139*, 231 (1996).

79. K. G. Mooney, M. A. Mintun, K. J. Himmelstein, and V. J. Stella, *J. Pharm. Sci., 70*, 13 (1981).

80. T. Higuchi, S. Dayal, and I. H. Pitman, *J. Pharm. Sci., 61*, 695 (1972).

81. C. Doherty and P. York, *Int. J. Pharm., 47*, 141 (1988).

82. T. Uchida, E. Yonemochi, T. Oguchi, K. Terada, K. Yamamoto, and Y. Nakai, *Chem. Pharm. Bull., 41*, 1632 (1993).

83. R. Li, P. T. Mayer, J. S. Trivedi, and J. J. Fort, *J. Pharm. Sci., 85*, 773 (1996).

84. M. C. Martínez-Ohárriz, C. Martín, M. M. Goñi, C. Rodríguez-Espinosa, M. C. Tros de Ilarduya-Apaolaza, and M. Sánchez, *J. Pharm. Sci., 83*, 174 (1994).

85. N. Laihanen, E. Muttonen, and M. Laaksonen, *Pharm. Dev. Tech., 1*, 373 (1996).

86. M. Lagas and C. F. Lerk, *Int. J. Pharm., 8*, 11 (1981).

87. W. I. Higuchi, P. D. Bernardo, and S. C. Mehta, *J. Pharm. Sci., 56*, 200 (1967).

88. A. J. Aguiar and J. E. Zelmer, *J. Pharm. Sci., 58*, 983 (1969).

89. S. A. Botha and D. R. Flanagan, *Int. J. Pharm., 82*, 195 (1992).

90. H. G. Ibrahim, F. Pisano, and A. Bruno, *J. Pharm. Sci., 66*, 669 (1977).

91. S. A. Botha, M. R. Caira, J. K. Guillory, and A. P. Lötter, *J. Pharm. Sci., 78*, 28 (1989).

92. M. S. Suleiman and N. M. Najib, *Int. J. Pharm., 50*, 103 (1989).

93. M. C. Tros de Ilarduya, C. Martín, M. M. Goñi, and M. C. Martínez-Ohárriz, *Drug Dev. Indust. Pharm., 23*, 1095 (1997).

94. S. Ghosh, D. A. Adsmond, and D. J. W. Grant, *J. Pharm Sci., 84*, 568 (1995).

95. J. W. Poole, G. Owen, J. Silverio, J. N. Freyhof, and S. B. Rosenman, *Current Therap. Res., 10*, 292 (1968).

96. J. W. Poole and C. K. Bahal, *J. Pharm. Sci., 57*, 1945 (1968).

97. C. MacLeod, H. Rabin, J. Ruedy, M. Caron, D. Zarowny, and R. O. Davies, *Can. Med. Assoc. J., 107*, 203 (1972).

98. A. A. Ali and A. Farouk, *Int. J. Pharm., 9*, 239 (1981).

99. B. E. Cabana, L. E. Willhite, and M. E. Bierwagen, *Antimicrob. Agents Chemotherap., 9*, 35 (1969).

100. M. Mayersohn and L. Endrenyi, *Can. Med. Assoc. J., 109*, 989 (1973).

101. S. A. Hill, K. H. Jones, H. Seager, and C. B. Taskis, *J. Pharm. Pharmacol., 27*, 594 (1975).
102. A. J. Aguiar, J. Krc, A. W. Kinkel, and J. C. Samyn, *J. Pharm. Sci., 56*, 847 (1967).
103. C. M. Anderson, *Aust. J. Pharm., 47*, S44 (1966).
104. M. A. Moustafa, A. R. Ebian, S. A. Khalil, and M. M. Motawi, *J. Pharm. Pharmacol., 23*, 868 (1971).
105. S. A. Khalil, M. A. Moustafa, A. R. Ebian, and M. M. Motawi, *J. Pharm. Sci., 61*, 1615 (1972).
106. N. Khalafallah, S. A. Khalil, and M. A. Moustafa, *J. Pharm. Sci., 63*, 861 (1974).
107. J. K. Haleblian, R. T. Koda, and J. A. Biles, *J. Pharm. Sci., 60*, 1485 (1971).
108. J. K. Haleblian, R. T. Koda, and J. A. Biles, *J. Pharm. Sci., 60*, 1488 (1971).
109. P. V. Allen, P. D. Rahn, A. C. Sarapu, and A. J. Vanderwielen, *J. Pharm. Sci., 67*, 1087 (1978).
110. Y. Fukumori, T. Fukuda, Y. Yamamoto, Y. Shigitani, Y. Hanyu, Y. Takeuchi, and N. Sato, *Chem. Pharm. Bull., 93*, 4029 (1983).
111. K. S. Murthy, N. A. Turner, R. U. Nesbitt, and M. B. Fawzi, *Drug Dev. Ind. Pharm., 12*, 665 (1986).
112. E. Laine, P. Kahela, R. Rajala, T. Heikkilä, K. Saarnivaara, and I. Piippo, *Int. J. Pharm, 38*, 33 (1987).
113. G. N. Kalinkova and Sv. Stoeva, *Int. J. Pharm, 135*, 111 (1996).

8

Effects of Pharmaceutical Processing on Drug Polymorphs and Solvates

Harry G. Brittain

Discovery Laboratories, Inc.
Milford, New Jersey

Eugene F. Fiese

Pfizer Central Research
Groton, Connecticut

I. INTRODUCTION

In the previous chapters, the structural origin, energetics, and thermodynamics of polymorphs and solvates have been largely described for pure chemical entities. In most of the studies reported, the compounds were intentionally converted among various polymorphic forms for the purpose of study. In the present chapter, we will discuss the *unintentional* conversion of polymorphs and the desolvation of hydrates upon exposure to the energetics of pharmaceutical processing. Environments as harsh as 80°C and 100% RH for up to 6 h are not unusual during the routine manufacture of dosage forms. As previously noted, the various crystalline polymorphs frequently differ in their heats of fusion by as little as 1 kcal/mol, with the transition temperature being well below the boiling point of water. In the case of hydrates, removal of water from the crystal lattice requires more energy but is very much dependent on the temperature and humidity history of the sample.

In this chapter we will discuss the effects of pharmaceutical processing upon the crystalline state of polymorphic and solvate systems. Given the degree of attention lavished on drug substances that is required by solid-state pharmaceutical [1] and regulatory [2] concerns, it is only logical that an equivalent amount of attention be paid to processing issues. A variety of phase conversions are possible upon exposure to the energetic steps of bulk material storage, drying, milling, wet granulation, oven drying, and compaction. In this setting, an environment as harsh as 80°C and 100% RH for up to 12 h is not unusual, and the mobility of water among the various components must be considered.

II. PRODUCTION AND STORAGE OF BULK DRUG SUBSTANCE

The first processing opportunity to effect a change in polymorphic form or solvate nature is with the final crystallization step in the synthesis of the bulk drug substance. Crystallization is thought to occur by first forming hydrogen-bonded aggregates in the solution state, followed by the buildup of molecules to produce a crystal nucleus. A number of parameters are known to affect the crystallization process, including solvent composition and polarity, drug concentration and degree of supersaturation, temperature and cooling rate during the crystallization process, presence of seed crystals and/or nucleation sites, additives that influence crystal habit or add strain to the crystal lattice, agitation, pH, and the presence of a salt-forming molecule. It is evident that ample opportunity exists for the appearance of a polymorphic change when a process is scaled up, or moved to a new site, or run by a new operator.

Since the discovery chemist would have been able to make gram quantities of a quasi-crystalline drug substance, logic holds that the process chemist should be able to make kilograms of the same substance in a GMP manufacturing setting. While Mother Nature and equilibrium may have been fooled at the bench-top (where reaction steps are short and yield is improved through the use of anti solvents), longer processing times and improvements in purity usually mean that thermodynamic equilibrium will be achieved for the first time at the scale-up stage. On the other hand, the need for high yield frequently is the chief motivating force early in a development program, so even the first scale-up phase often fails to produce the thermodynamically preferred polymorph. Eventually, either at the first scale-up site or when the process is moved to a new site, the thermodynamically preferred form will appear. This observation has been attributed to Gay-Lussac, who noted that unstable forms are frequently obtained first, and that these subsequently transform to stable forms.

Eventually, either at the first scale-up site or when the process is moved to a new site, the thermodynamically preferred form will appear.

When this change occurs late in development, retesting of the substance is required, to correct the analytical profile of the compound, as well as expensive clinical or toxicology testing. In the competitive environment of today, time is money, and delaying an NDA can be as costly as the expense of additional clinical or toxicological testing. It is the appearance of a thermodynamically preferred polymorph late in the development cycle that often requires the use of seed crystals during processing. One occasionally runs into legends of "whiffle dust," or seed crystals circulating in the air-handling system, which lead to the occurrence of the disappearing polymorphs discussed earlier in Chapter 1. Experienced process chemists who have tried to generate an unstable polymorph for an analytical standard generally support the tenets of this legend and contribute to its dissemination.

III. EFFECTS OF PARTICLE SIZE REDUCTION

The last processing step in the production of bulk drug substances usually involves milling to reduce the particle size distribution of the material. This is ordinarily conducted using the mildest conditions possible to render a sample homogeneous, or through the use of more rigorous milling to reduce the primary particle size in an effort to improve formulation homogeneity or dissolution and bioavailability. In the latter process, a substantial amount of energy is used to process the substance, which can result either in a polymorphic conversion or in the generation of an amorphous substance. The formation of an amorphous material is highly undesirable, since it will often be hygroscopic and more water soluble than any of the crystalline forms. The amorphous material must also be considered as being a thermodynamically metastable state of high energy, and a variety of pathways exist that can result in a back-conversion to a crystalline form of the material. This will certainly alter the dissolution characteristics and possibly even the bioavailability of the drug. In any case, the use of any rigorous milling process requires careful analysis for consequent changes in crystallinity or crystal form.

Hancock and Zografi have reviewed the characteristics of the amorphous state and methods whereby this form of a given material may be obtained [4], and this information is also covered in the final chapter of this book. Amorphous materials are disordered, have the highest free energy content, have the highest water solubility, and are usually hygroscopic. The effect of amorphous materials or amorphous regions within pharmaceutical solids is to convert the normal solid properties of high elasticity and brittleness to varying degrees of visco-elasticity that allow the materials to flow under the mechanical stress of milling or tableting. Upon heating, amorphous materials pass through a glass transition temperature and convert to a crystalline state. The effect of adsorbed moisture from the atmosphere or excipients is to act as a plasticizer, lowering the glass transition temperature, increasing molecular mobility in the solid, and allowing crystallization to occur at a lower temperature. Thus any rigorous milling of a drug substance requires careful analysis and monitoring for subsequent changes in crystallinity.

Much of the literature on the milling of polymorphs involves long grinding or ball-milling times, which impart unusual thermodynamic stress to the material and possibly lead to confusion about the real effects of milling on polymorphism. In practice, drugs are typically milled in a Bantam Mill or a Fitzpatrick Mill (alone or with excipients) where the exposure time of the material to stress is very short. One would predict that the thermodynamic impact would be very small, and therefore one might expect to observe little or no change in polymorphism arising from effects of industrial-scale milling.

Nevertheless, one can glean an understanding as to the effect of milling on polymorphism from ball-milling studies, such as those conducted by Miyamae et al. on (E)-6-(3,4-dimethoxy-phenyl)-1-ethyl-4-mesitylimino-3-methyl-3,4-dihydro-2(1H)-pyrimidinone [5]. In this work, the title compound was milled for up to 60 min and then stored at ambient conditions for up to 2 months. It was found that the unstable Form A (melting point 118°C) was converted to a noncrystalline solid as a result of the milling process but then crystallized upon storage to the stable Form B (melting point 141°C).

Fostedil exists in two polymorphic forms, characterized by melting points of 95.3°C (Form I) and 96.4°C (Form II), a free energy difference of only 71.8 cal/mol at 37°C, and distinctly different infrared absorption and x-ray powder diffraction patterns [6]. Solubility studies suggested that Form I was more stable than Form II. Mechanical mixing in an automated mortar showed that complete conversion of Form II to Form I occurred in 2 h. Milling fostedil in an industrial fluid energy mill and a hammer mill resulted in a similar conversion of Form II to Form I, presumably through the generation of nuclei on the surface of the crystal. The hammer milled material was exposed to less energy, which resulted in fewer seed nuclei and therefore less conversion to Form I even when exposed to elevated temperature and humidity. In contrast, since the fluidized energy mill provides more energy, its use generated more surface defect nuclei and greater conversion to the Form I polymorph. The presence of excipients also was found to exert a strong perturbation on the kinetics of the phase transformation. As illustrated in Fig. 1, the presence of microcrystalline cellulose retarded the conversion of Form II into Form I with grinding.

Chloramphenicol palmitate Form B was found to transform to the less therapeutically desirable Form A during grinding [7]. The transformation could be accelerated by the presence of appropriate seed crystals, and it was also found that the least stable Form C could be progressively converted from Form B and then to Form A if the grinding times were sufficiently long. In another study, it was shown that the temperature increases that accompanied the grinding process could accelerate

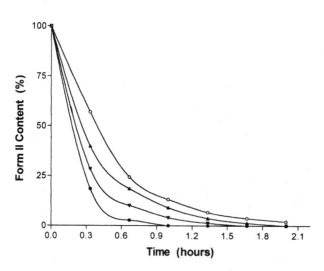

Fig. 1.   Effect of grinding time on the Form-II-to-Form-I phase transformation composition of pure fostedil (○), and also for fostedil/microcrystalline cellulose mixtures having ratios of 3:1 (▲), 1:1 (▼), and 1:3 (●). (The figure is adapted from data presented in Ref. 6.)

the phase conversion [8]. In fact, when one grinds chloramphenicol stearate for a sufficiently long period of time, Form III first turns into Form I, and ultimately into an amorphous state [9]. One interesting note of this latter study was the finding that the rate of phase conversion was accelerated by the presence of microcrystalline cellulose.

Clearly, the temperature maintained during the milling process can influence the degree of any polymorphic transitions. The α-and γ-forms of indomethacin could be converted to an amorphous solid during grinding at 4°C, but at 30°C the γ-form converted to the α-form, which could not be further transformed [10,11]. The authors concluded that although the noncrystalline form was stable at 4°C, it evidently was unstable with respect to crystallization at 30°C. An illustration of the crystal form present in samples of indomethacin as a function of grinding time is provided in Fig. 2.

The transformation behavior of phenylbutazone polymorphs during grinding at 4 and 35°C, and the solid-state stability and dissolution behavior of the ground materials, was investigated [12]. The α, β, and δ forms were transformed to the new ζ-form, which in turn was transformed to the ε-form (which was stable at 4°C). On the other hand,

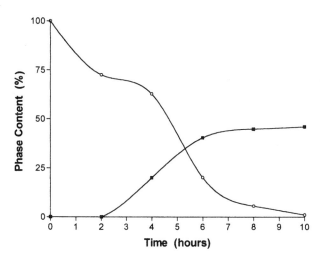

Fig. 2. Effect of grinding time (conducted at 30°C) on the phase composition of γ-indomethacin (○), showing also the α-phase (■) formed. (The figure is adapted from data presented in Ref. 11.)

during grinding at 35°C, the δ-form was not changed, while the α-form was transformed to the δ-form by way of the ζ-form. The β-form was apparently transformed directly to the δ-form. Once obtained by grinding, the ε-form was transformed to the δ-form under 0% relative humidity at various storage temperatures after an induction time of a few hours.

A common occurrence during grinding is the formation of an amorphous, noncrystalline phase. For instance, the grinding of cephalexin has been found to decrease the degree of crystallinity [13] as well as a wide range of physical properties that depend on the crystalline content of the material [14]. When the material becomes less crystalline, its stability decreases, and the hydrate forms become easier to dehydrate. Similar conclusions were reached with respect to the strength of the crystal lattice when the effect of grinding on the stability of cefixime trihydrate was studied [15]. The degree of interaction between the waters of hydration and the cefixime molecules was weakened by grinding, negatively affecting the solid-state stability.

Four modifications of cimetidine were prepared as phase-pure materials, and each was found to be stable during dry storage [16]. The milling process enabled the conversions of Forms B and C to Form A, while Form A transformed into Form D only upon nucleation. In all cases, the particle size reduction step resulted in substantial formation of the amorphous phase. Compression of the various forms into compacts did not yield any evidence for phase conversion.

The pitfalls that can accompany the development of the amorphous form of a drug substance were shown for the capsule formulation of (3R,4S)-1,4-bis(4-methoxy-phenyl)-3-(3-phenylpropyl)-2-azetidinone [17]. A variety of spectroscopic techniques were used to study the amorphous-to-crystalline phase transition, and it was determined that problems encountered with dissolution rates could be related to the amount of crystalline material generated in the formulation.

One general finding of milling studies conducted on hydrate species is that the grinding process serves to lower the dehydration temperatures of ground materials, facilitating the removal of lattice water and formation of an amorphous product. This behavior has been noted for cephalexin [13,14], cefixime [15], cyclophosphamide [18], 2-[(2-methylimidazoyl-1-yl)-methyl]-benzo[f]thiochromen-1-one [19], and lactitol [20], to name a few representative examples.

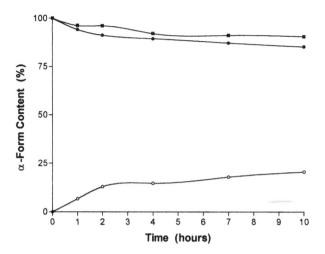

Fig. 3. Changes in α-form content for the α-monohydrate (■), α-anhydrate (●), and β-anhydrate (○) modifications of lactose with grinding time. (The figure is adapted from data presented in Ref. 22.)

Although most studies have been concerned with the state of the drug entity in either bulk or dosage form, milling can also affect the crystalline state of excipients. For example, isomerization was found to take place during the grinding of the α-monohydrate, α-anhydrate, and β-anhydrate forms of lactose [21]. The crystalline materials were all transformed into amorphous phases upon grinding, but these materials quickly re-sorbed water to approach the equilibrium composition ordinarily obtained in aqueous solutions. This behavior has been illustrated in Fig. 3. In a subsequent study [22], these workers determined the degree of crystallinity during various stages of the grinding process and concluded that while the isomerization rate of the α-monohydrate phase depended on the crystallinity, the rates for the two anhydrate phases depended on the content of absorbed water.

IV. EFFECTS DUE TO GRANULATION

When the requirements of a given formulation require a granulation step, this will represent an opportunity for a change in crystal form of the drug to take place. The problem will be most acute when a wet granulation step is used, since a potential transforming solvent is added to the blend of drug and excipients. The mixture is physically processed in the wet stage until homogeneity is obtained and then dried at a fairly significant temperature. The rationale for including a granulation step is to improve flow and blend homogeneity, permitting the high-speed compression of tablets. The granulation process is used to maintain the distribution of the drug throughout the formulation even during free flow, thus blend homogeneity.

Clearly, one can consider a wet granulation to be equivalent to a suspension of the drug entity in a mixture of solvent and excipients. Since the usual solvent is water, one can encounter a variety of interconversions between anhydrates and hydrates, or between hydrates and hydrates, which are mediated by the presence of the solvent. It is equally clear that one should not expect to be able to wet-granulate the metastable phase of a particular compound if that metastable phase is capable of transforming into a more stable form. A discussion of solvent-mediated phase transformations has been given in an earlier chapter and need not be repeated here.

It has been noted that chlorpromazine hydrochloride Form II exhibits severe lamination and capping when compressed, and that wet granulation with ethanol and water significantly improves the tableting characteristics [23]. The reason for this process advantage was shown to entail a phase change of the initial Form II to the more stable Form I during the granulation step. It was concluded that the improvements in tabletability and tablet strength that followed the use of wet granulation were due to changes in lattice structure that facilitated interparticulate bonding on compaction.

The effects of granulation solvent on the bulk properties of granules prepared using different polymorphic forms of carbamazepine have been studied [24]. It was found that although the anhydrate phase of the drug transformed to the dihydrate phase in the presence of 50% aqueous ethanol, the phase transformation did not take place when either pure water or pure ethanol was used as the granulating solution. This unusual finding was attributed to the fact that the solubility of the drug in 50% ethanol was 37 times higher than that in pure water. Granules processed using 50% ethanol were harder than those processed with pure solvents and led to the production of superior tablets. This work demonstrates the extreme need to evaluate the robustness of the granulation step during process development.

The crystal characteristics of a drug substance can undergo a sequence of reactions during wet granulation and subsequent processing. When subjected to a wet granulation step where 22% w/w water was added to dry solids, and the resulting mass dried at 55°C, an amorphous form of (S)-4-[[[1-(4-fluorophenyl)-3-(1-methylethyl)-1H-indol-2-yl]-ethynyl]-hydroxyphosphinyl]-3]-hydroxybutanoic acid, disodium salt, was obtained [25]. However, this amorphous gradually became crystalline upon exposure to humidity conditions between 33 and 75% RH. Depending on the exact value of the applied humidity, different hydrates of the drug substance could be obtained in the formulation.

V. EFFECTS DUE TO DRYING

Whenever a wet suspension containing a drug substance is dried, the possibility exists that a change in crystal state will take place. It has been amply shown in earlier chapters that anhydrates or amorphates can be produced by the simple desolvation of a solvate species, and such reactions are always possible during any drying step. The production of anhydrate phases by simple drying is one example of a reaction that can accompany the drying process, and this reaction type has been fully discussed in an earlier chapter.

In the simplest type of drying procedure, moist material from the wet granulation step is placed in coated trays and the trays are placed in drying cabinets with a circulating air current and thermostatic heat control. Historically, the tray drying method was the most widely used method, but more recently, fluid-bed drying is now widely used. In drying tablet granulations by fluidization, the material is suspended and agitated in a warm air stream. The chief advantages of the fluid-bed approach are the speed of drying and the degree of control that can be exerted over the process. The wet granulations are not dried to zero moisture; for a variety of reasons one seeks to maintain a residual amount of moisture in the granulation. It is evident from these simple considerations that the combination of solvent and drying conditions provides a suitable environment for the generation of new polymorphs or solvates when such conversion routes are available.

During ordinary manufacturing processes, crystals of the drug substance are dried under vacuum or through the circulation of warm dry air. Evaluation of wet and dried samples of the bulk substance by thermogravimetric analysis and hygroscopicity studies usually yields the necessary insight into the limiting parameters of the drying process. Since drying times can be relatively long and the environmental temperature mild (such as 24 to 48 h at 60°C), partial or total polymorphic conversion is possible. The likelihood that phase conversion will take place becomes increasingly less as the solvent is removed during drying. This area has been thoroughly covered earlier in Chapter 5, so only a few illustrative examples will be quoted at this time.

Shefter et al. used x-ray diffraction analysis to show that over-drying ampicillin trihydrate resulted in an unstable amorphous product, and that the drying of theophylline hydrate yielded a crystalline anhydrous form [26]. Kitamura et al. dehydrated (with vacuum) cefixime trihydrate to obtain a crystalline anhydrous form, which also proved to be unstable [27].

Otsuka et al. studied the effect of humidity and drying on the polymorphic transformation rate at 45°C for six forms of phenobarbital [28]. Figure 4 shows that the monohydrate (Form C) and the hemihydrate (Form E) slowly convert to the anhydrous Form B under high-humidity stress conditions (45°C and 75% RH), while the conversion is very fast at low humidity (45°C and 0% RH). This is reasonable, since the driving force to form anhydrous material is a dry environment. The key point of Figure 4 is the rapid conversion of phenobarbitol in less than 12 at 45°C and 0% RH, which represents very reasonable processing conditions for an industrial drying oven.

The sodium salt of 3-(((3-(2-(7-chloro-2- quinolinyl)ethenyl)phe-nyl)((3-dimethylamino)-3-oxo-propyl)thio)propanoic acid represents an unusual case of a hygroscopic drug substance, where the surface-active substance (in both crystalline and lyophilized forms) produced nonflowing, semisolid masses with exposure to increasing relative humidity [29]. The substance was shown to form first an amorphous material and then a mesomorphic phase, as the moisture sorption increased with exposure to relative humidity.

The favorite endpoint of the process chemist is to "dry to constant weight," but this laudable goal can lead to the production of a desol-

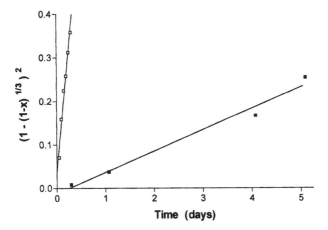

Fig. 4. Jander plots of one polymorphic transformation of phenobarbital. Illustrated are data for the conversion at 45°C of Form C (monohydrate phase) to Form B (anhydrate phase). The 0% relative humidity atmosphere points are given by the open symbols, and the 75% relative humidity atmosphere points are given by the closed symbols. (The figure is adapted from data presented in Ref. 28.)

vated product. Without doubt, the overdrying process has produced many anhydrous lots of bulk drug substance, which unfortunately are eventually found to be unstable.

Since the effects of simple drying have been more fully discussed in detail in Chapter 5 of this book, we will turn to case studies that illustrate how the use of two specialized drying methods can lead to phase interconversions.

A. Changes in Crystalline Form Accompanying the Spray-Drying Process

The spray-drying process first requires the formation of a slurry to be sprayed, which can be a concentrated solution of the agent to be dried or a dispersion of the agent into a suitable nondissolving medium. The dispersion is then atomized into droplets, which are exposed to a heated atmosphere to effect the drying process. Completion of the process yields a dry, freeflowing powder that ordinarily consists of spherical particles of relatively uniform particle size distribution. For instance, the morphology of spray-dried lactose has been contrasted with that of the monohydrate and anhydrous phases of lactose, with significant differences being reported [30].

One interesting feature of spray-drying is the possibility of controlling the final crystal form through manipulation of the drying conditions. For instance, three different crystalline forms of phenylbutazone were prepared from methylene chloride solution by varying the drying temperature of the atomized droplets between 30 and 120°C [31]. One of the isolated polymorphs could not be obtained by any crystallization procedure and could only be produced using extremely slow solvent evaporation rates.

Using phenobartitone and hydroflumethiazide as examples, it has been shown that high-energy solids can be produced through the use of spray-drying [32]. Commercially available phenobartitone is obtained as Form II, but with spray-drying the authors were able to obtain Form III having a large specific surface area. The analogous spray-drying of hydroflumethiazide yielded an amorphous solid. The spray drying of indomethacin produced a viscous glassy phase, which was found to be physically unstable with time [33]. Upon storage, the glassy solid converted into a mixture of crystalline Forms I and II. Representative x-ray powder diffraction patterns of these materials are shown in Fig. 5.

Spray-drying is often used to encapsulate drug substances into excipient matrices, but phase interconversions are still possible in such situations. Sulfamethoxazole Form I was microencapsulated with cellulose acetate phthalate and talc, colloidal silica, or montmorillonite clay by a spray-drying technique, and the drug was found to convert to Form II during the process [34]. Increasing the amount of cellulose acetate phthalate in the formulation led to increased amorphous drug content, but increasing the talc level yielded more polymorphic conversion. In a subsequent study, the authors showed that no phase conversion took place in formulations containing colloidal silica [35].

Frequently, the spray drying process yields amorphous materials, which undergo crystallization upon storage or exposure to suitable environmental conditions. The effect of various additives on the recrystallization of amorphous spray-dried lactose has been studied [36]. The

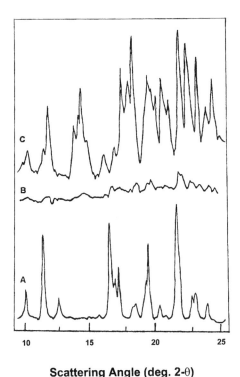

Scattering Angle (deg. 2-θ)

Fig. 5. X-ray powder diffraction patterns for (A) crystalline indomethacin, (B) spray-dried indomethacin, and (C) indomethacin spray-dried with 5% polyvinylpyrrolidone and stored for two months. (The figure is adapted from data presented in Ref. 32.)

presence of magnesium stearate was found to inhibit the recrystallization process, while layers of microcrystalline cellulose in the spray-dried products could result in long lag times prior to recrystallization. It was deduced that the onset of crystallization was critically related to the mobility of water in the formulations.

B. Changes in the Crystalline State of Lyophilized Products

Another drying procedure that can yield changes in drug crystal form is that of freeze-drying, or lyophilization. In this approach, the material to be dried is prepared as an aqueous solution or suspension and then frozen rapidly and cooled to a temperature below its eutectic point. The frozen formulation is then exposed to vacuum and the ice removed by sublimation. Since lyophilized products are often dried to moisture levels less than 1%, the materials are ordinarily hygroscopic and must be protected from adventitious water to prevent any unwanted phase transformation steps. In usual practice, the product is produced in its amorphous state, which is favorable for its subsequent solubilization at the time of its intended use. Consequently, the study of any possible moisture-induced crystallization is important to the characterization of a lyophilized product.

The effect of storage conditions on the crystalline nature of lyophilized ethacrynate sodium has been reported [37]. Samples of the drug substance were dried to various moisture levels and then stored at either 60°C or 30°C/75% RH. It was found that crystalline drug substance was obtained after short periods of time as long as sufficient levels of water were either present in the initial lyophilized solid or introduced by adsorption of humidity. The crystalline form of the drug was identified as a monohydrate phase.

Cefazolin sodium is capable of existing in a number of crystalline modifications, but the amorphous state is produced by the lyophilization process [38]. The amorphous form retained its nature at relative humidities less than 56% but converted completely to the pentahydrate above 75% RH. Interestingly, the hygroscopicity of the amorphous form was essentially similar to that exhibited by the dehydrated monohydrate or dehydrated α-form. Lamotrigine mesylate was also found to exhibit a moisture-dependent amorphous-to-crystalline transition [39]. The transition was found to be independent of the presence of mannitol, whether the amorphous solid was produced by lyophilization or by spray-drying.

Excipients in a lyophilized product can also undergo amorphous-to-crystalline phase transitions. The transformation of amorphous sucrose to its crystalline phase was studied as a function of applied relative humidity [40]. As illustrated in Fig. 6, the amorphous material gains up to 6% water when exposed to a relative humidity of 33%, but continued storage leads to crystallization of the anhydrous crystalline phase and a consequent expulsion of the initially adsorbed water. It

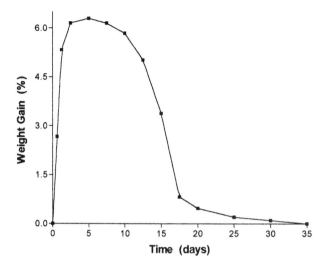

Fig. 6. Moisture update curve for amorphous sucrose, obtained at 23°C and a relative humidity of 33%. (The figure is adapted from data presented in Ref. 40.)

was hypothesized that the lag time that accompanied the amorphous-to-crystalline phase transition was required for the random motion of molecules to rotate and translate into the proper orientation to form a nucleus of sufficient size to allow crystallization.

Mannitol is commonly used as a bulking agent in lyophilization, owing to its tendency to freeze-dry into a mixture of crystalline and amorphous solids [41]. However, since mannitol can exist in a number of crystalline modifications, slight variations in the exact details of the lyophilization procedure can lead to the production of differing solids [42]. The effect of surface-active agents on the composition of lyophilized mannitol formulations has been studied, and it was reported that polysorbate 80 affected both the polymorphic state and the degree of crystallinity of freeze-dried mannitol [43]. When lyophilizing pure mannitol, one obtains a mixture consisting of the α, β, and δ forms in various ratios. In the presence of polysorbate 80, however, production of the δ-phase is favored.

VI. EFFECTS DUE TO COMPRESSION

It is often assumed that the crystalline form of a drug entity will remain unchanged during the compression steps associated with the manufacture of tablets. The reason for this line of thought is that under the applied load of tableting, the softer excipients should deform preferentially to the relatively hard crystalline drug substance. As tableting speeds increase towards commercial production, exposure times to stress decrease and one would anticipate even less chance for crystalline conversion. For the production of many substances, this situation is certainly true. There are numerous examples, however, which do not follow the general rule, and where changes in polymorph or solvate form accompany the compression step.

Effects due to compression can be further divided two categories. One of these concerns the phase transformations that accompany the compaction and consolidation steps of tablet manufacture. The other relates to the effect on tablet properties that arise from the use of different crystal forms in a direct compression tablet formulation. Case studies pertaining to each of these categories will be discussed in turn.

A. Changes in Crystal Form Effected by Compaction

In one study of wide scope, 32 drugs known to exist in different polymorphic states were subjected to a trituration test for possible phase transitions [44]. Out of 11 transforming substances, detailed studies of tableting were conducted on caffeine, sulfabenzamide, and maprotiline hydrochloride. Tablets were produced and sectioned, so that thermal analysis could be used to investigate polymorphic changes at the upper and lower surfaces, middle region, and sides of the compacts. Relationships were examined between the extent of transformation and the compression pressure and energy, as well as the effect of drug particle size. It was deduced that the form with the lowest melting point would have the least degree of intermolecular attractive forces and probably the least yield stress values at a given temperature. This conclusion was supported by the observation that the compression of metastable phases brought about a transformation to the most stable phase. As illustrated for the particular example of caffeine in Fig. 7, the extent of transforma-

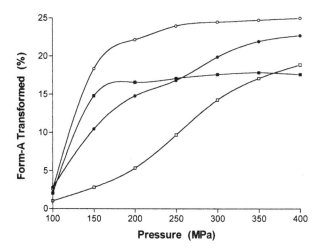

Fig. 7. Percentage of caffeine Form A transformed into Form B as a function of applied pressure. Phase composition data are shown for the upper surface (○), lower surface (□), middle region (■), and side (●) of compressed tablets. (The figure is adapted from data presented in Ref. 44.)

tion was found to depend on the zone of the tablet, the pressure applied, and the particle size of the powder.

Carbamazepine is a drug where irregular plasma levels can be obtained owing to differences in dissolution rates that are associated with the production of differing polymorphs by tablet manufacturing operations. The behavior of the α, β, and dihydrate crystalline forms during compression has been investigated, as well as the possibility of crystalline structural changes accompanying grinding and tableting conditions [45]. Grinding was performed in a ball mill for 15 and 60 minutes, while compression was carried out using an instrumented single punch machine. It was found that although the dihydrate form exhibited the best compressibility characteristics, it was not phase-stable with respect to compression. The α-form exhibited the best stability, but the use of this phase induces sticking in the tablets. It appears that the commercially available β-form represents the best compromise between performance and stability, and it was found to remain stable under normal conditions of fabrication and storage.

The effect of tablet compression mechanical energy on the polymorphic transformation of chlorpropamide has been examined [46]. A single-punch eccentric tableting machine with a load cell and a noncontact displacement transducer was used to measure compression stress, distance, and energy. Both the stable Form A and the metastable Form C were compressed at a stress of 196 MPa at room temperature. The compression cycle was repeated from 1 to 30 times, and the phase composition of deagglomerated tablets was measured using powder x-ray diffraction. It was found that each form could be transformed into the other by the mechanical energy input during tableting, with the composition reaching equilibrium above 100 J/g of compression energy after more than 10 cycles. It was hypothesized that the phase transformation required the conversion of the crystalline forms to an amorphous intermediate, which then transformed into the equilibrium ratio of forms A and C.

The effect of compressional force on the polymorphic transition in piroxicam has been examined for phase-pure materials, where the needle-like α-phase was found to convert to the cubic β-phase during compression [47]. The phase transformation took place only at higher applied forces, such that formation of the α-phase could only be observed after compression to a tablet hardness of 9 kg/cm^2. Compression of the β-phase yielded no changes in its crystalline form.

The extent of the polymorphic transformation of anhydrous caffeine has been studied as a function of grinding time and compression pressure [48]. It was found that both grinding and compression induced the transformation from the metastable Form I into the stable Form II. The transformation could be observed even after only 1 min of grinding, or by using a tableting compression pressure of about 50 MPa. Quantitative measurements of phase conversion indicated that the degree of transformation was greater near the surface of the tablet than in its interior.

The effect of temperature on polymorphic transformations that take place during compression has been studied. Tableting temperature was found to exert a definite effect on the polymorphic transformation of chlorpropamide during compression, and on the physical properties of the produced tablets [49]. Compression stress, distance, and energy were measured using a noncontact displacement transducer mounted within a temperature-controlled single punch eccentric tableting machine with two load cells (upper and lower punches). The phase composition of deagglomerated compacts was subsequently measured using powder x-ray diffraction to calculate the polymorphic content. It was learned that the amount of Form C produced at 45°C by the transformation of Form A was about twice that produced at 0°C using the same compression energy. The amount of Form A transformed from Form C by compression at 45°C was almost the same as that at 0°C. This suggested that the compressional effect on Form A depended on the compression temperature, while the effect on Form C was independent of temperature. The crushing strength of tablets produced using Form A was about twice that obtained for tablets produced from Form C, even when they were compressed at the same porosity.

A study of the effect of temperature on the polymorphic transformation and compression of chlorpropamide Forms A and C during tableting has been reported [50]. With the aid of apparatus similar to that previously described [49], these authors used multitableting at room temperature and single tableting at 0–45°C to effect compression. In the first method, the stable Form A or the metastable Form C was loaded into the die and the sample was compressed with a compression stress of 196 MPa. It was found that both forms were mutually transformed, and as evident in Fig. 8, an equilibrium ratio of Forms A and C was attained above roughly 200 J/g of compression energy. In the second method, the amount of Form C transformed from Form A at 45°C was about two times larger than that at 0°C at the same compression energy. The crushing strength of tablets prepared from Form A was about twice that of tablets manufactured using Form C as the source of drug substance.

A further study was conducted to study the effect of compression temperature on the consolidation mechanism of the polymorphs of chlorpropamide [51]. The effect of environmental temperature on the compression mechanism of the A and C forms of chlorpropamide was investigated with an eccentric type tableting machine with two load cells and a noncontact displacement transducer. The temperature of the sample powders was controlled at 0 and 45°C, and these were compressed at almost 230 MPa. The tableting dynamics were evaluated by Cooper and modified Heckel analyses. The results suggest that particle

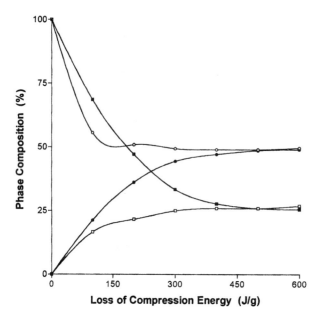

Fig. 8. Relationship between phase composition and compression energy during the compaction of different polymorphs of chlorpropamide. Shown are data for the residual content of Form A during the compression of Form A (○), the amount of Form C formed during the compression of Form A (□), the residual content of Form C during the compression of Form C (■), and the amount of Form A formed during the compression of Form C (●). (The figure is adapted from data presented in Ref. 50.)

brittleness or plasticity was affected by compression at different temperatures. The higher tablet hardness of Form A tablets produced at 45°C was thought to be caused by the increased plasticity of primary particles, whereas the hardness of tablets containing Form C (and also produced at 45°C) was ascribed to the decreased size of the secondary particles. Part of the reasoning behind this conclusion was the observation that Form A melted approximately 8°C lower than did Form C and would therefore be more plastic with respect to deformation.

Sebhatu et al. studied the tableting characteristics of spray-dried lactose [52], which typically contains 15% amorphous material. The spray-dried material is known to absorb moisture when exposed to high levels of relative humidity, which lowers the glass transition temperature of dry material from 104°C to 37°C (at moisture content 7.17%). It was found that tablets made from spray-dried lactose that had been stored for less than 4 h at 57% RH before compaction showed an increased in postcompaction tablet hardness when stored at 57% RH. This effect was ascribed to crystallization of the amorphous regions brought into close contact by compression. On the other hand, spray-dried lactose stored at 57% RH for more than 6 h prior to compaction showed no change in postcompression hardness for tablets stored at 57% RH because the amorphous regions had completed crystallization prior to compaction. One concludes that any tableted product initially containing a large portion of amorphous material (either active drug or inactive excipient) may be subject to changes in tablet hardness upon storage in high humidity environments.

B. Effects on Tablet Properties Associated with the Use of Different Crystal Forms

The dissolution rate of phenylbutazone from disintegrated tablets has been used to determine whether the drug particles underwent crushing or bonding during compression [53]. By using two polymorphic forms of the drug substance, it was shown that the predominant effect during compression in formulations containing high drug concentrations was dependent upon the original particle size of the active and on its polymorphic state. In formulations having a low drug concentration, the excipients effectively prevented the drug particles from bonding together. Lactose was found to exert a more abrasive action (relative to Avicel) on Form A but had little effect on the more ductile Form B. This effect is illustrated in Fig. 9, where the dissolution rates of tablets produced from the two polymorphs are contrasted. The effect of using differing particle size ranges of input material is particularly evident in the dissolution data of tablets prepared using the finest grade of drug substance. A second aspect of this work concerns the analysis of how the rate of compression during tableting affects the dissolution time. For larger particle sizes of raw material, no difference was observed on passing from 0.26 to 29 of compression time. For the smallest grade of input material, a significant difference was observed. The authors

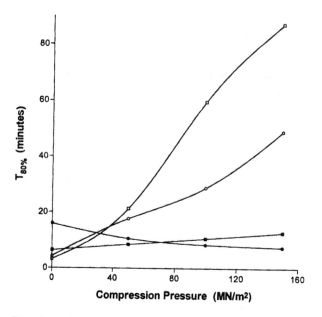

Fig. 9. Change in dissolution rate of disintegrated tablets of phenylbutazone as a function of particle size, compression pressure, and crystal form. Shown are the profiles for tablets prepared from Form A of 6 μm particle size (○), tablets prepared from Form B of 6 μm particle size (□), tablets prepared from Form A of 137 μm particle size (●), and tablets prepared from Form B of 146 μm particle size (■). (The figure is adapted from data presented in Ref. 53.)

concluded that a decrease in the rate of compression from 29 s to 0.26 s resulted in less bond formation within the tablet matrix and a more rapid dissolution.

The work of Tuladhar et al. [53] is significant because industrial tableting machines operate typically at 200 revolutions per minute. This means 0.3 seconds per revolution (i.e., the exposure time to compression), which is far less than the prolonged exposure on a manual Carver Press (on the order of minutes) or on some of the single-station tablet machines presented earlier (3 revolutions per minute). Prolonged exposure during compression may lead to erroneous conclusions about the significance of polymorphic transformation during tableting. Nevertheless, some interesting information can be derived from such studies if they are of proper design.

Others have sought to develop new methods for a better evaluation of the relation between the mechanical strength of solids and the polymorphic form used to produce these [54]. In order to evaluate the role played by polymorphism in the mechanical strength of solid dosage forms, and to minimize the influence of other factors, a melted disc technology was proposed. Tablet-shaped discs of zero porosity were prepared by melting powder and subsequent crystallization in the desired modification. Using phenobartitone as a model drug, discs consisting of Forms I, II, and III, as well as the amorphous form, were prepared. An evaluation of the mechanical properties of the discs was conducted, primarily through determinations of bending strength. It was found that the amorphous form and Form III yielded the strongest discs and would therefore probably be the most suitable materials for tablet manufacture.

The poor compression of paracetamol was addressed through the preparation of a new orthorhombic form [55]. Owing to its well-known poor compressional properties, commercially available paracetamol materials for direct compression are compounds of paracetamol with gelatin, polyvinylpyrrolidone, starch, or starch derivatives. Since a chemically pure paracetamol that could be used for direct compression would constitute a better compendial article, a new polymorph was produced. The new form was recrystallized from dioxane, and its crystals were found to consist of sliding planes that led to good compressibility. However, the orthorhombic form is metastable with respect to the monoclinic form; phase conversion was observed if the raw material contained greater than 20% of the monoclinic phase. However, the dissolution rates for the two forms were found to be similar, and therefore any questions as to the relative bioavailability of the two forms would probably be meaningless.

Compressive deformation studies were conducted on single crystals of α-lactose, using mechanical strength and acoustic emission analyses to deduce any possible differences in the deformation behavior of α-lactose monohydrate and anhydrous α-lactose [56]. The α-lactose monohydrate crystals were found to exhibit greater mechanical strength than did the anhydrous α-lactose crystals. The acoustic emission data showed that the fragmentation process of the monohydrate phase was acoustically more active and energetic. Amplitude distribution analysis of the acoustic signals further confirmed that the nature of fragmentation during the deformation of the two types of lactose was different. This was attributed to fundamental differences in the internal crystal structure of the two lactose types. This work showed that mechanical strength and acoustic emission analyses are able to provide significant insights into the fundamental deformation characteristics of different polymorphic or solvate species.

The compressional behavior of four polymorphs of mannitol (the α, β, and δ forms, as well as one unidentified phase) has been studied [57]. It was found that the compressibility of the α-phase is superior, and fortunately this phase is the major form in most commercial products. The particle shape was found to exert an influence upon the compressibility properties; granulated powders show better behavior than native crystalline powders. It was reported that no polymorphic transitions took place under the compression stresses used during the tableting process.

VII. SUMMARY

The key parameters to consider at the bulk drug substance production stage are prolonged processing times due to batch size, improved product purity, a lengthy final crystallization step to improve purity, improve yield, or increase particle size, drying conditions for the product, and milling to improve homogeneity or reduce particle size. In all these steps, the small energy barriers between crystalline forms are easily overcome, resulting in the wrong polymorphic form, a desolvated product, or an amorphous product.

Once the kilogram bulk has been produced in its preferred polymorphic form, or salt form, or solvate form, it will proceed to the pharmaceutics department, where it will be mixed with excipients, exposed to processing conditions, and converted to a marketable dosage form. The pharmaceutical scientist must formulate this material into a dosage form that is homogeneous, scalable, stable, and bioavailable. Since most dosage forms are solid dosage forms, capsules or tablets, this effort will focus primarily on those processing parameters that are operative in the production of solid dosage forms.

Solution dosage forms are ordinarily independent of polymorphic problems, but they require one warning. If there should happen to exist a less soluble form, it will appear upon temperature cycling stability testing. The reason for this solution state warning is that temperature cycling is the most severe challenge to solubility, and if one should generate seed crystals of a less soluble form of a compound during cooling, then equilibrium will rapidly be established. The final stage of this scenario will be precipitation or crystal growth. The classic cases in the industry are the soluble anhydrous material converting to an insoluble hydrate upon stability testing, or moisture leakage into a delivery system.

In solid dosage forms, the first opportunity for a change in crystallinity will be in the blending and milling of drug with excipients to produce a homogeneous blend. The excipients in a formulation can exert a strong influence on polymorphic conversion and may create new pathways that did not exist for the pure drug substance. One should expect polymorphs with similar thermodynamic energies to be prone to substantial conversion during a milling operation.

The key parameters that cause polymorphic conversion or dehydration of hydrates at the formulation stage are milling and granulation wetting. Other factors that can be important are drying, tableting, and drying of the film coating. In addition, one can have migration of water between drug substance, excipients, and capsule shells. All these parameters can become more important during prolonged processing times should the batch size become sufficiently large.

During mechanical treatment, the most likely mechanism for polymorphic interconversion is that of nucleation and growth of a second phase within the original phase. Nucleation proceeds from dislocation sites in a crystal, since their higher free energy ensures that the energy needed for transformation is lower at these sites. In the majority of cases, metastable phases transform to the most stable form. The degree of polymorphic conversion will depend on the relative stability of the phases in question, and on the type and degree of mechanical processing applied. These phase conversions are generally modulated by the presence of excipients in the formulation, and consequently they must be investigated in the dosage form as well as in the bulk drug substance.

The goal in any industrial pharmaceutical organization is to have the thermodynamically preferred polymorph or solvate present in the first scaled-up batch of drug substance. If this situation is achieved, then all toxicology, pharmacokinetic, and clinical studies will be conducted with the crystalline form that is likely to be the commercial form of the drug substance. This will eliminate expensive retesting should a more stable but previously unknown polymorph appear. Prudent drug development programs will identify the preferred crystalline form early in development, with a polymorph/salt group working in close conjunction with process chemists to make the thermodynamically preferred polymorphic form the initial kilogram-scale batch. They will also identify and define the physicochemical boundaries of that polymorph.

In the end, however, there is no substitute for multidisciplinary studies whose goal is to determine the likelihood of polymorphic interconversions at any time during the handling of the drug substance. Fortunately, the wide variety of available characterization methods makes it possible to detect virtually any problem that could be encountered during the course of drug development. The successful approach to problem solving will require the full collaboration of process, formulation, and analytical scientists and the complete sharing of information. Although problems in development invariably occur at the worst possible moment, the judicious use of information previously gathered from properly designed studies should minimize their impact and lead to ready deductions of methods for their elimination.

REFERENCES

1. S. R. Byrn, R. R. Pfeiffer, G. Stephenson, D. J. W. Grant, and W. B. Gleason, *Chem. Materials, 6,* 1148 (1994).
2. S. R. Byrn, R. R. Pfeiffer, M. Ganey, C. Hoiberg, and G. Poochikian, *Pharm. Res., 12,* 945 (1995).
3. K. C. Johnson and A. C. Swindell, *Pharm. Res., 13,* 1795 (1996).

4. B. Hancock and G. Zografi, *J. Pharm. Sci., 80*, 1 (1997).
5. A. Miyamae, S. Kitamura, To. Tada, S. Koda, and T. Yasuda, *J. Pharm. Sci., 80*, 995 (1991).
6. Y. Takahashi, K. Nakashima, T. Ishihara, H. Nakagawa, and I. Sugimoto, *Drug Dev. Ind. Pharm., 11*, 1543 (1985).
7. M. Otsuka and N. Kaneniwa, *J. Pharm. Sci., 75*, 506 (1986).
8. M. M. De Villiers, J. G. van der Watt, and A. P. Lötter, *Drug Dev. Indust. Pharm., 17*, 1295 (1991).
9. F. Forni, G. Coppi, V. Iannuccelli, M. A. Vandelli, and M. T. Bernabei, *Drug Dev. Indust. Pharm., 14*, 633 (1988).
10. M. Otsuka, T. Masumoto, and N. Kaneniwa, *Chem. Pharm. Bull., 34*, 1784 (1986).
11. M. Otsuka, K. Otsuka, and N. Kaneniwa, *Drug Dev. Indust. Pharm., 20*, 1649 (1994).
12. T. Masumoto, J.-I. Ichikawa, N. Kaneniwa, and M. Otsuka, *Chem. Pharm. Bull., 36*, 1074 (1988).
13. M. Otsuka and N. Kaneniwa, *Chem. Pharm. Bull., 31*, 4489 (1983).
14. M. Otsuka and N. Kaneniwa, *Chem. Pharm. Bull., 32*, 1071 (1984).
15. S. Kitamura, A. Miyamae, S. Koda, and Y. Morimoto, *Int. J. Pharm., 56*, 125 (1989).
16. A. Bauer-Brandl, *Int. J. Pharm., 140*, 195 (1996).
17. R. J. Markovich, C. A. Evans, C. B. Coscolluela, S. A. Zibas, and J. Rosen, *J. Pharm. Biomed. Anal., 16*, 661 (1997).
18. J. Ketolainen, A. Poso, V. Viitasaari, J. Gynther, J. Pirttimäki, E. Laine, and P. Parroen, *Pharm. Res., 12*, 299 (1995).
19. S. Ito, Y. Kobayashi, M. Nishmura, K. Masumoto, S. Itai, and K. Yamamoto, *J. Pharm. Sci., 85*, 1117 (1996).
20. K. Yajima, A. Okhira, and M. Hoshino, *Chem. Pharm. Bull., 45*, 1677 (1997).
21. M. Otsuka, H. Ohtani, N. Kaneniwa, and S. Higuchi, *J. Pharm. Pharmacol., 43*, 148 (1991).
22. M. Otsuka, H. Ohtani, K. Otsuka, and N. Kaneniwa, *J. Pharm. Pharmacol., 45*, 2 (1993).
23. M. W. Y. Wong and A. G. Mitchell, *Int. J. Pharm., 88*, 261 (1992).
24. M. Otsuka, H. Hasegawa, and Y. Matsuda, *Chem. Pharm. Bull., 45*, 894 (1997).
25. K. R. Morris, A. W. Newman, D. E. Bugay, S. A. Ranadive, A. K. Singh, M. Szyper, S. A. Varia, H. G. Brittain, and A. T. M. Serajuddin, *Int. J. Pharm., 108*, 195 (1994).
26. E. Shefter, H.-L. Fung, and O. Mok, *J. Pharm. Sci., 62*, 791 (1973).

27. S. Kitamura, S. Koda, A. Miyamae, T. Yasuda, and Y. Morimoto, *Int. J. Pharm., 59,* 217 (1990).
28. M. Otsuka, M. Onoe, and Y. Matsuda, *Pharm. Res., 10,* 577 (1993).
29. E. B. Vadas, P. Toma, and G. Zografi, *Pharm. Res., 8,* 148 (1991).
30. H. G. Brittain, S. J. Bogdanowich, D. E. Bugay, J. DeVincentis, G. Lewen, and A. W. Newman, *Pharm. Res., 8,* 963 (1991).
31. Y. Matsuda, S. Kawaguchi, H. Kobayashi, and J. Nishijo, *J. Pharm. Sci., 73,* 173 (1984).
32. O. I. Corrigan, K. Sabra, and E. M. Holohan, *Drug Dev. Indust. Pharm., 9,* 1 (1983).
33. O. I. Corrigan, E. M. Holohan, and M. R. Reilly, *Drug Dev. Indust. Pharm., 11,* 677 (1985).
34. H. Takenaka, Y. Kawashima, and S. Y. Lin, *J. Pharm. Sci., 70,* 1256 (1981).
35. Y. Kawashima, S. Y. Lin, and H. Takenaka, *Drug Dev. Indust. Pharm., 9,* 1445 (1983).
36. G. Buckton and P. Darcy, *Int. J. Pharm., 121,* 81 (1995).
37. R. J. Yarwood and J. H. Collett, *Drug Dev. Indust. Pharm., 11,* 461 (1985).
38. T. Osawa, M. S. Kamat, and P. DeLuca, *Pharm. Res., 5,* 421 (1988).
39. E. Schmitt, C. W. Davis, and S. T. Long, *J. Pharm. Sci., 85,* 1215 (1996).
40. J. T. Carstensen and K. van Scoik, *Pharm. Res., 7,* 1278 (1990).
41. R. H. M. Hatley, *Dev. Biol. Stand., 74,* 105 (1991).
42. N. A. Williams and T. Dean, *J. Parenter. Sci. Tech., 45,* 94 (1991).
43. R. Haikala, R. Eerola, V. P. Tanninen, J. Yliruusi, *PDA J. Pharm. Sci. Tech., 51,* 96 (1997).
44. H. K. Chan and E. Doelker, *Drug Dev. Indust. Pharm., 11,* 315 (1985).
45. C. Lefebvre, A. M. Guyot-Hermann, M. Draguet-Bruchmans, and R. Bouché, *Drug Dev. Indust. Pharm., 12,* 1913 (1986).
46. M. Otsuka, T. Matsumoto, and N. Kaneniwa, *J. Pharm. Pharmacol., 41,* 665 (1989).
47. G. A. Ghan and J. K. Lalla, *J. Pharm. Pharmacol., 44,* 678 (1991).
48. J. Pirttimäki, E. Laine, J. Ketolainen, and P. Paronen, *Int. J. Pharm., 95,* 93 (1993).
49. T. Matsumoto, N. Kaneniwa, S. Higuchi, and M. Otsuka, *J. Pharm. Pharmacol., 43,* 74 (1991).
50. M. Otsuka and Y. Matsuda, *Drug Dev. Indust. Pharm., 19,* 2241 (1993).
51. M. Otsuka, T. Matsumoto, S. Higuchi, K. Otsuka, and N. Kaneniwa, *J. Pharm. Sci., 84,* 614 (1995).
52. T. Sebhatu, A. A. Elamin, and C. Ahlneck, *Pharm. Res., 11,* 1233 (1994).

53. M. D. Tuladhar, J. E. Carless, and M. P. Summers, *J. Pharm. Pharmacol., 35*, 269 (1983).
54. S. Kopp, C. Beyer, E. Graf, F. Kubelt, and E. Doelker, *J. Pharm. Pharmacol., 41*, 79 (1989).
55. P. Di Martino, A.-M. Guyot-Hermann, P. Conflant, M. Drache, and J.-C. Guyot, *Int. J. Pharm., 128*, 1 (1996).
56. D. Y. T. Wong, M. J. Waring, P. Wright, and M. E. Aulton, *Int. J. Pharm., 72*, 233 (1991).
57. B. Debord, C. Lefebvre, A. M. Guyot-Hermann, J. Hubert, R. Bouché, and J. C. Guyot, *Drug Dev. Indust. Pharm., 13*, 1533 (1987).

9

Structural Aspects of Molecular Dissymmetry

Harry G. Brittain

Discovery Laboratories, Inc.
Milford, New Jersey

I. INTRODUCTION

The molecular chirality associated with an optically active molecule is ordinarily manifested in the crystallography of the compound [1]. Since the historical development of optical activity was greatly aided by systematic studies of the habits of enantiomorphic crystals, the concepts of molecular dissymmetry, crystallography, and chirality are inexorably linked. As with any other solid, these materials can be characterized on the basis of their crystal structures and through an understanding of their melting point phase diagrams.

It has become abundantly clear that the stereoselective actions associated with the enantiomeric constituents of a racemic drug can differ markedly in their pharmacodynamic or pharmacokinetic properties [2–5]. These factors can lead to much concern, especially if a drug containing a potentially resolvable center is marketed as a racemic mixture. This situation is not invariably bad, but it is clear that a racemate should not be administered when a clear-cut advantage exists with the use of a resolved enantiomer [6]. An excellent discussion regarding the possible selection of a resolved enantiomer over a racemate, from both a practical and a regulatory viewpoint, has been provided by De Camp [7].

When the mirror images of a compound are not superimposable, these mirror images are called enantiomers. Numerous techniques exist that permit the physical separation of enantiomers in a mixture [8], but a discussion of these is outside the scope of this work. Individual enantiomeric molecules are completely equivalent in their molecular properties, with the exception of their interaction with circularly polarized light. An equimolar mixture of two enantiomers is termed a racemic mixture. The generally accepted configurational nomenclature for tetrahedral carbon enantiomers was devised by Cahn, Ingold, and Prelog and is based on sequencing rules [9]. Enantiomers are identified as being either R or S, depending on the direction (clockwise or counterclockwise) of substituents after they have been arranged according to increasing atomic mass. Compounds containing more than one center of dissymmetry are identified as diastereomers, and in compounds containing n dissymmetric centers the number of diastereomers will equal 2^n.

Although they do not fit the formal definition of a polymorphic solid, the relationship between racemic mixtures and resolved enantiomers presents an intriguing parallel. By virtue of symmetry constraints, a resolved enantiomer must crystallize in a noncentrosymmetric space group. Racemic mixtures are under no analogous constraint, but over 90% of all racemic mixtures (which are merely equimolar mixtures of the enantiomers) are found to crystallize in a centrosymmetric space group [10]. Thus we immediately reach the intriguing situation where differing crystal structures can be obtained for the same chemical compound, depending only on the degree of resolution. Since this type of behavior falls within the confines of this book, we will therefore explore the relation between molecular and crystallographic chirality and examine some of the consequences of these relationships.

II. ENANTIOMER AND RACEMATE CRYSTAL STRUCTURES

A relatively small number of the possible crystallographic lattice symmetries are available for the crystals of separated enantiomers, since these must crystallize in a lattice structure totally devoid of inverse elements of symmetry. Out of the 230 possible space groups belonging to the 32 crystal classes, only 66 space groups within 11 crystal classes are noncentrosymmetric and can accommodate homochiral sets of enantiomers. Racemates are permitted to crystallize in any of the 230 space groups but are not restricted to crystallizing in a centrosymmetric group. It is found that most racemates crystallize into a group that possesses some elements of inverse symmetry. As will be discussed later, racemic mixtures occasionally crystallize in an enantiomorphic lattice system and consequently spontaneously resolve the enantiomers.

A symmetry element is defined as an operation that when performed on an object results in a new orientation of that object that is indistinguishable from and superimposable on the original. There are five main classes of symmetry operations: (a) the identity operation (an operation that places the object back into its original orientation), (b) proper rotation (rotation of an object about an axis by some angle), (c) reflection plane (reflection of each part of an object through a plane bisecting the object), (d) center of inversion (reflection of every part of an object through a point at the center of the object), and (e) improper rotation (a proper rotation combined with either an inversion center or a reflection plane). Every object possesses some element or elements of symmetry, even if this is only the identity operation.

The rigorous group theoretical requirement for the existence of chirality in a crystal or a molecule is that no improper rotation elements be present. This definition is often trivialized to require the absence of either a reflection plane or a center of inversion in an object, but these two operations are actually the two simplest improper rotation symmetry elements. It is important to note that a chiral object need not be totally devoid of symmetry (i.e., asymmetric); it must merely be dissymmetric (containing no improper rotation symmetry elements). The tetrahedral carbon atom bound to four different substituents may be asymmetric, but the reason it represents a site of chirality is by virtue of its dissymmetry.

Jacques et al. have evaluated the compilations published by various authors and reported that 70–90% of homochiral enantiomers crystallize in the $P2_12_12_1$ or $P2_1$ space groups [10]. The most frequently encountered chiral space group, $P2_12_12_1$, is orthorhombic, which on a macroscopic level translates to a system consisting of three orthogonal binary axes. The unit cell commonly consists of four homochiral molecules that are related to each other by three binary screw axes. Space group $P2_1$ belongs to the monoclinic system and is characterized by a plane of symmetry and a twofold axis. The cell generally contains two molecules related by a binary screw axis.

Jacques et al. have also concluded that among the 164 space groups possessing at least one element of inverse symmetry, 60–80% of racemic compounds crystallize in either the $P2_1c$, $C2/c$, or $P\bar{1}$ space groups [10]. The most common group is $P2_1c$, which belongs to the monoclinic system, and whose unit cell contains two each of the opposite enantiomers related to one another by a center of symmetry and a binary screw axis.

A chirality classification of crystal structures that distinguishes between homochiral (type A), heterochiral (type B), and achiral (type C) lattice types has been provided by Zorkii, Razumaeva, and Belsky [11] and expounded by Mason [12]. In the type A structure, the molecules occupy a homochiral system, or a system of equivalent lattice positions. Secondary symmetry elements (e.g., inversion centers, mirror or glide planes, or higher-order inversion axes) are precluded in type A lattices. In the racemic type B lattice, the molecules occupy heterochiral systems of equivalent positions, and opposite enantiomers are related by secondary lattice symmetry operations. In type C structures, the molecules occupy achiral systems of equivalent positions, and each molecule is located on an inversion center, on a mirror plane, or on a special position of a higher-order inversion axis. If there are two or more independent sets of equivalent positions in a crystal lattice, the type D lattice becomes feasible. This structure consists of one set of type B and another of type C, but it is rare. Of the 5,000 crystal structures studied, 28.4% belong to type A, 55.6% are of type B, 15.7% belong to type C, and only 0.3% are considered as type D.

A detailed discussion of crystal packing and the resulting space groups has been given by Kitaigorodskii [13]. This approach assumes that molecular crystals are assemblages for which compactness tends toward the maximum that is compatible with the molecular geometry. He defined a packing coefficient as vZ/V, where v is the volume of the molecule, V is the volume of the cell, and Z is the number of molecules in the unit cell. In this way, the space-filling or packing coefficient in crystals always lies between 0.65 and 0.77, which is of the same order as the regular packing of spheres or ellipsoids. Molecules that owing to peculiarities of their geometry cannot attain a packing coefficient at least equal to 0.6 do not crystallize and can only form glasses from the melt. In order to fill space in the most compact manner with objects of indeterminate geometry, planes must be filled in a compact fashion. There are only a limited number of two-dimensional arrays in which an object can reside in contact with six neighbors, which is necessary for the optimal packing of a molecular assembly. It has been concluded that the binary screw axis is highly conducive to efficient packing. In the final analysis, the limited number of possible combinations of stacking leads to a limited number of space groups they fulfill the requirements of three-dimensional close-packing [13].

The close-packing criterion for crystal stability generally yields lower free energies for a structure composed of a racemic assembly over that composed of homochiral molecules. This view has been expressed in the empirical rule of Wallach [14], which states that the combination of two opposite enantiomers to form a racemate is accompanied by a volume contraction. The many exceptions to this rule have been discussed in a systematic manner [15]. The modification to Wallach's rule contributed by Walden [16] is more generally valid; it states that if an enantiomer has a lower melting point than its corresponding racemate, then the crystals of the latter will have the higher density.

III. RELATIONSHIP BETWEEN MACROSCOPIC AND MICROSCOPIC STRUCTURES AS ILLUSTRATED BY QUARTZ AND ITS NONSUPERIMPOSABLE HEMIHEDRAL CRYSTAL FACES

The most appropriate point to begin a discussion of molecular and crystallographic chirality is to discuss the case of quartz. The holohedral morphology of a quartz crystal consists of a hexagonal prism, capped at each end by a hexagonal pyramid. This geometrical figure is highly symmetrical, possessing 24 symmetry elements of various types. It was noted by Hauy at the beginning of the nineteenth century that quartz crystals contained many small facets that considerably reduced the overall symmetry [17]. These facets were found only on alternate corners of the crystal and were therefore described as being hemihedral. As shown in Fig. 1, the distribution of these hemihedral faces gives rise to two forms of quartz, identified at the time as being either left- or right-handed [18]. It was noted that the two forms of quartz were mirror images of each other, and that no amount of positioning or orientation permitted the superimposition of one form onto the other. For instance, on a right-handed crystal the *s* trigonal pyramid lies to the right of the *m* face, which is below the predominating positive rhombo-

(a) **(b)**

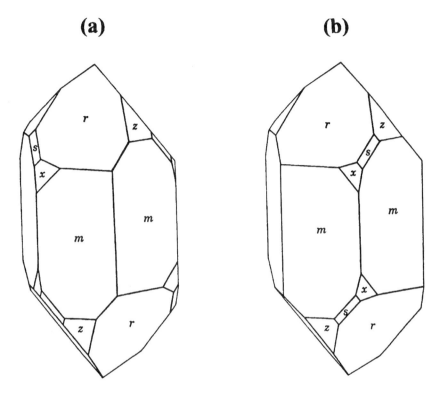

Fig. 1. Crystal morphologies of (a) left-handed quartz and (b) right-handed quartz. These are readily distinguished by the relative placement of the hemihedral facets. (The figure was adapted from Ref. 11.)

hedron r. On a left-handed crystal, the s trigonal pyramid lies to the left of the m face below the r rhombohedron.

The first complete determinations of the simple β-quartz structure were made by Bragg and Gibbs [19] and by Wyckoff [20], who reported that each silicon atom was surrounded by four oxygen atoms, two of which were above and two of which were below the silicon atom. The tetrahedral groups were found to arrange in spirals, each of which twisted in a definite manner. Thus all the spirals of left-handed quartz twist in one way, and all the spirals of right-handed quartz twist in the other way. The left-handed modification belongs to the enantiomorphic space group $C3_12$, while the right-handed modification belongs to the mirror image space group $C3_22$.

Herschel investigated the optical properties of quartz slabs, which were cut perpendicular to the long crystal axis [17]. He found that crystals cut from left-handed quartz would invariably rotate the plane of linearly polarized light in a clockwise fashion, while crystals cut from right-handed quartz rotated the plane of linearly polarized light in a counterclockwise direction. Elsewhere Biot was performing extensive studies on the optical rotatory properties of certain compounds dissolved in fluid solutions, and it was at this time that the connection between optical activity and crystallographic properties was made.

Observations of this type were extremely important to Pasteur in that they permitted him to deduce requirements for chirality. He connected the concept of nonsuperimposable mirror images with the existence of chirality. Since quartz loses its optical rotatory power when dissolved or melted, he inferred that it was a helical arrangement of molecules in this particular solid that conveyed the given properties [17]. He also understood that solids could possess certain elements of symmetry and still exhibit optical activity (i.e., chirality) and therefore coined the term ''dissymmetry'' to describe materials whose mirror images were not superimposable on each other.

The twinning of quartz crystals can yield interesting structures. In the extreme case, the complete interpenetration of left- and right-handed structures can yield a crystal whose morphology is devoid of nonsuperimposable hemihedral faces (see Fig. 2a). As would be expected, no optical activity is observable from crystal slabs cut from this type of quartz structure [21]. Complete interpenetration would yield equal numbers of left-handed and right-handed spirals within the same crystal, with the optical effect of one type canceling the effect of the other. A more common occurrence is the so-called Dauphiné twin, which consists of a partial crystal interpenetration of differently handed quartz crystals (see Fig. 2b). In this latter instance one can find separate regions of opposite crystal chirality existing within the same twinned structure. In one sense, the instance of twinned quartz crystals can be considered to be a prototypical racemic mixture. The understanding of optical effects in quartz permitted Pasteur to develop impor-

(a)   (b)

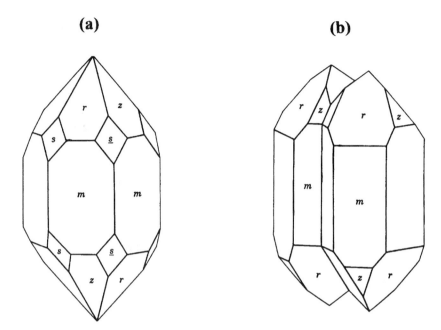

Fig. 2. Examples of twinned quartz crystals, illustrating (a) the lack of hemi-hedral facets in a completely interpenetrating crystal, and (b) the existence of oppositely handed quartz crystals in a single twinned structure. (The figure was adapted from Ref. 14.)

tant conclusions about the chirality of other crystal systems and in turn about individual molecules.

IV. RELATION OF MOLECULAR DISSYMMETRY AND THE CRYSTALLOGRAPHY OF TARTRATES

Crystalline quartz is one of those materials whose fundamental molecular units are achiral, but which are forced to adopt a chiral configuration by virtue of the crystal structure. The proof of this lies in the observation that quartz, when dissolved or melted, loses its ability to rotate the plane of linearly polarized light. The fusion/solubilization process serves to destroy the spiral arrangement of the crystal, and the isotropic fluid that results contains no residual optical activity.

Early in the study of optical activity, it was learned that many organic compounds did not lose their ability to interact with polarized light even after being melted or dissolved in a solvent. A significant number of these optically active materials could be obtained as crystals that exhibited nonsuperimposable hemihedral crystal facets. Pasteur was the first to use the term "dissymmetry" to describe situations distinguished by the existence of nonsuperimposable mirror images. He understood that in the case of quartz, the dissymmetry was a result solely attributable to the nature of the crystal structure and that destruction of the structure eliminated the dissymmetry. He realized, however, that organic molecules that remained optically active even after dissolution or fusion must possess an inherent molecular dissymmetry. The remarkable nature of these postulates is that Pasteur made his proposals prior to the general acceptance of the tetrahedral nature of carbon valency.

The development of molecular dissymmetry as an understood property was inescapably linked to optical and structural crystallography [1]; tartaric acid and its salts provide the best illustrations. Tartaric acid was discovered by Scheele in 1769 in conjunction with wine-making, and numerous studies on its optical activity were carried out by Biot [17]. Perplexing the scientists of that era was a compound that became named racemic acid, whose name was derived from the Latin *racemus*, or grape. This compound was shown by Gay Lussac to be identical in composition with tartaric acid, though it differed significantly in its physical properties. For instance, the melting point of tartaric acid is 168–170°C, while the melting point of racemic acid is 206°C. The solid density of tartaric acid (1.760 g/mL) exceeds that of racemic acid (1.697 g/mL), as does the water solubility (1390 g/L for tartaric acid as compared to 206 g/L for racemic acid). Most important, Biot conclusively demonstrated that racemic acid was not optically active [17].

Pasteur studied the crystals of both tartaric and racemic acids and found that while tartrate crystals contained nonsuperimposable hemihedral facets, racemate crystals did not. Examples of the ideal crystal

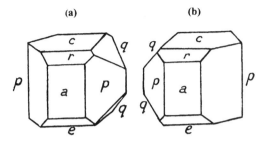

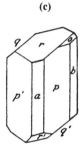

Fig. 3. Crystals of tartaric acid, obtained under conditions yielding the hemi-hedral facets distinctive of the (a) dextrorotatory and (b) levorotatory forms. Also shown is the (c) morphology associated with the holohedral modification of racemic acid. (The figure was adapted from Ref. 17.)

morphologies of these crystal forms are shown in Fig. 3. Pasteur subsequently proved the relationship between tartaric and racemic acids by mixing equal amounts of tartrate crystals having left-handed and right-handed hemihedrism and crystallizing out a tartrate substance whose crystal morphology was completely indistinguishable from that of racemate crystals isolated in the same manner [17].

While Pasteur was able to deduce many important conclusions using optical crystallography as his main tool, structural crystallography has the ability to go much further. It has been established that tartaric acid contains two centers of dissymmetry, with the naturally obtained optically active acid having the (R,R)-configuration [23]. Enantiomerically pure tartaric acid crystallizes as an anhydrate phase in the monoclinic class (space group $P2_1$), with the unit cell containing two molecules [24,25]. The reported crystallographic data are $a = 7.715$ Å, $b = 6.004$ Å, $c = 6.231$ Å, and $\beta = 100.1°$ [25].

Racemic acid has now been shown to consist of an equimolar mixture of (R,R)-tartaric acid and the unnatural (S,S)-enantiomer and is obtained as a monohydrate species. The compound crystallizes in the triclinic class (space group P1̄), with the unit cell consisting of one molecule of (R,R)-tartaric acid, one molecule of (S,S)-tartaric acid, and two water molecules [26]. This heterochiral unit cell is significantly different from the homochiral unit cell; $a = 8.06$ Å, $b = 9.60$ Å, $c = 4.85$ Å, $\alpha = 70.4°$, $\beta = 97.2°$, and $\gamma = 112.5°$. The unit cell of racemic acid contains a center of symmetry, while that of tartaric acid does not.

The molecular conformation of the tartaric acid is not significantly different in the tartrate and racemate structures, but the different modes of molecular packing lead to the existence of the two very different crystal structures (see Fig. 4 for a comparison). For the tartaric acid anhydrate crystal, the molecular planes are held directly together by a complicated network of intermolecular hydrogen bonds [25]. The hydrogen bonding pattern in the racemic acid monohydrate crystal is very different: the molecules are bound into columns that are linked up to form sheets [26].

Another acid was occasionally encountered in the French winemaking industry that was chemically identical to tartaric and racemic acids but exhibited a completely different range of physical properties. It did not exhibit optical activity, nor could it be resolved into components that were themselves optically active [17]. The solid exhibited a melting point of 159–160°C, a density of 1.737 g/mL, and a solubility of 1250 g/L [22]. Pasteur was ultimately able to obtain quantities of this acid as a by-product from the heating of tartaric acid in the presence of basic substances.

It was learned that this nonresolvable, optically inactive form of tartaric acid was the meso form and that it contained two centers of opposite-sense dissymmetry in the same molecule. This compound has been found to crystallize in four distinct modifications, two of which are polymorphic anhydrates and two of which are polymorphic mono-

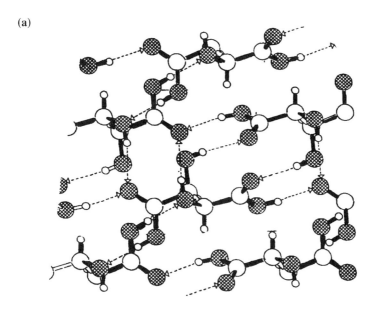

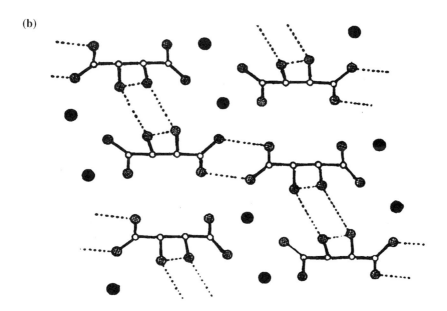

Fig. 4. Comparison of the molecular packing in crystals of (a) tartaric and (b) racemic acids. (The figure was adapted from data contained in Refs. 25 and 26.)

hydrates [27]. Triclinic *meso*-tartaric acid monohydrate crystallizes in the P$\bar{1}$ space group with two molecules in the unit cell; a = 5.516 Å, b = 9.220 Å, c = 7.330 Å, α = 115.11°, β = 93.62°, and γ = 93.64°. The monoclinic monohydrate contains four molecules in the unit cell and crystallizes in the P2_1c space group; a = 5.215 Å, b = 5.019 Å, c = 25.92 Å, and β = 99.77.2°. The triclinic anhydrate crystallizes in the P$\bar{1}$ space group with two molecules in the unit cell; a = 9.459 Å, b = 6.464 Å, c = 5.396 Å, α = 68.99°, β = 76.36°, and γ = 75.77°. Finally the orthorhombic anhydrate crystallizes in the Pbn2_1 space group with 16 molecules in the unit cell; a = 19.05 Å, b = 9.88 Å, and c = 12.16 Å.

In the triclinic *meso*-tartaric acid structures, the relative positions of most of the nonhydrogen atoms of the molecule are approximately equivalent to those of the corresponding atoms of racemic acid. As mentioned earlier, one of the important linkages in the racemic acid structure is a square system of hydrogen bonds between hydroxyl groups on four molecules, which forms columns parallel to the *c*-axis [26]. Analogous squares are formed in the *meso*-tartaric acid structures. In the structure of racemic acid, one carboxyl group of the molecule is directly linked with a carboxyl group of a centrosymmetrically related molecule, and the other carboxyl group is indirectly connected with a counterpart through water molecules. The first type of linkage is found at one side of the molecule in the triclinic anhydrate form of *meso*-tartaric acid, while the second type is found at one side of the molecule in the triclinic hydrate structure. In the monoclinic hydrate, one side of the molecule takes part in a carboxylic acid dimer configuration, but the linkage through water molecules at the other side of the molecule is a less common mutation of the hydrated links found in other cases. Evidently the geometry of *meso*-tartaric acid does not permit an ideal three-dimensional arrangement to exist with both types of intermolecular linkages present.

Undoubtedly, Pasteur's most widely known work in the area of molecular dissymmetry and crystallography came about during his study of the salts of tartaric acid [17]. In studies concerning the mixed 1:1 sodium/ammonium salts of tartaric and racemic acids, Mitscherlich had previously reported that both salts exhibited completely identical properties. This finding contradicted the growing body of evidence that optically active and inactive forms of the same chemical species should exhibit different physical properties. As he expected, Pasteur found that sodium ammonium tartrate crystallized with the presence of nonsuperimposable hemihedral facets, but he was surprised to learn that crystals of sodium ammonium racemate contained nonsuperimposable hemihedral facets as well. Upon closer examination, Pasteur found that half of the crystals of sodium ammonium racemate exhibited hemihedral facets that spiraled to the left, while the other half exhibited hemihedral facets that spiraled to the right. He was able to separate manually the left-handed crystals from the right-handed ones and found that these separated forms were optically active upon dissolution. Drawings of these crystal habits are provided in Fig. 5.

The sodium ammonium salt crystallized from racemic tartaric acid has been found to crystallize in the orthorhombic $P2_12_12_1$ space group and contains four molecules in the unit cell [28]. This particular crystal class is noncentrosymmetric, and as a result individual crystals will be optically active. In fact, efficient growth of this tartrate salt only takes place if all the (R,R)-tartrate molecules crystallize in one ensemble of crystals, and if all the (S,S)-tartrate molecules crystallize in another ensemble. When formed below a temperature of 26°C, the preferred molecular packing does not permit the intermingling of the enantiomers to yield a true racemic crystal. The crystallization of sodium ammonium tartrate below 26°C results in a spontaneous resolution of the substance into physically separable enantiomers. Interestingly, a different polymorph forms above 26°C that requires a completely different packing pattern that allows for the formation of a racemic modification of sodium ammonium tartrate.

A series of half-neutralized salts of tartaric acid are also found to crystallize in the orthorhombic $P2_12_12_1$, space group (all containing four molecules in the unit cell) and are thus potentially resolvable by a mechanical separation. Ammonium hydrogen tartrate is found to crystallize with $a = 7.648$ Å, $b = 11.066$ Å, and $c = 7.843$ Å [29]. Sodium hydrogen tartrate crystallizes with $a = 8.663$ Å, $b = 10.583$ Å, and $c = 7.228$ Å, while potassium hydrogen tartrate crystallizes with $a = 7.782$ Å, $b = 10.643$ Å, and $c = 7.608$ Å [29]. Finally, rubidium hydrogen tartrate is found to crystallize with $a = 7.923$ Å, $b = 10.988$ Å, and $c = 7.653$ Å [29].

(a)   **(b)**

(c)

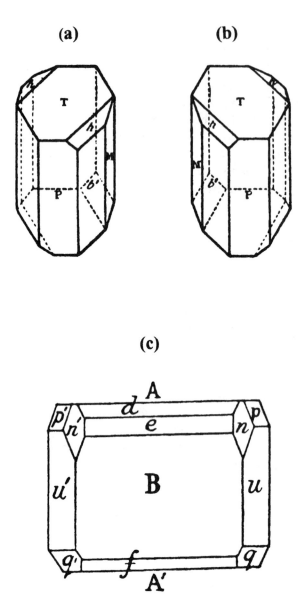

Fig. 5. Crystals of sodium ammonium tartrate, obtained under conditions yielding the hemihedral facets distinctive of the (a) dextrorotatory and (b) levorotatory forms. Also shown is the (c) morphology associated with the holohedral modification of sodium ammonium racemate. (The figure was adapted from Ref. 17.)

V. PHYSICAL CHARACTERISTICS OF RACEMIC MIXTURES OF DISSYMMETRIC SOLIDS

The early work of Pasteur clearly demonstrated the possibility that racemic compounds were mixtures that could be separated into their component enantiomers. It was found that while some compounds could be separated by direct crystallization, the majority required some type of chemical means to effect the separation. Compounds separable by the former method were termed *conglomerates*, while compounds separable only by the latter were termed *true racemates*. Much work has gone into the investigation of these two types of dissymmetric solids, and we will now consider some of the structural aspects of these systems.

A. Conglomerate Systems

Conglomerate solids are characterized by the presence of a single enantiomer within the unit cell of the crystal, even when the solid is obtained through crystallization of a racemic or partially resolved mixture. These solids consist of separate crystals, each of which consists entirely of one enantiomer or the other, which may be separated entirely on the basis of their physical properties. Compounds known to crystallize as conglomerates can be quite easily resolved, since the resolution step takes place spontaneously upon crystallization. The key to a successful resolution by direct crystallization lies in the means used to separate physically the crystals containing the opposite enantiomers. Jacques and coworkers have provided extensive summaries of the methods whereby direct crystallization can be used to effect the resolution of a racemic mixture [10,32].

The first method is the classical technique of Pasteur, which entails the mechanical separation of enantiomorphic crystals formed simultaneously while the mother liquor remains racemic. Enantiomer separation by this particular method is extremely time-consuming and is really impossible unless the crystals form with well-defined hemihedral faces. Nevertheless, it is often the method of choice to obtain the seed crystals required for other direct crystallization procedures. When a particular system has been shown to be a conglomerate, and the crystals are not sufficiently distinct so as to be separated, polarimetry can often be used to establish the chirality of the enantiomer. Even a few seed crystals, mechanically separated, can be used to produce larger quantities of resolved enantiomerically pure material.

A second method of resolution by direct crystallization involves the localized crystallization of each enantiomer from a racemic, supersaturated solution. Oppositely handed, enantiomerically pure seed crystals of the compound are placed in geographically distant locations in the crystallization vessel, and they then serve as nuclei for the further crystallization of the like enantiomer. This procedure has been used to obtain both enantiomers of methadone, where approximately 50% total yield of enantiomerically pure material can be obtained [33]. An apparatus has been described that permits the automated use of localized crystallization, and it has successfully been used to separate the enantiomers of hydrobenzoin [34].

Enantiomer separation may be practiced on the large industrial scale using the procedure known as resolution by entrainment [35]. The method is based on the condition that the solubility of a given enantiomer be less than that of the corresponding racemate. To begin, a solution is prepared that contains a slight excess of one enantiomer. Crystallization is induced (usually with the aid of appropriate seed crystals), whereupon the desired enantiomer is obtained as a solid and the mother liquor is enriched in the other isomer. In a second crystallization step, the other enantiomer is obtained. The method can be applied to any racemic mixture that crystallizes as a conglomerate, and the main complication that can arise is when the compound exhibits polymorphism. In that case, the entrainment procedure must be carefully designed to generate only the desired crystal form.

Resolution by entrainment is best illustrated through the use of an example. Consider the bench scale resolution of hydrobenzoin [32]. 1100 mg of racemic material was dissolved along with 370 mg of (−)-hydrobenzoin in 85 g of 95% ethanol, and the solution was cooled to 15°C. 10 mg of the (−)-isomer was added as seeds, and crystals were allowed to form. After 20 minutes, 870 mg of (−)-hydrobenzoin was recovered. 870 mg of racemic hydrobenzoin was then dissolved with heating; the resulting solution was cooled to 15°C and seeded with 10 mg of the (+)-isomer. 900 mg of (+)-hydrobenzoin was recovered at this time. The process was cycled 15 times and ultimately yielded 6.5 g of (−)-hydrobenzoin and 5.7 g of (+)-hydrobenzoin. Each isomer was obtained as approximately 97% enantiomerically pure.

Jacques and coworkers have provided compilations of the known conglomerate systems, for which approximately 250 organic compounds have been identified [36]. This relatively small number represents only about 10% of the number of chiral compounds for which resolutions have been published and this small fraction is seen as the major limitation to enantiomer resolution by direct crystallization. Nevertheless, it has been recognized that the formation of salts can greatly improve the odds of obtaining a conglomerate system, since one of the numerous possible salts of a racemate might exhibit the proper solid-state characteristics [37]. Substitution onto other portions of a compound can sometimes also yield a conglomerate system. For example, although mandelic acid and many of its derivatives crystallize as racemic mixtures, *ortho*-chloromandelic acid is obtained as a conglomerate [38]. Interestingly, both the *meta*- and *para*-chloromandelic acids crystallize as true racemates.

This seeming randomness of conglomerate formation has been explained through examinations of the thermodynamics of these systems. The stability of true racemic solids is defined by the free energy change associated with the process of combining the *R*-enantiomer with the *S*-enantiomer to produce the *R,S*-enantiomeric solid. This has been calculated to be in the range of 0 to −2 kcal/mol and is roughly proportional to the difference in melting points between the racemate and the resolved enantiomers [39]. In most cases, the free energy associated with the formation of racemates is exothermic, owing to the positive nature of the enthalpies and entropies of formation. In those cases where the melting point of the racemic mixture is at least 20 degrees lower than the melting points of the separated enantiomers, one generally obtains a conglomerate system.

Since conglomerate systems consist of totally independently formed enantiomer crystals and are therefore mere physical mixtures of the enantiomer components, these constitute a binary system. Such binary mixtures are easily described by the phase rule and can be profitably characterized by their melting point phase diagrams [10]. Since the components of a conglomerate racemate will melt independently, the material behaves as if it were a pure substance. The melting point phase diagram will therefore exhibit a eutectic in its melting point phase diagram at the racemic composition. For a partially resolved conglomerate system, one can separate out the enantiomer present in excess simply by heating the sample to a temperature just above the melting point of the racemate and then collecting the crystalline excess of the residual enantiomer.

The liquidus line of a phase diagram can be calculated using the Schröder–Van Laar equation:

$$\ln X_i = \left(\frac{\Delta H^F}{R} \right) \left(\frac{1}{T_P} - \frac{1}{T_i} \right)$$

where X_i is the ith mole fraction of one of the enantiomers, ΔH^F is the molar heat of fusion of the pure enantiomer, T_P is the melting point of the pure enantiomer, and T_i is the melting point of the substance having mole fraction X_i.

An example of the type of melting point phase diagram that can be obtained for a conglomerate system is shown in Fig. 6, which illustrates one-half of the phase diagram reported for 4,4′-dimethyl-8,9,10-trinor-spiro-2,2′-bornane [40]. Below the eutectic temperature of 67°C, the system exists as a mixture of solid D-enantiomer and L-enantiomer. At the exact composition of the racemic mixture ($X = 0.5$), the system will exist entirely in the liquid phase above the eutectic temperature. At mole fractions where the amount of L-enantiomer exceeds that of the D-enantiomer, the system will exist as an equilibrium mixture of racemic liquid and solid L-enantiomer. As required by the phase diagram of a conglomerate, the eutectic temperature is the lowest temperature attainable at which any liquid phase can exist in equilibrium with any solid phase. An excellent fit of the data according to the Schröder–Van Laar equation was obtained, with $\Delta H^F = 5.9$ kcal/mol and $T_P = 95°C$.

An alternative approach for the characterization of conglomerate systems is through the use of ternary phase diagrams, where one component is the solvent and the compound solubility is used as the observable parameter [10,32]. The racemic mixture is found to be more solu-

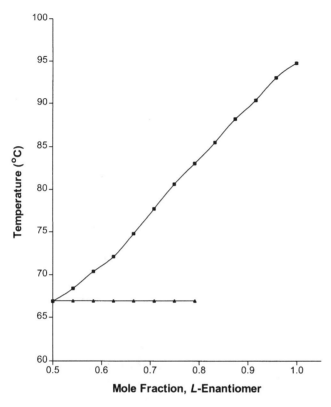

Fig. 6. Portion of the melting point phase diagram obtained for the crystalline conglomerate formed by 4,4′-dimethyl-8,9,10-trinor-spiro-2,2′-bornane. (The figure was adapted from data contained in Ref. 40.)

ble than are the separated enantiomers, and the rule of Meyerhoffer states that a conglomerate will have a solubility twice that of the resolved enantiomer [41]. This situation arises since ideally the mole fraction of an independent constituent in a liquid depends only on the enthalpy of fusion and melting point of the substance, and since according to the Schröder–Van Laar equation the solubility of one enantiomer cannot affect the solubility of the other. Jacques and Gabard have examined the solubilities of a number of conglomerate systems and have largely confirmed the double solubility rule [42]. A selection of their results obtained for substances that are not dissociated in solution is presented in Table 1.

B. True Racemate Systems

True racemate solids are characterized by the presence of equimolar amounts of both enantiomers within the unit cell of the crystal. Materials of this type will crystallize within a centrosymmetric space group, which will necessarily be different from the space group of the separated enantiomers. As a result, the range of physical properties associated with a heterochiral solid must be completely different from those of the homochiral solid. For instance, the melting point of the exact racemic mixture may be greater than or less than that of the separated enantiomers; no general rule is available for a reliable prediction of behavior.

When a compound forms a true racemate, resolution by direct crystallization cannot be used to separate the enantiomers. Whether one works with melting point or solubility phase diagrams, one invariably finds that these systems yield diagrams that contain two minima. True racemates can therefore be separated after the performance of a derivatization reaction that alters the thermodynamics of the system so that the phase diagram now contains only a single minimum. In general, this minimum will not be located at the exact racemic composition but will instead be observed at some other concentration value. In fact, for the p-salt to be separated from the n-salt in a single crystallization step, the position of the eutectic should be substantially removed from the equimolar point. However, if the eutectic point should happen by accident to be close to the equimolar point, then the phase diagram would closely resemble that of a conglomerate. In that particular instance, it would be possible to devise an entrainment procedure for the separation of the p- and n-salts. This possibility has been demonstrated by the preferential crystallization of 4-methylpiperidinium hydrogen (R,S)-succinate [43].

The procedure most commonly used to alter the crystallization thermodynamics of true racemate systems involves the formation of dissociable diastereomer species [44]. These are most often simple salts

Table 1 Solubilities of Separated Enantiomers and Racemic Mixtures of Conglomerate Materials

Compound	System	Solubility of separated enantiomer (g/100 mL)	Solubility of racemic mixture (g/100 mL)	Ratio of racemate solubility to enantiomer solubility
Asparagine	water (25°C)	2.69	5.61	2.16
(α-Naphtoxy)-2-propionamide	acetone (25°C)	1.38	2.24	2.08
p-Nitrophenylamino propanediol	methanol (25°C)	1.63	3.35	2.14
N-Acetyl-leucine	acetone (25°C)	1.86	4.12	2.13
N-Acetyl-glutamic acid	water (25°C)	4.14	8.28	2.12
Diacetyldiamide	water (35.5°C)	12.4	23.7	2.14
3,5-Dinitrobenzoate-lysine	water (30°C)	6.96	13.17	1.99

formed between electron-pair donors and electron-pair acceptors. For instance, the first resolving agents introduced for acidic enantiomers were alkaloids, and hydroxy acids were used for the resolution of bases. The procedure has been known since the time of Pasteur, and extensive tables of resolving agents and procedures are available [45,46].

As an example, consider the case where a racemic acid is to be resolved through the use of a basic resolving agent. The first step of the resolution procedure involves formation of diastereomeric salts as in the table.

Racemic mixture	Resolving agent $\rightarrow$	Diastereomer salts
(+)R-COOH	+ 2NH$_2$-R'(+)	$\rightarrow$ (+)R-COO-NH$_3$-R'(+)
(−)R-COOH	+ 2NH$_2$-R'(+)	$\rightarrow$ (−)R-COO-NH$_3$-R'(+)

According to convention [47], the (R,R) and (S,S) diastereomers are termed the **p**-salts, and the (R,S) and (S,R) diastereomers are identified as the **n** salts. In the example used above, the (+)R-COO-NH$_3$-R'(+) diastereomer would correspond to the p-salt, and the (−) R-COO-NH$_3$ -R'(+) diastereomer is the n-salt. In the usual practice, the p- and n-salts are separated by fractional crystallization, and the success of the resolution process is critically related to this crystallization step.

As mentioned earlier, the desirable effect of derivatization is to transform the undesirable thermodynamics of a true racemate system into those more amenable for direct separation of the diastereomers. For example, the enantiomers of phenylsuccinic acid can be separated after formation of the proline salt [48], and an example of the solubility phase diagram reported for the 1:2 phenylsuccinate/proline salt is shown in Fig. 7. In this work, it was found that the degree of resolution for the salt having the 1:2 stoichiometry was only minimally affected by temperature, but that separation of the 1:1 stoichiometric phenylsuccinate/proline salt was strongly affected by temperature. These findings illustrate the need for obtaining the phase diagram when designing enantiomeric separations, since the efficiency of the process can be strongly affected by environmental factors.

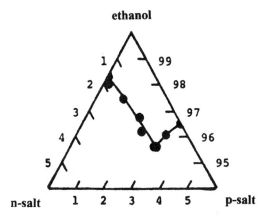

Fig. 7. Solubility phase diagram reported for the diastereomer system (1:2 stoichiometry) formed by (±)-phenylsuccinic acid and (−)-proline, using ethanol as the solvent system. (The figure was adapted from data contained in Ref. 47.)

For instance, the aqueous solubility of the n-salt of α-methylbenzylamine mandelate has been reported as 49.1 g/L, and the solubility of the p-salt as 180 g/L [49]. However, the relative solubilities of the p- and n-salt pairs is greatly affected by the choice of temperature or solvent system, and the pathway of the diastereomer separation can easily be manipulated by variations in experimental conditions. In fact, it is possible preferentially to crystallize either the p- or the n-salt depending on the range of parameters used.

The chiral discrimination existing in the diastereomer systems formed by ephedrine and various mandelic acids has been studied in great detail, using a variety of spectroscopic and structural tools. Although significant differences in the solubilities of the n- and p-salts obtained after reaction of ephedrine with mandelic acid were known to exist, both salts were found to crystallize in the same monoclinic space group with crystal structures that were isosteric [50]. In subsequent work, it was shown that the crystalline diastereomer salts formed by R-mandelic acid formed a more compact repeating pattern (with higher melting points and enthalpies of fusion) than did analogous salts formed by S-mandelic acid [51]. It was deduced that subtle chiral discriminations led to the existence of different hydrogen-bonding modes, which in turn became manifest in a variety of other physical properties.

Distinctions in hydrogen-bonding networks were also noted during studies of the structures formed by α-amino acids and S-phenylethanesulfonic acid [52]. The differences noted for the two salts formed between p-hydroxyphenylglycine and the resolving agent are shown in Fig. 8. In another work, the diastereomers formed during the resolution of R,S-mandelic acid with R-2-*tert*-butyl-3-methylimidazoldin-4-one were characterized by varying networks of hydrogen-bonding, while the conformations of the molecular species were relatively equivalent [53].

Depending on the details of the lattice dynamics, it is certainly possible that the p-salt and the n-salt can crystallize in different space groups. For example, in the diastereomeric system formed by α-methylbenzylamine and hydratropic acid, the p-salt was found to crystallize in the $P2_1$ space group, while the n-salt crystallized in the $P2_12_12_1$ space group [54]. Although the molecular conformations in the two crystal forms were reported to be similar, the mode of packing to generate the individual crystals was different.

This latter behavior differs from that observed in the crystalline diastereomers of mandelic acid with 1-phenylethylamine [55]. Here the n-salt crystallizes in the triclinic P1 space group, while the p-salt crystallizes in the monoclinic $P2_1$ space group. The crystallographic structures of the two diastereomers reveal the existence of fairly equivalent hydrogen-bonding but substantially different conformations for the two molecular ions making up the donor-acceptor complex. These structural differences are manifest in the relative solubilities of the two diastereomers, where the aqueous solubility of the p-salt is much less than that of the n-salt.

Once either the p-salt or the n-salt (or both) is successfully crystallized, the diastereomer salt is usually dissociated and the resolving agent separated. In the specific instance of acid resolution, the diastereomeric salt is efficiently cleaved by means of a hydrolysis reaction as in the table.

p-salt

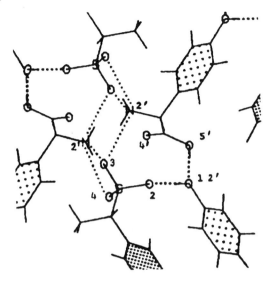

n-salt

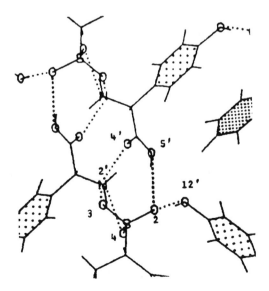

Fig. 8. Details of the hydrogen bonding existing in crystals of the two salts formed between *p*-hydroxyphenylglycine (HPG) and *S*-phenylethanesulfonic acid (PES). (The figure was adapted from data contained in Ref. 52.)

Separated diastereomers	Hydrolysis agent		Separated enantiomers
(+)R-COO-NH$_3$-R′(+)	+	HCl $\rightarrow$	(+)R-COOH + (+)R′-NH$_3$Cl
(−)R-COO-NH$_3$-R′(+)	+	HCl $\rightarrow$	(−)R-COOH + (+)R′-NH$_3$Cl

The diastereomer cleavage must be simple and selective, must take place in quantitative yield, must not racemize the resolved compound, and must leave the resolving agent in a form that is easily recovered. The recovery is usually effected by precipitation or extraction of the resolving agent. One important criterion in the choice of a resolving agent is the ease by which it can be dissociated and removed from the compound being resolved [44].

VI. SUMMARY

Observations on the chirality of crystals made it possible for Pasteur and others to identify dissymmetry as the true origin of optical activity. It became quickly evident that the molecular chirality associated with a given compound could be directly evident in the bulk crystallography of that compound. This in turn led to observable differences in a variety of physical properties, such as the melting point and the solubility of such species. Many chiral molecules have been observed to resolve spontaneously upon crystallization, forming enantiomorphic crystals that can be physically separated. Others can only be resolved through the formation and separation of diastereomeric species.

Technically, the crystallographic differences that exist between enantiomers and diastereomers do not fit the formal definition of polymorphism, since one may consider molecules of opposite dissymmetric senses (or mixtures of these) to be distinct chemical species. Nevertheless, such systems contain a number of features that are common to genuine polymorphic systems, and many of the same characterization techniques discussed in Chapter 5 of this book are equally appropriate here.

With more and more therapeutic agents being administered as

resolved enantiomers, methods for isolation of the desired dissymmetric species will be in constant development. Determination of the melting point or solubility phase diagram becomes a key element to the development process, since one can use this information to identify the agent in question as being either a conglomerate or a true racemate. Should the compound happen to crystallize as a conglomerate, then enantiomer separation by direct crystallization could represent the cost-effective route. Compounds identified as true racemates would by necessity need to be separated by the formation of dissociable diastereomers. Of course, when the compound in question can be synthesized directly with the proper enantiomeric composition, the need to conduct a resolution procedure must be critically examined. One should realize, however, that even when asymmetric synthesis is the chosen route, an investigation into the relationship between chirality and crystallography is essential to having an understanding of the molecular system.

REFERENCES

1. H. G. Brittain, *Pharm. Res.*, *7*, 683 (1990).
2. I. W. Wainer, and D. E. Drayer, eds., *Drug Stereochemistry: Analytical Methods and Pharmacology*, Marcel Dekker, New York, 1988.
3. B. Testa, *Trends Pharmacol. Sci.*, *7*, 60 (1986).
4. E. J. Ariens, E. W. Wuis, and E. J. Veringa, *Biochem. Biopharmacol.*, *37*, 9 (1988).
5. F. Jamail, R. Mehvar, and F. M. Pasutto, *J. Pharm. Sci.*, *78*, 695 (1989).
6. R. L. Smith, and J. Caldwell, *Trends Pharmacol. Sci.*, *9*, 75 (1988).
7. W. H. De Camp, *Chirality*, *1*, 2 (1989).
8. P. Newman, ed., *Optical Resolution Procedures for Chemical Compounds*, Vols. 1–3, Manhattan College Press, New York, 1984.
9. R. S. Cahn, C. K. Ingold, and V. Prelog, *Angew. Chem., Int. Ed. Engl.*, *5*, 385 (1966).
10. J. Jacques, A. Collet, and S. H. Wilen, *Enantiomers, Racemates, and Resolutions*, John Wiley, New York, 1981.
11. P. M. Zorkii, A. E. Razumaeva, and V. K. Belsky, *Acta Cryst.*, *A33*, 1001 (1977).
12. S. F. Mason, *Molecular Optical Activity and the Chiral Discriminations*, Cambridge Univ. Press, Cambridge, 1982, pp. 165–166.

13. A. I. Kitaigorodskii, *Molecular Crystals and Molecules*, Academic Press, New York, 1973.

14. G. Wallach, *Annalen*, *286*, 140 (1895).

15. C. P. Brock, W. B. Schweizer, and J. D. Dunitz, *J. Am. Chem. Soc.*, *113*, 9811 (1991).

16. P. Walden, *Chem. Ber.*, *29*, 1692 (1896).

17. For an excellent introduction into the early work involving chirality and crystallography, see T. M. Lowry, *Optical Rotatory Power*, Longmans, Green, London, 1935, pp. 25–36.

18. E. S. Dana, and W. E. Ford, *A Textbook of Mineralogy*, John Wiley, New York, 1922, p. 470.

19. W. H. Bragg, and R. E. Gibbs, *Proc. Roy. Soc.*, *A109*, 405 (1925).

20. R. W. G. Wyckoff, *Am. J. Sci.*, *11*, 101 (1926).

21. C. Frondel, *The System of Mineralogy*, Vol. 3, John Wiley, New York, 1962, pp. 75–98.

22. *The Merck Index*, 11th ed. (S. Budavari, M. J. O'Neil, A. Smith, and P. E. Heckelman, eds), Merck, Rahway, NJ, 1989, pp. 1432–1433.

23. W. Klyne, and J. Buckingham, *Atlas of Stereochemistry*, 2d ed., Oxford Univ. Press, New York, 1978, p. 6.

24. F. Stern, and C. A. Beevers, *Acta Cryst.*, *3*, 341 (1950).

25. Y. Okaya, R. R. Stemple, and M. I. Kay, *Acta Cryst.*, *21*, 237 (1966).

26. G. S. Parry, *Acta Cryst.*, *4*, 131 (1951).

27. G. A. Bootsma, and J. C. Schoone, *Acta Cryst.*, *22*, 522 (1967).

28. Z. Brozek, and K. Stadnicka, *Acta Cryst.*, *B50*, 59 (1994).

29. A. J. van Bommel, and J. M. Bijvoet, *Acta Cryst.*, *11*, 61 (1958).

30. Y. Kubozone, A. Hirano, S. Nagasawa, H. Maeda and S. Kashino, *Bull. Chem. Soc. Jap.*, *66*, 2166 (1993).

31. L. K. Templeton, and D. H. Templeton, *Acta Cryst.*, *C45*, 675 (1989).

32. A. Collet, M.-J. Brienne, and J. Jacques, *Chem. Rev.*, *80*, 215 (1980).

33. H. E. Zaugg, *J. Am. Chem. Soc.*, *77*, 2910 (1955).

34. J. Brugidou, H. Christol, and R. Sales, *Bull. Soc. Chim. France*, 2033 (1974).

35. G. Amiard, *Bull. Soc. Chim. France*, 447 (1956).

36. A. Collet, M.-J. Brienne, and J. Jacques, *Bull. Soc. Chim. France*, 127 (1972); 494 (1977).

37. J. Jacques, M. Leclercq, and M.-J. Brienne, *Tetrahedron*, *37*, 1727 (1981).

38. A. Collet, and J. Jacques, *Bull. Soc. Chim. France*, 3330 (1973).

39. M. Leclercq, A. Collet, and J. Jacques, *Tetrahedron*, *32*, 821 (1976).

40. J. Lajzerowicz-Bonneteau, J. Lajzerowicz, and D. Bordeaux, *Phys. Rev.*, *34*, 6453 (1986).

41. W. Meyerhoffer, *Berichte, 37*, 2604 (1904).

42. J. Jacques, and J. Gabard, *Bull. Soc. Chim. France*, 342 (1972).

43. T. Shiraiwa, Y. Ohki, Y. Sado, H. Miyazaki, and H. Kurokawa, *Chirality*, *6*, 202 (1994).

44. S. H. Wilen, *Topics in Stereochemistry, 6*, 107 (1971).

45. S. H. Wilen, *Tables of Resolving Agents and Optical Resolutions*, Univ. of Notre Dame Press, South Bend, IN, 1972.

46. P. Newman, *Optical Resolution Procedures for Chemical Compounds*, 3 vols., Optical Resolution Information Center, Riverdale, NY, 1984.

47. I. Ugi, *Z. Naturforsch., 20B*, 405 (1965).

48. T. Shiraiwa, Y. Sado, S. Fujii, M. Nakamura, and H. Kurokawa, *Bull. Chem. Soc. Jap.*, *60*, 824 (1987).

49. A. W. Ingersoll, S. H. Babcock, and F. B. Burns, *J. Am. Chem. Soc.*, *55*, 411 (1933).

50. E. J. Valente, J. Zubkowski, and D. S. Egglestron, *Chirality*, *4*, 494 (1992).

51. E. J. Valente, C. W. Miller, J. Zubkowski, D. S. Egglestron, and X. Shui, *Chirality*, *7*, 652 (1995).

52. R. Yoshioka, O. Ohtsuki, T. Da-Te, K. Okamura, and M. Senuma, *Bull. Chem. Soc. Jap. 67*, 3012 (1994).

53. M. Acs, E. Novotny-Bregger, K. Simon, and G. Argay, *J. Chem. Soc. Perkin Trans., 2.*, 2011 (1992).

54. M. C. Brianso, *Acta Cryst., B32*, 3040 (1976).

55. H. Lopez de Diego, *Acta Chem. Scand., 48*, 306 (1994).

10

Impact of Polymorphism on the Quality of Lyophilized Products

Michael J. Pikal

University of Connecticut
Storrs, Connecticut

I. INTRODUCTION

Polymorphism in solids is a well-known phenomenon that, at least in principle, may dramatically impact both manufacturability and performance of a dosage form. Variation in the structure of the solid will cause variation in physical and chemical properties of the dosage form, and while such variations are not always of practical significance, a development scientist does need to be aware of the potential for polymorphism and the possible consequences. Awareness is generally high when dealing with tablet and/or capsule dosage forms, where the focus is often on manufacturability or on the impact of dissolution variations on bioavailability. The potential for significant polymorphism issues with lyophilized (freeze-dried) dosage forms is frequently ignored. It is not uncommon to see the procedure for preparation of a freeze-dried product described by one word, "lyophilized." In addition, the solid-state characterization of the freeze-dried product is frequently also described by the same word, "lyophilized." However, the implication that "lyophilized" is a satisfactory description of both the process and the resulting product is a gross oversimplification. The "phase chemistry" in freeze-drying is exceedingly complex, and stability problems are common (particularly with protein products). Depending on the formulation, process, and details of storage, changes in structure of the solid occur, often with dramatic consequences. Furthermore, the freeze-drying process is often long and expensive, and the impact of crystallization may change the processing time by an order of magnitude.

While polymorphism issues conventionally focus on different crystal forms of the same compound, the major question for freeze-dried products is normally whether the solid phases are crystalline or amorphous. The question of which crystalline polymorph of a given compound forms is usually of secondary importance.

While crystallization during freeze-drying may be of no practical consequence, there are a number of scenarios for which crystallization can have major impact on design of the process and/or performance of the drug. Altered stability is perhaps the most common result of crystallization. Crystallization of the drug will nearly always produce a significant stability enhancement, often by more than an order of magnitude [1,2]. Conversely, crystallization of one of the excipients will often result in a major loss of stability, particularly when crystallization of buffer and/or crystallization of a stabilizer takes place in a protein formulation [3]. Since the amorphous form has greater solubility than any of its crystalline forms, crystallization of a sparingly soluble drug may prolong reconstitution time or even lead to incomplete dissolution at the intended reconstitution volume.

Crystallization can also have a major impact on design of the process. The impact may be positive or negative, depending on what crystallizes and when crystallization occurs. Since the glass transition temperature T_g' of the freeze concentrate depends upon the composition of the freeze concentrate, crystallization of one or more formulation components alters the composition of the solute phase(s) and thereby changes the value of T_g'. An increase of T_g', as would be produced by crystallization of most inorganic salts, would allow the positive result of freeze-drying at a higher temperature. Conversely, a decrease in T_g' would require freeze-drying at a lower temperature, which would mean longer processing times and a negative outcome. Mannitol, perhaps the most commonly used excipient, is a useful bulking agent, but it can lead to vial breakage problems upon crystallization. While a bulking agent should crystallize during the freeze-drying process, mannitol crystallization should take place during freezing (not during the primary drying), or else significant vial breakage may occur [4]

The objective of this chapter is to analyze situations where crystallization may occur, and where one would anticipate a high probability of negative impact on product quality and/or process design. Following a summary of methodology for detection of crystallinity and a brief review of the freeze-drying process, a number of case histories will be described to illustrate the major generalizations that can be made regarding crystallization during freeze-drying and the impact of such crystallization.

II. DETECTION OF CRYSTALLINITY

Crystallinity in the freeze-dried product can normally be detected by polarized light microscopy, through the presence of birefringence in the sample [1]. Although microscopy is simple, quick, and sensitive to small levels of crystallinity, it does have several disadvantages. Cubic crystals are isotropic and therefore nonbirefringent. Very small crystals may escape detection, and liquid crystal phases may be mistaken for three-dimensional crystallinity.

Differential scanning calorimetry (DSC) can also be used effectively in crystallinity studies. A sharp endotherm near the known melting point of a given polymorph is a clear signal of crystallinity arising from that polymorph. Further, when suitable calibration standards exist, the degree of crystallinity in a one-component system can be estimated from the areas under the melting endotherms (i.e., the heats of fusion) for the unknown and the standard. Conversely, a DSC scan of a sample containing a glassy phase produces a sharp increase in heat capacity at the glass transition. While enthalpy recovery effects may obscure the glass transition [5,6], enthalpy recovery can be eliminated by scanning slightly past the thermal event, cooling, and immediately rescanning. Enthalpy relaxation effects that produce the enthalpy recovery endotherm are irreversible on the time scale of the DSC scan, and the second scan will ordinarily reveal the glass transition. However, newer DSC techniques, such as "modulated DSC" [7] or "dynamic DSC" [8], have the potential for more efficient separation of enthalpy recovery effects from the glass transition. By superimposing a nonlinear temperature variation on the linear temperature scan rate, these methods can separate the reversible increase in heat capacity at the glass transition from pure kinetic effects, such as enthalpy recovery.

Degrees of crystallinity can also be evaluated from heat of solution measurements that compare the sample to be characterized with known crystalline and amorphous standards [1]. The heat of solution technique is particularly useful for materials that decompose upon melting. The method sensitivity depends upon the difference in heat of solution between crystalline and amorphous phases, and it can be better than ±1%, allowing detection of small differences in highly crystalline samples [1].

X-ray powder diffraction is perhaps the "gold standard" for the qualitative determination of crystallinity. Not only can the presence of a crystalline phase be confirmed, but since each polymorph produces a unique diffraction pattern, the question of which polymorph crystallized can be addressed. While x-ray powder diffraction requires expensive specialized equipment not likely to reside in a freeze-drying laboratory, such equipment is widely available in materials characterization laboratories. The experiments can be done quickly, and the results are normally unambiguous. One significant disadvantage is that very low degrees of crystallinity (i.e., much less than 10%) are frequently not detectable. An ensemble of techniques is often needed, such as the combination of x-ray powder diffraction and polarized light microscopy or DSC.

The preceding discussion was directed at the determination of crystallinity in a freeze-dried product, which from a quality control viewpoint is the most important material to characterize. However, from a research viewpoint it is often desirable to determine just where in the process crystallization does occur. Crystallization during freezing can easily be studied using variations of several of the techniques described above. Polarized light microscopy studies can be carried out with a freeze-drying microscope apparatus [9,10]. DSC studies on frozen solutions are routine, and at least at high solute concentration, glass transitions from an amorphous phase are easily recognized. In addition, a crystalline solute provides a eutectic melting endotherm that, if well removed from the main ice melting endotherm, is easily detected. Recently, x-ray diffraction has been used to characterize crystallinity in frozen systems [11]. This extension of conventional x-ray powder diffraction has great promise in studying complex crystallization problems directly relevant to freeze drying. However, all the methods for studying crystallization during freezing do have one limitation. It is well known that crystallization phenomena are sensitive to the thermal history of the sample (i.e., the cooling rate) and the volume of the sample. None of these laboratory techniques mimic either the thermal history or the sample volume of a product freezing in a freeze-dryer. Freezing in a vial is slower, it frequently involves less supercooling of water, and the volumes normally employed are far greater. In general, solute crystallization proceeds more readily in a vial. While special thermal treatments applied after the first freezing process (i.e., annealing) can be employed in an attempt to mimic freezing in a vial, some caution must be exercised in the extrapolation of the specialized studies of crystallization to behavior in a vial.

III. THE FREEZE-DRYING PROCESS: THE IMPORTANCE OF TEMPERATURE CONTROL

Freeze-Drying is a drying process by which a solution (usually aqueous) is first frozen and then subjected to removal of the ice by direct sublimation during the stage of the process called primary drying. As a sample is cooled during the freezing process, the solution normally supercools to a temperature about 10–20°C below the equilibrium freezing point where ice nucleates and crystallizes. Once crystallization begins, the product temperature rises rapidly to near the equilibrium freezing point, decreases slowly until most of the water has crystallized, and then decreases sharply to approach the shelf temperature. The process of ice formation is accompanied by an increase in solute concentration between the growing ice crystals and an increase in solution viscosity. Solutes that tend to crystallize easily from aqueous solution, such as sodium chloride, buffer salts, and some low molecular weight drugs, may crystallize. Generally the drug itself and most carbohydrate excipients do not crystallize. Rather, they remain amorphous, and at the end of the freezing process they exist in a glassy state containing a relatively large amount of unfrozen water (i.e., approximately 20% w/w). The water that does not freeze is removed by desorption during the secondary drying stage of the process. Both primary drying and secondary drying are conducted at low pressure, typically in the range of 50–300 m Torr. Processing temperatures are low for primary drying (−10 to −40°C) and normally above ambient in the secondary drying step (25°C to 50°C). The objective is to produce a solid that will lead to the observation of a satisfactory storage stability for labile drugs. The resulting product is normally a parenteral product and may be a small molecule or biopolymer. Since the physical state of the product is sensitive to the product temperature history, it follows that product temperature control is critical, particularly in freezing and primary drying.

The rate of primary drying increases sharply with increasing product temperature (i.e., roughly by a factor of 2 for a 5°C increase in product temperature). Therefore an efficient process is one that runs at a high product temperature. However, the temperature must not be too high or the product quality can be compromised. Ideally, the freeze-dried solid should be a porous cakelike structure with dimensions roughly equal to the frozen solution. Another way of stating this is that freeze-drying should proceed with retention of structure. Freeze-drying with retention of structure demands that one freeze-dry from a solid system. In a system where at least one solute component is crystalline, primary drying must be conducted below the eutectic temperature. As most drugs and excipients either have high eutectic temperatures or do not crystallize, eutectic melt is normally not a practical issue. Systems containing high levels of NaCl are an exception, as this salt exhibits a eutectic temperature of −21°C. In a system where the solute does not crystallize, primary drying must be carried out below the collapse temperature. Collapse is the amorphous system analog of eutectic melting, which is caused by the product temperature exceeding that of the glass transition temperature of the solute phase. If a product is being dried significantly above the glass transition temperature of the solute phase, the solute has sufficient fluidity to undergo viscous flow when ice is removed from that region of the sample. This phenomenon will lead to a loss of cake structure. Collapse of the cake means a loss of pharmaceutical elegance, higher residual moisture at the end of the freeze-drying process, possible increase of the reconstitution time, and possible crystallization of some components. Loss of activity can also occur with some protein formulations [12].

IV. CRYSTALLIZATION CASE HISTORIES

A. Impact of Crystallization on Process Design

1. Crystallization and Collapse

The collapse temperature depends upon the composition of the amorphous phase, and it has been found that crystallization of one or more components can significantly alter the collapse temperature. For example, human growth hormone (hGH) formulated with glycine and mannitol in a hGH: glycine:mannitol weight ratio of 1:1:5 can form a completely amorphous system if frozen very quickly. Here, the collapse temperature is $-24°C$. However, a slower freezing rate permits crystallization of most of the mannitol, which results in a collapse temperature greater than $-5°C$ [13]. Another example is provided by the glycine: sucrose system [14–16], where glycine is present in excess relative to sucrose. If frozen quickly, glycine remains amorphous, and the glycine: sucrose freeze concentrate has a very low collapse temperature (roughly $-45°C$, depending upon the exact composition). However, if glycine is allowed to crystallize, the glass transition temperature is essentially that of the sucrose freeze concentrate (about $-34°C$). Further, since the structure is maintained by the crystalline glycine, exceeding T_g' does not result in macroscopic collapse, and the system can be freeze-dried without apparent collapse (even at temperatures exceeding $-10°C$). Thus crystallization can transform a formulation from one that is nearly impossible to freeze-dry in a commercial operation to one where freeze-drying is relatively easy.

2. Mannitol Crystallization and Vial Breakage

Vial breakage is a common occurrence when freeze-drying mannitol solutions. This situation is particularly troublesome with high fill volumes, high mannitol concentrations, and low levels of other solutes. While in principle the stress created by a very rapid temperature change may be the cause, the fact that mannitol solutions are much more problematic than other formulations suggests that other factors dominate. It appears [4,17] that vials typically break during the early portion of primary drying as a result of mannitol crystallization. That is, when amorphous mannitol is created during freezing, the shelf temperature increase during early primary drying increases the product temperature above the T_g' (approximately $-30°C$), allowing rapid crystallization of mannitol at about $-26°C$ and making the water that was part of the freeze concentrate free to crystallize. The resulting rapid volume increase is the stress that breaks the vials. As illustrated in Fig. 1, the probability of vial breakage depends significantly upon both the fill volume and the freezing procedure. A fast freeze to $-50°C$ (which in reality is a typical commercial freezing procedure) results in about 40% of the batch being lost to vial breakage with the higher fill volume (5 mL). However, with a slow freeze to $-18°C$, no vial breakage is ob-

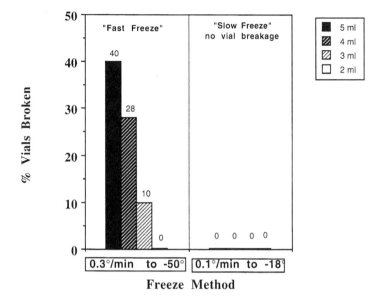

Fig. 1. Impact of process on crystallization: mannitol crystallization and vial breakage. Aqueous mannitol (5%) freeze-dried at a product temperature of −15°C during primary drying. (Data taken from Ref. 4.)

served. Vial breakage is also sensitive to the fill volume. No vial breakage is observed with the fast freeze procedure when the fill volume is only 2 mL.

The greater breakage with higher fill volumes is likely a direct result of mechanical effects. The key factor, however, is mannitol crystallization during freezing. If mannitol is induced to crystallize during freezing, mannitol crystallization occurs slowly from a fluid system without undue mechanical stress being developed. Using a slow cooling rate and limiting the product temperature to only −18°C during freezing (which is well above the T_g' temperature) produces crystalline mannitol at the end of the freezing process, thereby eliminating the rapid crystallization during early primary drying that is responsible for vial breakage.

3. Impact of Drying Process on Solubility

In some cases, a primary consideration in freeze drying is the production of the amorphous phase of the drug so that rapid and complete solubilization is allowed. Dobutamine hydrochloride is an example where the room temperature equilibrium solubility of the crystalline form is marginal under the recommended reconstitution conditions. For this reason, its reconstitution is slow and often incomplete. Thus the drug is formulated with mannitol in a 1:1 weight ratio to produce highly soluble amorphous dobutamine hydrochloride. However, during early process development studies [18], it was noted that some batches produced a significant fraction of vials in which dissolution was incomplete. This finding was attributed to the presence of residual crystalline drug in batches that dissolved poorly. These batches also contained crystalline mannitol and crystalline drug in the solid state.

The results of one series of laboratory studies designed to explore the impact of process variations on crystallization are summarized in Table 1. In this work, the two process variations studied were stopper position and the criterion for judging the end of primary drying. Stopper position was judged a relevant variable due to the unfortunate fact that the 13-mm stoppers being used would frequently be seated too deeply into the vial, thereby restricting the vapor flow, raising the product temperature, and prolonging primary drying. The criterion for judging the end of primary drying was based upon product temperature response. The product temperature during primary drying normally varied from about $-25°C$ early in primary drying to about $-15°C$ at the end of primary drying. Therefore, once the product temperature sig-

Table 1 Impact of Process on Crystallization: The Effect of Drying Cycle on Crystallization of Dobutamine HCl

Stopper position	Product temperature when shelf→+40°C	#experiments (total # samples)	Percent crystallized
Normal	+10°C to +20°C	3 (27)	0
1/2 to 3/4 closed	+10°C to +20°C	2 (13)	38
Normal	≤ +5°C	2 (13)	23
1/2 to 3/4 closed	≤ +5°C	3 (22)	64

The formulation is 1:1 drug:mannitol. The freeze drying conditions are 100 mTorr chamber pressure and shelf temperature +15°C, followed by +40°C when the product reaches the indicated temperature. Stoppers are 13 mm finish stoppers.
Source: Data taken from Ref. 18.

nificantly exceeded $-15°C$, the product temperature would rise sharply, thereby signaling the end of primary drying in that vial. In theory, the shelf temperature could be increased for secondary drying at this point without the risk of collapse. However, to allow for variation in primary drying times, and/or to allow for reduction of residual water in the "dry" product at low temperature, the shelf temperature is commonly held at the primary drying setting beyond the time at which all vials containing temperature sensors have completed primary drying. That is, only after this "operational" end of primary drying criterion has been met is the shelf temperature increased for secondary drying. Selection of this operational criterion is somewhat arbitrary. A process that specifies a longer hold time and/or a higher product temperature is more conservative in the sense that at the point of shelf temperature increase, the probability of a vial still containing ice is smaller and the residual water in the "dry" samples is less. Obviously, the more conservative process has the disadvantage of adding time to the process. Two operational criteria for the end of primary drying were used. The more conservative process yielded product temperatures in the $+10°C$ to $+20°C$ range before the shelf temperature was increased to $40°C$. The more aggressive process yielded product temperatures less than $5°C$ at the point of shelf temperature increase.

The data shown in Table 1 clearly demonstrate the probability of crystallization and resulting insolubility increases as stoppers are positioned too deep into a vial. The " 1/2 to 3/4 closed" term approximately represents a position such that 1/2 to 3/4 of the areas of the openings are closed. The operational criterion for the end of primary drying is also important, as the probability of crystallization is significantly higher for the more aggressive criterion of increasing the shelf temperature (i.e., a hold time giving measured product temperatures less than or equal to $5°C$). It seems obvious that the higher product temperature and/or the prolonged primary drying time caused by improper stopper position increases the probability of crystallization. It is possible that the aggressive criterion for ending primary drying resulted in some vials still containing ice when the shelf temperature was increased to $40°C$, thereby yielding collapse and subsequent crystallization. However, the collapse temperature is relatively high ($-10°C$), and collapse was not visually apparent. It is more likely that crystallization occurred in product devoid of ice when the product temperature exceeded the glass transition temperature of the moisture-rich solute early during secondary drying.

B. Buffer Crystallization and pH Shifts

It is ironic that while the function of a buffer system is to maintain constant pH, massive pH shifts during freezing may result from selective crystallization of one of the buffer components. In principle, and in practice, pH shifts often lead to degradation [19]. Under equilibrium conditions attained by seeding, the sodium phosphate buffer system shows a dramatic decrease of about 4 pH units due to crystallization of the basic buffer component, $Na_2HPO_4 \cdot 2H_2O$ [20]. Conversely, the potassium phosphate system shows only a modest increase in pH of about 0.8 pH unit. Under nonequilibrium conditions (i.e., no seeding) and with lower buffer concentrations, the degree of crystallization is less, and the resulting pH shifts are moderated [21]. Table 2 shows data accumulated during freezing of phosphate buffer solutions in large volumes at cooling rates intended to mimic freezing in vials [22,23]. For the concentrated buffer solutions (100 mM), the frozen pH values were close to the equilibrium values. However, lowering the buffer concentration by an order of magnitude considerably reduced the pH shift observed during freezing. It should also be noted that under some

Table 2 Shifts in pH During Nonequilibrium
Freezing with Phosphate Buffer Systems

Concentration,mM	Initial pH	Frozen pH	ΔpH
Sodium phosphate buffer			
100	7.5	4.1	-3.4
8	7.5	5.1	-2.4
Potassium phosphate buffer			
100	7.0	8.7	$+1.7$
100	5.5	8.6	$+3.1$
10	5.5	6.6	$+1.1$

Source: Data taken from Refs. 22 and 23.

conditions, potassium phosphate buffers also exhibited large pH shifts during freezing. As shown in Table 2, if the initial pH is 5.5, the 100 mM potassium phosphate buffer increases in pH by 3.1 units during freezing.

Clearly, if drug stability is sensitive to changes in the solution pH, buffer crystallization must be avoided. In the author's experience, the best solution is to formulate so that the weight ratio of buffer to other solutes is very low [24,25]. The impact of other components on buffer crystallization and resulting pH shift is illustrated by the data in Table 3 [26]. As the weight ratio of sucrose to buffer increases, the pH shift on freezing moderates. At a weight ratio of 6:1, pH in the frozen material is identical to the starting solution pH. X-ray powder diffraction studies conducted on freeze-dried samples corresponding to the compositions given in Table 2 showed no evidence of crystallinity in the 6:1 weight ratio sample, which was consistent with the lack of a pH shift on freezing the 6:1 formulation.

Table 3 Impact of Formulation on Crystallization: The Effect of Sucrose on the pH Shift in Sodium Phosphate Buffer Systems During Nonequilibrium Freezing

Weight ratio, sucrose:buffer	Frozen pH (at $-10°C$)
0.0	4.1
0.36	4.2
1.0	4.6
2.0	5.0
6.0	7.5[1]

[1] No crystallization of buffer observed in freeze-dried sample (x-ray powder diffraction).
The buffer concentration is 100 mM with an initial pH of 7.5.
Source: Data taken from Ref. 26.

C. Crystallization During Storage: Formulation, Water Content, and the Glass Transition

At least for materials that can crystallize, an amorphous form represents a metastable state, so thermodynamics requires that crystallization eventually occur. In general, however, crystallization of most materials is very slow until the storage temperature approaches or exceeds the glass transition temperature. However, crystallization well below the glass transition temperature does occasionally occur. Indomethacin crystallizes from the glassy state at a significant rate [27], and observations from the author's laboratory indicate that sodium cephalothin (a cephalosporin antibacterial) crystallizes very slowly at temperatures more than 100°C below the glass transition temperature. The degree of crystallinity was found to increase for relatively dry (approximately 1% water) freeze-dried sodium cephalothin (initially with partial crystallinity) after storage at −20°C after nearly 20 years. This behavior was observed even through the glass transition for "dry" sodium cephalothin is over 100°C.

Residual moisture has a profound influence on the rate of crystallization from the amorphous state [28–30]. As the moisture content increases, the glass transition temperature decreases sharply, and particularly for accelerated stability testing, it is not unusual for the glass transition temperature at modest levels of residual moisture to be less than the storage temperature. In such cases, increased molecular mobility in the amorphous material allows crystallization to take place on the timescale of the experiment (i.e., during storage). This phenomenon is illustrated by the water sorption isotherms shown in Fig. 2. As the relative humidity increases, the equilibrium water content increases smoothly until a moisture content is reached at which the moisture content suddenly decreases due to crystallization of the sugar. The moisture content at which the glass transition temperature of the sugar has been reduced to the sample temperature (25°C in these experiments) is denoted "Wg" on the plot. Note that crystallization does not occur on the rapid timescale of the experiment (i.e., several hours) until the moisture content is well above Wg. This indicates that rapid crystallization does not occur until the glass transition temperature has been reduced to well below the sample temperature.

While a well-designed freeze drying process can produce a prod-

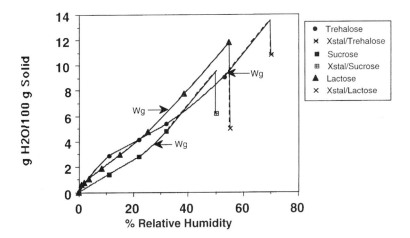

Fig. 2. The effect of water content on crystallization from the amorphous state: water sorption isotherms and crystallization of lactose, sucrose, and trehalose at 25°C. The equilibration time scale is hours. (Data taken from unpublished observations from this laboratory.)

uct of very low water content, and crystallization from the glassy state is not normally a practical problem at these low water contents, one often finds that the water content increases significantly upon storage. This can facilitate crystallization of one or more of the formulation components. The cause of the moisture increase is not moisture transmission from the ambient environment through the stopper but rather transfer of moisture from the stopper to the product [31]. Evidently, moisture is "pumped" into the stopper during steam sterilization, and while vacuum drying after steam sterilization and during the freeze-drying process removes water from the surface region of the stopper, water deeper in the interior is not completely removed. During storage, water slowly diffuses out of the stopper and absorbs into the dry product until equilibrium is reached. For the gray butyl rubber stopper studied, increases in water content of over 3% w/w were observed [31]. The level of water in the product is extremely sensitive to the method of vacuum drying the stoppers after their steam sterilization, with longer drying times leading to much reduced levels of moisture transfer. In addition, the equilibrium water content in the product depends on the mass of amorphous material in the vial, so that a small mass leads to a high water content at equilibrium. Equilibrium is reached slowly at normal storage temperatures, and equilibration times are about 2 weeks at 40°C and 4–6 months at 25°C. While the moisture transfer problem can be moderated by heroic drying procedures after sterilization, a better solution is to use stoppers that do not release large amounts of water to the product. Such low-moisture stoppers are now available from some vendors.

The author has had experience with Daikyo Flurotech stoppers (available from the West Company), which produce a moisture increase of only about 0.1% under conditions where the butyl rubber stoppers produced an increase of over 1%. It is possible that the Flurotech coating (a fluorocarbon polymer coating) is responsible for the low moisture release properties. However, we have also found that Teflon-coated butyl rubber stoppers release moisture essentially in an equivalent fashion to stoppers of the same rubber formulation without the coating. Thus the low moisture properties of the Flurotech stoppers are more likely due to the rubber formulation.

It is important to note that, consistent with the impact of additional formulation components on buffer crystallization during freezing (Table 2), crystallization during storage is also retarded by the presence of additional solute components. The induction time for sucrose crystallization (Fig. 3) at 30°C (and 30% relative humidity) was studied as a function of level and type of saccharide additive [30]. Raffinose, lactose, and trehalose all significantly extend the induction time for crystallization, even at the relatively low levels investigated (less than 10%), with raffinose being the most effective of the three sugars studied. Since these additive sugars all exhibit glass transition temperatures in excess of 100°C when dry (compared with 75°C for sucrose), it is tempting to conclude that the mechanism involves an increase in the glass transition temperature. However, the actual mechanism may be more complex [30].

D. Stability Consequences of Crystallization

1. Crystallization of Drug

For nearly all compounds studied, the crystalline form has far greater stability than the corresponding amorphous form. This result is likely due to the crystalline form being able to provide an ''inert'' and rigid

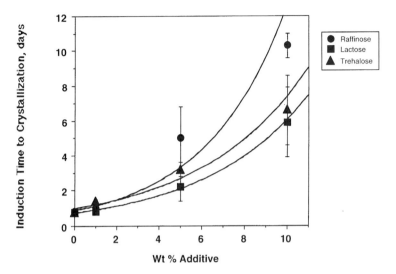

Fig. 3.   Effect of formulation on crystallization: impact of saccharides on the crystallization of sucrose from the amorphous state. (Data taken from Ref. 30.)

environment for the reactive groups of a molecule. The few known "small molecule" examples of greater reactivity in the crystalline form arise from unusual details of crystal geometry where two reactive groups are forced in close proximity so that reaction requires little movement [32]. Insulin is the only known pharmaceutical example where greater reactivity is observed in the crystalline state relative to its amorphous phase [33].

The qualitative trend of degradation rate with crystallinity shown in Fig. 4 illustrates more typical behavior [1]. Here, the stability of cephalothin sodium clearly increases dramatically as the crystallinity increases. For this work, the degree of crystallinity was evaluated from the heat of solution of a given sample relative to the heats of solution of crystalline and amorphous standards. Although the instability of "hydrated" samples (i.e., in equilibrium with 31% relative humidity) is far greater than for corresponding dry samples, the qualitative trend of degradation rate with degree of crystallinity is the same. Many of the partially crystalline samples represented in Fig. 4 were prepared by freeze-drying, where significant differences in crystallinity arose

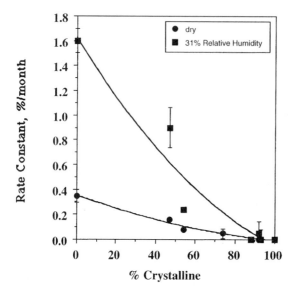

Fig. 4. Impact of drug crystallization on storage stability: effect of degree of crystallinity on the storage stability of cephalothin sodium at 50°C. The degrees of crystallinity were estimated from heat of solution data. (Data taken from Ref. 1.)

from both variations in process and inter-vial variations in crystallinity produced by nominally the same process [1]. Clearly, even with a given compound, crystallinity variations and corresponding stability variations can be quite striking.

2. Crystallization of Stabilizer

While crystallization of the drug increases stability, crystallization of an excipient intended to function as a stabilizer will decrease drug stability. To function as a stabilizer, the excipient must be in the same phase as the drug [24]. Complete crystallization of the stabilizer will lead to the existence of a simple physical mixture of an excipient and a drug (such as crystalline sucrose and glassy protein) and will not allow either molecular dispersion of the drug or a stabilizing interaction of drug and excipient. Thus the stability of human growth hormone in a system of amorphous glycine and mannitol is superior when compared to a formulation with crystalline glycine or mannitol [24].

An excellent example of the impact of stabilizer crystallization on protein stability during freeze-drying is provided by the series of mannitol/buffer formulations of β-galactosidase studied by Izutsu and coworkers [34]. Figure 5 illustrates the relative activity after freeze-drying as a function of the weight ratio of mannitol to phosphate buffer. As the weight ratio of mannitol to buffer increases, the stability also increases, and it reaches essentially 100% at a weight ratio of about 4.0. Interestingly, the stability decreased at higher weight ratios. It was noted that many of the systems contained crystalline material. Since systems without mannitol (where buffer crystallization might be expected) were amorphous, it appears that the crystalline phase was mannitol. In the figure, the data points referring to samples that were at least partially crystalline are given filled symbols. The vertical line at a weight ratio of 4.0 clearly separates the totally amorphous samples from the samples with crystallinity, and it also marks the region of maximum stability. As the mannitol content increases, stability in-

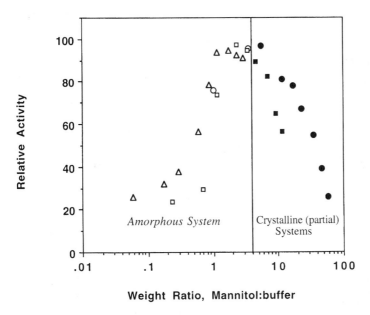

Fig. 5. Crystallization of stabilizer system and impact on enzyme stability during freeze drying: residual activity of β-galactosidase in mannitol:sodium phosphate formulations after freeze-drying. (Data taken from Ref. 34.)

creases as long as the mannitol remains amorphous. However, at sufficiently high amounts of mannitol, the substance crystallizes and removes much of the mannitol from the protein phase, thereby decreasing stability.

If a stabilizer crystallizes during storage, not only will the stabilizer be removed from the drug phase but also any moisture associated with the portion of the amorphous phase that crystallizes to an anhydrate will be shifted to the remaining amorphous phase (i.e., the drug phase), thereby increasing the moisture content of the drug and decreasing stability. Figure 6 illustrates the possible impact of this effect on stability. The data shown in Fig. 6 give estimated degradation rates at 5°C in a vinca alkaloid: monoclonal antibody conjugate formulated with a 1:1:1 weight ratio of conjugate/glycine/mannitol [25]. This formulation tends to crystallize during storage, producing a significant decrease in the glass transition temperature of the drug phase. Degradation rates, at least in systems above the glass transition temperature, were found to be strongly correlated with the difference between storage temperature and glass transition temperature. Degradation rates via hydrolysis of the vinca-antibody linkage (free vinca generation) and through vinca decomposition (largely oxidation) are shown as a function of water content for formulations that remain amorphous during storage and for samples where the glycine and mannitol excipients crystallize during storage. These data were projected from experimental observations at 25°C and 40°C at water contents of about 1% and higher. Due to sorption of water from the stopper, very low levels of moisture could not be maintained during storage. In Fig. 6, the circles with corresponding error bars represent direct experimental observation on samples where crystallization had occurred during storage. The close agreement between the projected data and the direct experimental observations provides some support for the accuracy of the extrapolation procedure used [25]. However, it must be recognized that the very sharp decrease in reaction rates projected below 1% water and the roughly fivefold increase in reaction rate upon crystallization are based upon the validity of correlation between reaction rate and T-T_g. Assuming that the projected data are at least semiquantitative, it is obvious that crystallization of excipient during storage has a serious stability consequence.

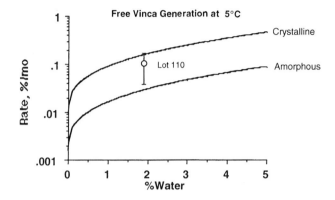

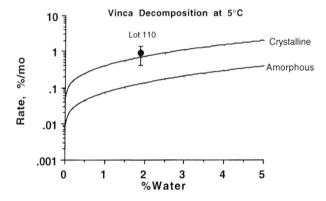

Fig. 6. Crystallization of stabilizer system and impact on storage stability: stability at 5°C of a KS1/4 monoclonal antibody:vinca alkaloid conjugate formulated with mannitol and glycine in a 1:1:1 weight ratio of conjugate:mannitol:glycine. Both mannitol and glycine crystallize early in storage. The smooth lines are data projected from 40°C and 25°C data for crystalline and amorphous excipient systems. The circles represent direct experimental observations at 5°C for "lot #110", which was "mostly crystalline" after storage. (Data taken from Ref. 25.)

A complex but informative example of the impact of crystallization on storage stability is provided by the effect of sucrose on the
storage stability of ribonuclease A [35–37]. Figure 7 gives the percentage of initial enzyme activity remaining after storage for 105 days at
45°C for sucrose and Ficoll formulations of ribonuclease A (RNASE)
buffered with 100 mM sodium phosphate (pH 6.4). With no excipient,
the residual activity is about 60%, and the addition of increasing levels
of Ficoll enhance stability to about 90% of initial. However, at low
levels of added sucrose, the stability is sharply decreased, and only
about 15% of initial activity remains at a level of 0.5% sucrose; but
at higher sucrose levels, sucrose stabilizes at least as well as Ficoll. It
was noticed that only the low sucrose content samples were brown in
color, and they were similar in appearance to sucrose formulations that
were freeze-dried from solutions at pH 3. The instability and color of
the pH 3 samples were correlated with the onset of sucrose hydrolysis
to the constituent reducing sugars. The interpretation of the data seems
clear. At low sucrose levels, the sodium phosphate buffer crystallized,

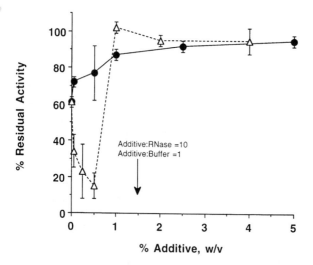

Fig. 7. The effect of formulation on storage stability of freeze-dried ribonuclease A: residual activity after storage for 105 days at 45°C. All formulations
were freeze-dried from formulations containing 100 mM sodium phosphate
buffer at pH 6.4. (Data taken from Ref. 35–37.)

producing a significant reduction in pH, thereby hydrolyzing the sucrose. The reducing sugars then qualitatively produced the same instability as was found with samples freeze-dried from pH 3 solutions. However, at higher levels of sucrose, buffer crystallization was reduced or eliminated, thereby circumventing sucrose hydrolysis.

V. SUMMARY AND CONCLUSIONS

A number of factors are critical for crystallization during or after freeze-drying. First, the nature of the solute is of obvious importance. Unless the solute will readily crystallize from an aqueous system at ambient temperatures, crystallization during freeze-drying is unlikely. Our experience suggests that if an aqueous solution crystallizes upon air drying on the laboratory bench, crystallization is possible in a freeze-drying process. Crystallization is obviously favored by higher concentrations of crystallizable solute. However, and most important, the presence of high levels of other solutes that will remain amorphous will generally impede or prevent crystallization of the potentially crystallizable solute. For example, while mannitol easily crystallizes from systems that are nearly pure mannitol, mannitol crystallization during freeze-drying is unlikely if it is present as a minor component. Crystallization during storage is sensitive to these factors, and it is also extremely sensitive to both residual water content and storage temperature. As a general rule, crystallization is rapid if the system is stored above the glass transition temperature. Nevertheless, with prolonged storage times, crystallization in some systems is possible even at temperatures well below the glass transition temperature.

Polymorphism may well be a significant problem with freeze-dried products. However, the issue of practical significance lies normally along the ''amorphous vs. crystalline'' pathway, rather than being concerned about which particular crystalline polymorph forms. In general, a crystalline drug implies better stability, but if crystallization of an excipient occurs, the result is normally inferior stability. Of course, if the drug is intrinsically stable in any form, then crystallization will have no stability implication. Finally, it should be emphasized poor inter- or intra-batch uniformity can result in systems where crystallization is either erratic or partially present, leading to particular concerns for quality control groups. Again, the lack of uniformity in crystallization is only an issue if crystallization impacts some aspect of product quality.

REFERENCES

1. M. J. Pikal, A. L. Lukes, J. E. Lang, and K. Gaines, *J. Pharm. Sci.*, *67*, 767 (1978).
2. M. J. Pikal, A. L. Lukes, and J. E. Lang, *J. Pharm. Sci.*, *66*, 1312 (1977).
3. M. J. Pikal, *Biopharm.*, *3*(9), 26–30 (1990).
4. N. A. Williams, Y. Lee, G. P. Polli, and T. A. Jennings, *J. Parent. Sci. Tech.*, *40*, 136 (1986).
5. A. T. M. Serajuddin, M. Rosoff, and D. Mufson, *J. Pharm. Pharmacol.*, *38*, 219–220 (1985).
6. B. C. Hancock, S. L. Shamblin, and G. Zografi, *Pharm. Res.*, *12*, 799–806 (1995).
7. M. Reading, A. Luget, and R. Wilson, *Thermochim. Acta*, *238*, 295–307 (1994).
8. M. P. DiVito, R. B. Cassel, M. Margulies, and S. Goodkowsky, *Am. Lab.*, August, 28–37 (1995).
9. M. J. Pikal, S. Shah, D. Senior, and J. E. Lang, *J. Pharm. Sci.*, *72*, 635–650 (1983).
10. S. L. Nail, L. M. Her, C. P. B. Proffitt, and L. L. Nail, *Pharm. Res.*, *11*, 1098–1100 (1994).
11. R. K. Cavatur, and R. Suryanarayanan, *Pharm. Res.*, *15*, 194–199 (1998).
12. F. Franks, *Cryo-Letters*, *11*, 93–110 (1990).
13. M. J. Pikal, K. M. Dellerman, M. L. Roy, and R. M. Riggin, *Pharm. Res.*, *8*, 427–436 (1991).
14. S. Chongprasert, and S. L. Nail, *Pharm. Res.*, *14*, S-409 (1997).
15. T. Suzuki, and F. Franks, *J. Chem. Soc. Far. Trans.*, *89*, 3283–3288 (1993).
16. E. Shalaev, and A. Kanev, *Cryobiology*, *31*, 374–382 (1994).
17. N. A. Williams, and J. Guglielmo, *J. Parent. Sci. Tech.*, *47*, 119–123 (1993).
18. M. J. Pikal, and S. Shah, Eli Lilly & Co., unpublished observations.
19. S. S. Larsen, *Dansk. Tidsskr. Farm.*, *45*, 307–316 (1971).

20. S. S. Larsen, *Arch. Pharm., Chem. Sci. Ed., 1,* 41–53 (1973).
21. N. Murase, and F. Franks, *Biophys. Chem., 34,* 293–300 (1989).
22. G. Gomez, N. Rodriguez-Hornedo, and M. J. Pikal, *Pharm. Res., 11,* S-265 (1994).
23. B. A. Szkudlarek, N. Rodriguez-Hornedo, and M. J. Pikal, 1994 Mid-Western AAPS meeting, Chicago, poster #24.
24. M. J. Pikal, K. M. Dellerman, M. L. Roy, and R. M. Riggin, *Pharm. Res., 8,* 427–436 (1991).
25. M. L. Roy, M. J. Pikal, E. C. Rickard, and A. M. Maloney, *Dev. Biol. Stand., 74,* 323–340 (1991).
26. G. Gomez, Ph.D. thesis, Univ. of Michigan, 1995.
27. M. Yoshioka, B. Hancock, and G. Zografi, *J. Pharm. Sci., 83,* 1700–1705 (1994).
28. J. T. Carstensen, and K. Van Scoik, *Pharm. Res., 7,* 1278–1281 (1990).
29. Y. Aso, S. Yoshioka, T. Otsuka, and S. Kojima, *Chem. Pharm. Bull., 43,* 300–303 (1995).
30. A. Saleki-Gerhardt, and G. Zografi, *Pharm. Res., 11,* 1166–1173 (1994).
31. M. J. Pikal, and S. Shah, *Dev. Biol. Stand., 74,* 165–179 (1991).
32. C. N. Sukenik, J. A. P. Bonapase, N. S. Mandel, R. G. Bergman, P. Lau, and G. Wood, *J. Am. Chem. Soc., 97,* 5290–5291 (1975).
33. M. J. Pikal, and D. R. Rigsbee, *Pharm. Res., 14,* 1379–1387 (1997).
34. K. Izutsu, S. Yoshioka, and T. Terao, *Pharm. Res., 10,* 1233–1238 (1993).
35. M. W. Townsend, and P. P. DeLuca, *J. Parent. Sci. Tech., 42,* 190–199 (1988).
36. M. W. Townsend, P. R. Byron, and P. P. DeLuca, *Pharm. Res., 7,* 1086–1091 (1990).
37. M. W. Townsend, and P. P. DeLuca, *J. Pharm. Sci., 80,* 63–66 (1991).

Index